STUDIES IN THE THEORY OF DESCENT

VOLS.
I & II

AMS PRESS
NEW YORK

Part I.

ON THE SEASONAL DIMORPHISM OF BUTTERFLIES.

STUDIES IN THE THEORY OF DESCENT

BY

DR. AUGUST WEISMANN

PROFESSOR IN THE UNIVERSITY OF FREIBURG

WITH NOTES AND ADDITIONS BY THE AUTHOR

TRANSLATED AND EDITED, WITH NOTES, BY

RAPHAEL MELDOLA, F.C.S.

LATE VICE-PRESIDENT OF THE ENTOMOLOGICAL SOCIETY OF LONDON

WITH A PREFATORY NOTICE BY

CHARLES DARWIN, LL.D., F.R.S.

Author of "The Origin of Species," &c.

VOL. I.

WITH EIGHT COLOURED PLATES

London

SAMPSON LOW, MARSTON, SEARLE, & RIVINGTON

CROWN BUILDINGS, 188, FLEET STREET

1882

Library of Congress Cataloging in Publication Data

Weismann, August, 1834-1914.
Studies in the theory of descent.

Translation of Studien zur Descendenz-Theorie.
Reprint of the 1882 ed. published by S. Low, Marston, Searle, & Rivington, London.
Includes bibliographical references and index.
1. Lepidoptera. 2. Insects—Evolution. 3. Mechanism (Philosophy). I. Title.
QL562.2.W4413 1975 595.7'8'0438 72-1661
ISBN 0-404-08192-4

Reprinted from an original copy in the collections of the University of Kansas Libraries

From the edition of 1882, London
First AMS edition published in 1975
Manufactured in the United States of America

AMS PRESS INC.
NEW YORK, N. Y. 10003

PREFATORY NOTICE.

THE present work by Professor Weismann, well known for his profound embryological investigations on the Diptera, will appear, I believe, to every naturalist extremely interesting and well deserving of careful study. Any one looking at the longitudinal and oblique stripes, often of various and bright colours, on the caterpillars of Sphinx-moths, would naturally be inclined to doubt whether these could be of the least use to the insect; in the olden time they would have been called freaks of Nature. But the present book shows that in most cases the colouring can hardly fail to be of high importance as a protection. This indeed was proved experimentally in one of the most curious instances described, in which the thickened anterior end of the caterpillar bears two large ocelli or eye-like spots, which give to the creature so formidable an appearance that birds were frightened away. But the mere explanation of the colouring of these caterpillars is but a very small part of the merit of the work. This mainly consists in the light thrown on the laws of variation and of inheritance by the facts given and discussed. There is also a valuable discussion on classification, as founded on characters displayed at different ages by animals belonging to the same group. Several distinguished naturalists maintain with much confidence that organic beings tend to vary and to rise in the scale, independently of the conditions to which they and their progenitors have been exposed; whilst others maintain that all variation is due to such exposure, though the manner in which the environment acts is as yet quite unknown. At the present time there is hardly any question in biology of more importance than this of the nature and causes of variability, and the reader will find in the present work an able discussion on the whole subject, which will probably lead him to pause before he admits the existence of an innate tendency to perfectibility. Finally, whoever compares the discussions in this volume with those published twenty years ago on any branch of Natural History, will see how wide and rich a field for study has been opened up through the principle of Evolution; and such fields, without the light shed on them by this principle, would for long or for ever have remained barren.

CHARLES DARWIN.

TRANSLATOR'S PREFACE.

IN offering to English readers this translation of Professor Weismann's well-known "Studies in the Theory of Descent," the main part of which is devoted to entomological subjects, I have been actuated by the desire of placing in the hands of English naturalists one of the most complete of recent contributions to the theory of Evolution as applied to the elucidation of certain interesting groups of facts offered by the insect world Although many, if not most, working naturalists are already familiar with the results of Dr. Weismann's researches, of which abstracts have from time to time appeared in English and American scientific journals, I nevertheless believe that a study of the complete work, by enabling the reader to follow closely the detailed lines of reasoning and methods of experiment employed by the author, will be found to be of considerable value to those biologists who have not been able to follow the somewhat difficult phraseology of the original. It is not my intention, nor would it be becoming in me to discuss here the merits of the results arrived at by the minute and laborious investigations with which Dr. Weismann has for many years occupied himself. I may however point out that before the appearance of the present work the author, in addition to his well-known papers on the embryology and development of insects, had published two valuable contributions to the theory of descent, viz. one entitled "Über die Berechtigung der Darwin'schen Theorie" (1868), and another "Über den Einfluss der Isolirung auf die Artbildung" (1872). These works, which are perhaps not so well known in this country as could be desired, might be advantageously studied in connection with the present volume wherein they are frequently referred to.

Since every new contribution to science is a fresh starting-point for future work, I may venture without any great breach of propriety to dwell briefly upon one or two of the main points which appear to me to be suggested by Prof. Weismann's investigations.

Although the causes of Glacial Epochs is a subject which has much occupied the attention of geologists and physiographers, the question is one of such great complexity that it cannot yet be regarded as finally settled. But apart from the question of causes—a most able discussion of which is given by the author of "Island Life"—there is not the least doubt that at no very distant geological period there occurred such an epoch, which, although intermittent, was of considerable duration. The last great geological event which our globe experienced was in fact this Ice Age, and the pure naturalist has not hitherto attributed in my opinion sufficient importance to the *direct* modifying effects of this prolonged period of cold. It is scarcely possible that such a vast climatic change as that which came on at the close of the Pliocene Period should have left no permanent effect upon our present fauna and flora, all the species of which have survived from the glacial age. The great principle of Natural Selection leads us to see how pre-glacial forms may have become adapted to the new climatic conditions (which came on gradually) by the "survival of the fittest" or "indirect equilibration." The influence of the last Glacial Epoch as a factor in determining the present geographical distribution of animals and plants has already been amply treated of by many writers since the broad paths were traced out by Darwin, Lyell, and Wallace. The last named author has indeed quite recently discussed this branch of the subject most exhaustively in his work on "Island Life" above mentioned. The reference of a particular group of phenomena—the seasonal dimorphism of butterflies—to the direct action of the Glacial Period and the subsequent influence of the ameliorating climate, was however the first step taken in this neglected field by the author of the present work in 1875. It is possible, and indeed probable, that future researches will show that other characters among existing species can be traced to the same causes.

The great generalizations of embryology, which science owes so largely to the researches of Karl Ernst von Baer, bear to the theory of descent the same relations that Kepler's laws bear to the theory of gravitation. These last-named laws are nothing more than generalized statements of the motions of the planets, which were devoid of meaning till the enunciation of the theory of gravitation. Similarly the generalized facts of embryology are meaningless except in the light of the theory of descent. It has now become a recognized principle in biology that animals in the course of their development from the ovum recapitulate more or less completely the phases through which their ancestors have passed. The practical application of this principle to the determination of the line of descent of any species or group of species is surrounded by difficulties, but attempts have been made of late years—as by Haeckel in his *Gastrula* theory—to push the law to its legitimate consequences. In this country Sir John Lubbock, in 1874, appealed to the embryonic characters of larvæ in support of his views on the origin of insects. To the author of this work (1876) is due the first application of the principle of Ontogeny as revealing the origin of the markings of caterpillars. A most valuable method of research is thus opened up, and entomologists should not be long in availing themselves of it. Our knowledge of the subject of larval development in Lepidoptera is still most imperfect, and it cannot as yet be foreseen to what extent the existing notions of classification in this much-studied order may have to be modified when a minute study of the Comparative Ontogeny of larval characters, worked out as completely as possible for each family, has enabled a true genealogical system to be drawn up. The extent to which such a larval genealogy would coincide with our present classification cannot now be decided, but he who approaches this fruitful line of inquiry in the true spirit of an investigator, will derive much instruction from Prof. Weismann's remarks on "Phyletic Parallelism in Metamorphic Species." The affinities of the larger groups among Lepidoptera would most probably be made out once and for ever if systematists would devote more time to observation in this field, and to the co-ordination and working up of the numerous data scattered throughout the vast number of entomological publications.

The doctrine of development by no means implies, as has sometimes been maintained, a continuous *advancement* in organization. Although the scale of organic nature has continued to rise as a whole, cases may occasionally occur where a *lower grade* of organization is better adapted to certain conditions of life. This principle of "degeneration" was recognized by Darwin as early as in the first edition of the "Origin of Species;" it was soon perceived to be applicable to the phenomenon of parasitism, and was first definitely formulated by Dr. Anton Dohrn in 1875. In a lecture delivered before the British Association at Sheffield in 1879, Prof. E. Ray Lankester ascribed to "degeneration" a distinct and well-defined function in the theory of descent. Dr. Weismann's explanation of the transformation of Axolotl given in the fourth essay of this work, may be regarded as a special contributiom to this phase of Darwinism. Whilst refuting the idea held by certain naturalists, that such cases are arguments against the origin of species by the accumulation of minute variations, and prove the possibility of development *per saltum*, the theory here advanced (that *Siredon* at a former period existed at a higher stage of development as *Amblystoma*, and that the observed cases of metamorphosis are but reversions to this lost higher stage) suggests the question whether there may not still be in existence many other degenerated forms quite unsuspected by naturalists.

Many of the opponents of Evolution have from time to time denounced this doctrine as leading to "pure materialism," a denunciation which may appear somewhat alarming to the uninitiated, but which may not seem fraught with any serious consequences to those who have followed the course of philosophical speculation during the last few years. Those who attack the doctrine on this ground will however do well to consider Prof. Weismann's views set forth in the last essay in this volume, before hastily assuming that the much dreaded "materialism" is incompatible with any other conception of Nature.

The small amount of leisure time which I have been able to devote to the translation of this volume has delayed its completion considerably beyond the anticipated time, and it was with a view to meeting this difficulty that I departed from the original form of the German edition and issued it in parts. Owing to the extremely idiomatic character of the German text, I have throughout endeavoured to preserve only the author's meaning, regardless of literal translation or of the construction of the original. In some few cases, however, I have intentionally adopted literal translations of certain technical expressions which might, I think, be advantageously introduced into our biological vocabularies. Some alterations have been made in the original text by the author for the present edition, and many new notes have been added. For those bearing my initials I am alone responsible.

It gives me much pleasure in conclusion to express my thanks to Dr. Weismann, not only for the readily given permission to publish an English translation of his work, but also for much valuable assistance during the execution of the task. The author has been good enough to superintend the drawing of the plates for this edition, and he has also read through the greater part of the manuscript. From Mr. Darwin also I have received much kindly encouragement, and among entomologists I am especially indebted to Mr. W. H. Edwards of West Virginia, for his valuable additions to the first part. To my friends Mr. A. G. Butler, Mr. Roland Trimen, and Mr. F. Moore, I owe acknowledgments for much useful information concerning the caterpillars of exotic *Sphingidæ*, which I have incorporated in the notes and appendices, and Mr. W. S. Simpson has given me occasional advice in the translation of some of the more difficult passages.

R. M.

London, November, 1881.

PREFACE TO THE ENGLISH EDITION.

WITH the appearance of Charles Darwin's work "On the Origin of Species," in the year 1858, there commenced a new era in biology. Weary of the philosophical speculations which, at the beginning of this century, had at first been started with moderation but had afterwards been pushed to excess, biologists had entirely let drop all general questions and confined themselves to special investigations. The consideration even of general questions had quite fallen into disuse, and the investigation of mere details had led to a state of intellectual shortsightedness, interest being shown only for that which was immediately in view. Immense numbers of detailed facts were thus accumulated, but they could not possibly be mastered; the intellectual bond which should have bound them together was wanting.

But all this was changed in a short time. At first only single and mostly the younger naturalists fell in with the new theory of development proclaimed by Darwin, but the conviction soon became general that this was the only scientifically justifiable hypothesis of the origin of the organic world.

The materials accumulated in all the provinces of biology now for the first time acquired a deeper meaning and significance; unexpected inter-relations revealed themselves as though spontaneously, and what formerly appeared as unanswerable enigmas now became clear and comprehensible. Since that time what a vast modification has the subject of animal embryology undergone; how full of meaning appear the youngest developmental stages, how important the larvæ; how significant are rudimentary organs; what department of biology has not in some measure become affected by the modifying influence of the new ideas!

But the doctrine of development not only enabled us to understand the facts already existing; it gave at the same time an impetus to the acquisition of unforeseen new ones. If at the present day we glance back at the development of the biological sciences within the last twenty years, we must be astonished both at the enormous array of new facts which have been evoked by the theory of development, and by the immense series of special investigations which have been called forth by this doctrine.

But while the development theory for by far the greater majority of these investigations served as a light which more and more illuminated the darkness of ignorance, there appeared at the same time some other researches in which this doctrine itself became the object of investigation, and which were undertaken with a view to establish it more securely.

To this latter class of work belong the "Studies" in the present volume.

It will perhaps be objected that the theory of descent has already been sufficiently established by Darwin and Wallace. It is true that their newly-discovered principle of selection is of the very greatest importance, since it solves the riddle as to how that which is useful can arise in a purely mechanical way. Nor can the transforming influence of direct action, as upheld by Lamarck, be called in question, although its extent cannot as yet be estimated with any certainty. The *secondary* modifications which Darwin regards as the consequence of a change in some other organ must also be conceded. But are these three factors actually competent to explain the complete transformation of one species into another? Can they transform more than mere single characters or groups of characters? Can we consider them as the sole causes of the regular phenomena of the development of the races of animals and plants? Is there not perhaps an unknown force underlying these numberless developmental series as the true motor power—a "developmental force" urging species to vary in certain directions and thus calling into existence the chief types and subtypes of the animal and vegetable kingdoms ?

At the time these "Studies" first appeared (1875) they had been preceded by a whole series of attempts to introduce into science such an unknown power. The botanists, Nägeli and Askenasy, had designated it the "perfecting principle" or the "fixed direction of variation;" Kolliker as the "law of creation;" the philosophers, Von Hartmann and Huber, as the "law of organic development," and also "the universal principle of organic nature."

It was thus not entirely superfluous to test the capabilities of the known factors of transformation. We had here before us a question of the highest importance—a question which entered deeply into all our general notions, not only of the organic world, but of the universe as a whole.

This question—does there exist a special "developmental force"?—obviously cannot be decided by mere speculation; it must also be attempted to approach it by the inductive method.

The five essays in this volume are attempts to arrive, from various sides, somewhat nearer at a solution of the problem indicated.

The first essay on the "Seasonal Dimorphism of Butterflies" is certainly but indirectly connected with the question; it is therein attempted to discover the causes of this remarkable dimorphism, and by this means to indicate at the same time the extent of one of the transforming factors with reference to a definite case. The experiments upon which I base my views are not as numerous as I could desire, and if I were now able to repeat them they would be carried out more exactly than was possible at that time, when an experimental basis had first to be established. In spite of this, the conclusions to which I was led appear to be on the whole correct. That admirable and most conscientious observer of the North American butterflies, Mr. W. H. Edwards, has for many years experimented with American species in a manner similar to that which I employed for European species, and his results, which are published here in Appendix II. to the first essay, contain nothing as far as I can see which is not in harmony with my views. Many new questions suggest themselves, however, and it would be a grateful task if some entomologist would go further into these investigations.

The second essay directly attacks the main problem above indicated. It treats of the "Origin of the Markings of Caterpillars," and is to some extent a test of the correctness and capabilities of the Darwinian principles; it attempts to trace the differences in form in a definite although small group entirely to known factors.

Why the markings of caterpillars have particularly been chosen for this purpose will appear for two reasons.

The action of Natural Selection, on account of the nature of this agency, can only be exerted on those characters which are of biological importance. As it was to be tested whether, besides Natural Selection and the direct action of external conditions, together with the correlative results of these two factors, there might not lie concealed in the organism some other unknown transforming power, it was desirable to select for the investigation a group of forms which, if not absolutely excluding, nevertheless appeared possibly to restrict, the action of one of the two known factors of transformation, that of Natural Selection; a group of forms consisting essentially of so-called "purely morphological" characters, and not of those the utility of which was obvious, and of which the origin by means of Natural Selection was both possible and probable *ab initio*. Now, although the *colouring* can readily be seen to be of value to the life of its possessors, this is not the case with the quite independent *markings* of caterpillars; excepting perhaps those occasional forms of marking which have been regarded as special cases of protective resemblance. The markings of caterpillars must in general be considered as "purely morphological" characters, *i. e.* as characters which we do not know to be of any importance to the life of the species, and which cannot therefore be referred to Natural Selection. The most plausible explanation of these markings might have been that they were to be regarded as ornaments, but this view precludes the possibility of referring them either to Natural Selection or to the influence of direct changes in the environment.

The markings of caterpillars offered also another advantage which cannot be lightly estimated; they precluded from the first any attempt at an explanation by means of Sexual Selection. Although I am strongly convinced of the activity and great importance of this last process of selection, its effects cannot be estimated in any particular case, and the origin of a cycle of forms could never be clearly traced to its various factors, if Sexual Selection had also to be taken into consideration. Thus, we may fairly suppose that many features in the markings of butterflies owe their origin to Sexual Selection, but we are, at least at present, quite in the dark as to how many and which of these characters can be traced to this factor.

An investigation such as that which has been kept in view in this second essay would have been impracticable in the case of butterflies, as well as in the analogous case of the colouring and marking of birds, because it would have always been doubtful whether a character which did not appear to be attributable to any of the other transforming factors, should not be referred to Sexual Selection. It would have been impossible either to exclude or to infer an unknown developmental force, since we should have had to deal with two unknowns which could in no way be kept separate.

We escape this dilemma in the markings of caterpillars, because the latter do not propagate in this state. If the phenomena are not here entirely referable to Natural Selection and the direct action of the environment—if there remains an inexplicable residue, this cannot be referred to Sexual Selection, but to some as yet unknown power.

But it is not only in this respect that caterpillars offer especial advantages. If it is to be attempted to trace transformations in form to the action of the environment, an exact knowledge of this environment is in the first place necessary, *i. e.* a precise acquaintance with the conditions of life under the influence of which the species concerned exist. With respect to caterpillars, our knowledge of the life conditions is certainly by no means as complete as might be supposed, when we consider that hundreds of Lepidopterists have constantly bred and observed them during a most extended period. Much may have been observed, but it has not been thought worthy of publication; much has also been published, but so scattered and disconnected and at the same time of such unequal credibility, that a lifetime would be required to sift and collect it. A comprehensive biology of caterpillars, based on a broad ground, is as yet wanting, although such a labour would be both most interesting and valuable. Nevertheless, we know considerably more of the life of caterpillars than of any other larvæ, and as we are also acquainted with an immense number of species and are able to compare their life and the phenomena of their development, the subject of the markings of caterpillars must from this side also appear as the most favourable for the problem set before us.

To this must be added as a last, though not as the least, valuable circumstance, that we have here preserved to us in the development of the individual a fragment of the history of the species, so that we thus have at hand a means of following the course which the characters to be traced to their causes—the forms of marking—have taken during the lapse of thousands of years.

If with reference to the question as to the precise conditions of life in caterpillars I was frequently driven to my own observations, it was because I found as good as no previous work bearing upon this subject. It was well known generally that many caterpillars were differently marked and coloured when young to what they were when old; in some very striking cases brief notices of this fact are to be found in the works,[1]

[1] A most minute and exact description of the newly hatched larva of *Chionobas Aëllo* is given by the American entomologist, Samuel H. Scudder. Ann. Soc. Ent. de Belgique, xvi., 1873.

more especially, of the older writers, and principally in that of the excellent observer Rösel von Rosenhof, the Nuremberg naturalist and miniature painter. In no single case, however, do the available materials suffice when we have to draw conclusions respecting the phyletic development. We distinctly see here how doubtful is the value of those observations which are made, so to speak, at random, *i. e.* without some definite object in view. Many of these observations may be both good and correct, but they are frequently wanting precisely in that which would make them available for scientific purposes. Thus everything had to be established *de novo*, and for this reason the investigations were extended over a considerable number of years, and had to be restricted to a small and as sharply defined a group as possible—a group which was easily surveyed, viz. that of the Hawk-moths or Sphinges.

Since the appearance of the German edition of this work many new observations respecting the markings of caterpillars have been published, such, for example, as those of W. H. Edwards and Fritz Müller. I have, however, made but little use of them here, as I had no intention of giving anything like a *complete* ontogeny of the markings in all caterpillars: larval markings were with me but means to an end, and I wished only to bring together such a number of facts as were necessary for drawing certain general conclusions. It would indeed be most interesting to extend such observations to other groups of Lepidoptera.

The third essay also, for similar reasons, is based essentially upon the same materials, viz. the Lepidoptera. It is therein attempted to approach the general problem—does there or does there not exist an internal transforming force?—from a quite different and, I may say, opposite point of view. The form-relationships of Lepidoptera in their two chief stages of development, imago and larva, are therein analysed, and by an examination of the respective forms it has been attempted to discover the nature of the causes which have led thereto.

I may be permitted to say that the fact here disclosed of a *different morphological*, with the *same genealogical* relationship, appears to me to be of decided importance. The agreement of the conclusions following therefrom with the results of the former investigation has, at least in my own mind, removed the last doubts as to the correctness of the latter.

The fourth and shortest essay on the "Transformation of the Axolotl into Amblystoma," starts primarily with the intention of showing that cases of sudden transformation are no proof of *per saltum* development. When this essay first appeared the view was still widely entertained that we had here a case proving *per saltum* development. That this explanation was erroneous is now generally admitted, but I believe that those who suppose that we have here to deal with some quite ordinary phenomenon which requires no explanation, now go too far towards the other extreme. The term "larval reproduction" is an *expression*, but no *explanation*; we have therefore to attempt to find out the true interpretation, but whether the one which I have given is correct must be judged of by others.

These four essays lead up to a fifth and concluding one "On the Mechanical Conception of Nature." Whilst the results obtained are here summed up, it is attempted to form them into a philosophical conception of Nature and of the Universe. It will be thought by many that this should have been left to professed philosophers, and I readily admit that I made this attempt with some misgiving. Two considerations, however, induced me to express here my own views. The first was that the facts of science are frequently misunderstood, or at any rate not estimated at their true value, by philosophers;[2] the second consideration was, that even certain naturalists and certainly very many non-naturalists, turn distrustfully from the results of science, because

[2] I am aware that this certainly cannot be said of philosophers like Lotze or Herbert Spencer; but these are at the same time both naturalists and philosophers.

they fear that these would infallibly lead to a view of the Universe which is to them unacceptable, viz. the materialistic view. With regard to the former I wished to show that the views of the development of organic Nature inaugurated by Darwin and defended in this work are certainly correctly designated *mechanical;* with reference to the latter I wished to prove that such a mechanical conception of the organic world and of Nature in general, by no means leads merely to one single philosophical conception of Nature, viz. to Materialism, but that on the contrary it rather admits of legitimate development in a quite different manner.

Thus in these last four essays much that appears heterogeneous will be found in close association, viz. scientific details and general philosophical ideas. In truth, however, these are most intimately connected, and the one cannot dispense with the other. As the detailed investigations of the three essays find their highest value in the general considerations of the fourth, and were indeed only possible by constantly keeping this end in view, so the general conclusions could only grow out of the results of the special investigations as out of a solid foundation. Had the new materials here brought together been already known, the reader would certainly have been spared the trouble of going into the details of special scientific research. But as matters stood it was indispensable that the facts should be examined into and established even down to the most trifling details. The essay "On the Origin of the Markings of Caterpillars" especially, had obviously to commence with the sifting and compilation of extensive morphological materials.

AUGUST WEISMANN.

Freiburg in Baden,
November, 1881.

CONTENTS.

Part I.

ON THE SEASONAL DIMORPHISM OF BUTTERFLIES.

I.

II.

III.

IV.

V.

VI.

Part II.

ON THE FINAL CAUSES OF TRANSFORMATION

I.

THE ORIGIN OF THE MARKINGS OF CATERPILLARS.

IV.

Biological Value of Colour, p. 289.

V.

Biological Value of special Markings, p 308.

VI.

Objections to a Phyletic Vital Force, p. 352.

VII.

Phyletic Development of the Markings of the Sphingidæ. Summary and Conclusion, p. 370.

II.

ON PHYLETIC PARALLELISM IN METAMORPHIC SPECIES.

Introduction, p. 390.

I.

Larva and Imago vary in Structure independently of each other, p. 401.

II.

Does the Form-relationship of the Larva coincide with that of the Imago? p. 432.

Part III.

ON THE FINAL CAUSES OF TRANSFORMATION (*continued*).

III.

THE TRANSFORMATION OF THE MEXICAN AXOLOTL INTO AMBLYSTOMA.

IV.

ON THE MECHANICAL CONCEPTION OF NATURE.

STUDIES IN THE THEORY OF DESCENT.

Part I.

ON THE SEASONAL DIMORPHISM OF BUTTERFLIES.

I.

THE ORIGIN AND SIGNIFICANCE OF SEASONAL DIMORPHISM.

THE phenomena here about to be subjected to a closer investigation have been known for a long period of time. About the year 1830 it was shown that the two forms of a butterfly (*Araschnia*) which had till that time been regarded as distinct, in spite of their different colouring and marking really belonged to the same species, the two forms of this dimorphic species not appearing simultaneously but at different seasons of the year, the one in early spring, the other in summer. To this phenomenon the term "seasonal dimorphism" was subsequently applied by Mr. A. R. Wallace, an expression of which the heterogeneous composition may arouse the horror of the philologist, but, as it is as concise and intelligible as possible, I propose to retain it in the present work.

The species of *Araschnia* through which the discovery of seasonal dimorphism was made, formerly bore the two specific names *A. Levana* and *A. Prorsa.* The latter is the summer and the former the winter form, the difference between the two being, to the uninitiated, so great that it is difficult to believe in their relationship. *A. Levana* (Figs. 1 and 2, Plate I.) is of a golden brown colour with black spots and dashes, while *A. Prorsa* (Figs. 5 and 6, Plate I.) is deep black with a broad white interrupted band across both wings. Notwithstanding this difference, it is an undoubted fact that both forms are merely the winter and summer generations of the same species. I have myself frequently bred the variety *Prorsa* from the eggs of *Levana,* and *vice versâ.*

Since the discovery of this last fact a considerable number of similar cases have been established. Thus P. C. Zeller[1] showed, by experiments made under confinement, that two butterflies belonging to the family of the 'Blues,' differing greatly in colour and marking, and especially in size, which had

[1] "Über die Artrechte des *Polyommatus Amyntas* und *Polysperchon.*" Stett. ent. Zeit. 1849. Vol. x. p. 177—182. [In Kirby's "Synonymic Catalogue of Diurnal Lepidoptera" *Plebeius Amyntas* is given as a synonym and *P. Polysperchon* as a var. of *P. Argiades* Pall. R.M.]

formerly been distinguished as *Plebeius* (*Lycæna*) *Polysperchon* and *P. Amyntas*, were merely winter and summer generations of the same species; and that excellent Lepidopterist, Dr. Staudinger, proved the same[2] with species belonging to the family of the 'Whites,' *Euchloe Belia* Esp. and *E. Ausonia* Hüb., which are found in the Mediterranean countries.

The instances are not numerous, however, in which the difference between the winter and summer forms of a species is so great as to cause them to be treated of in systematic work as distinct species. I know of only five of these cases. Lesser differences, having the systematic value of varieties, occur much more frequently. Thus, for instance, seasonal dimorphism has been proved to exist among many of our commonest butterflies belonging to the family of the 'Whites,' but the difference in their colour and marking can only be detected after some attention; while with other species, as for instance with the commonest of our small 'Blues,' *Plebeius Alexis* (= *Icarus*, Rott.), the difference is so slight that even the initiated must examine closely in order to recognize it. Indeed whole series of species might easily be grouped so as to show the transition from complete similarity of both generations, through scarcely

[2] "Die Arten der Lepidopteren-Gattung *Ino* Leach, nebst einigen Vorbemerkungen über Localvarietäten." Stett. ent. Zeit. 1862. Vol. xxiii. p. 342.

perceptible differences, to divergence to the extent of varieties, and finally to that of species.

Nor are the instances of lesser differences between the two generations very numerous. Among the European diurnal Lepidoptera I know of about twelve cases, although closer observation in this direction may possibly lead to further discoveries.[3] Seasonal dimorphism occurs also in moths, although I am not in a position to make a more precise statement on this subject,[4] as my own observations refer only to butterflies.

That other orders of insects do not present the same phenomenon depends essentially upon the fact that most of them produce only one generation in the year; but amongst the remaining orders there occur indeed changes of form which, although

[3] [Eng. ed. W. H. Edwards has since pointed out several beautiful cases of seasonal dimorphism in America. Thus *Plebeius Pseudargiolus* is the summer form of *P. Violacea*, and *Phyciodes Tharos* the summer form of *P. Marcia*. See Edwards' "Butterflies of North America," 1868-79.]

[4] [Eng. ed. I learn by a written communication from Dr. Speyer that two Geometræ, *Selenia Tetralunaria* and *S. Illunaria* Hüb., are seasonally dimorphic. In both species the winter form is much larger and darker.] [*Selenia Lunaria*, *S. Illustraria*, and some species of *Ephyra* (*E. Punctaria* and *E. Omicronaria*) are likewise seasonally dimorphic. For remarks on the case of *S. Illustraria* see Dr. Knaggs in Ent. Mo. Mag., vol. iii. p. 238, and p. 256. Some observations on *E. Punctaria* were communicated to the Entomological Society of London by Professor Westwood in 1877, on the authority of Mr. B. G. Cole. See Proc. Ent. Soc. 1877, pp. vi, vii. R.M.]

not capable of being regarded as pure seasonal dimorphism, may well have been produced in part by the same causes, as the subsequent investigation on the relation of seasonal dimorphism to alternation of generation and heterogenesis will more fully prove.

Now what are these causes?

Some years ago, when I imparted to a lepidopterist my intention of investigating the origin of this enigmatical dimorphism, in the hope of profiting for my inquiry from his large experience, I received the half-provoking reply: "But there is nothing to investigate: it is simply the specific character of this insect to appear in two forms; these two forms alternate with each other in regular succession according to a fixed law of Nature, and with this we must be satisfied." From his point of view the position was right; according to the old doctrine of species no question ought to be asked as to the causes of such phenomena in particular. I would not, however, allow myself to be thus discouraged, but undertook a series of investigations, the results of which I here submit to the reader.

The first conjecture was, that the differences in the imago might perhaps be of a secondary nature, and have their origin in the differences of the caterpillar, especially with those species which grow up during the spring or autumn and feed on different plants, thus assimilating different chemical substances, which might induce different deposits of colour in the wings of the perfect insect. This latter hypothesis was readily confuted by the fact, that the most strongly marked of the dimorphic species, *A. Levana*, fed exclusively on *Urtica major*. The caterpillar of this species certainly exhibits a well defined dimorphism, but it is not seasonal dimorphism : the two forms do not alternate with each other, but appear mixed in every brood.

I have repeatedly reared the rarer golden-brown variety of the caterpillar separately, but precisely the same forms of butterfly were developed as from black caterpillars bred at the same time under similar external conditions. The same experiment was performed, with a similar result, in the last century by Rösel, the celebrated miniature painter and observer of nature, and author of the well-known "Insect Diversions"—a work in use up to the present day.

The question next arises, as to whether the causes originating the phenomena are not the same as those to which we ascribe the change of winter and summer covering in so many mammalia and birds—whether the change of colour and marking does not depend, in this as in the other cases, upon the *indirect action* of external conditions of life, i. e., on adaptation through natural selection. We are certainly correct in ascribing white coloration to adaptation [5]—as with the ptarmigan, which is

[5] [In 1860 Andrew Murray directed attention to the dis-

white in winter and of a grey-brown in summer, both colours of the species being evidently of important use.

It might be imagined that analogous phenomena occur in butterflies, with the difference that the change of colour, instead of taking place in the same brood, alternates in different broods.[6] The nature of the difference which occurs in seasonal dimorphism, however, decidedly excludes this view; and moreover, the environment of butterflies presents such similar features, whether they emerge in spring or in summer, that all notions that we may be dealing with adaptational colours must be entirely abandoned.

I have elsewhere[7] endeavoured to show that butterflies in general are not coloured protectively during flight, for the double reason that the colour

guising colours of species which, like the Alpine hare, stoat, and ptarmigan, undergo seasonal variation of colour. See a paper "On the Disguises of Nature, being an inquiry into the laws which regulate external form and colour in plants and animals." Edinb. New Phil. Journ., Jan. 1860. In 1873 I attempted to show that these and other cases of "variable protective colouring" could be fairly attributed to natural selection. See Proc. Zoo. Soc., Feb. 4th, 1873, pp. 153—162. R.M.]

[6] [A phenomenon somewhat analogous to seasonal change of protecting colour does occur in some Lepidoptera, only the change, instead of occurring in the same individual, is displayed by the successive individuals of the same brood. See Dr. Wallace on *Bombyx Cynthia*, Trans. Ent. Soc. Vol. v. p. 485. R.M.]

[7] "Über den Einfluss der Isolirung auf die Artbildung." Leipzig, 1872, pp. 55—62.

of the background to which they are exposed continually changes, and because, even with the best adaptation to the background, the fluttering motion of the wings would betray them to the eyes of their enemies.[8] I attempted also to prove at the same time that the diurnal Lepidoptera of our temperate zone have few enemies which pursue them when on the wing, but that they are subject to many attacks during their period of repose.

In support of this last statement I may here adduce an instance. In the summer of 1869 I placed about seventy specimens of *Araschnia Prorsa* in a spacious case, plentifully supplied with flowers. Although the insects found themselves quite at home, and settled about the flowers in very fine weather (one pair copulated, and the female laid eggs), yet I found some dead and mangled every morning. This decimation con-

[8] [Mr. A. R. Wallace maintains that the obscurely coloured females of those butterflies which possess brightly coloured males have been rendered inconspicuous by natural selection, owing to the greater need of protection by the former sex. See "Contributions to the Theory of Natural Selection," London, 1870, pp. 112—114. It is now generally admitted that the underside of butterflies has undergone protectional adaptation; and many cases of local variation in the colour of the underside of the wings, in accordance with the nature of the soil, &c., are known. See, for instance, Mr. D. G. Rutherford on the colour-varieties of *Aterica Meleagris* (Proc. Ent. Soc. 1878, p. xlii.), and Mr. J. Jenner Weir on a similar phenomenon in *Hipparchia Semele* (*loc. cit.* p. xlix.) R.M.]

tinued—many disappearing entirely without my being able to find their remains—until after the ninth day, when they had all, with one exception, been slain by their nocturnal foes—probably spiders and *Opilionidæ*.

Diurnal Lepidoptera in a position of rest are especially exposed to hostile attacks. In this position, as is well known, their wings are closed upright, and it is evident that the adaptational colours on the under side are displayed, as is most clearly shown by many of our native species.[9]

Now, the differences in the most pronounced cases of seasonal dimorphism—for example, in *Araschnia Levana*—are much less manifest on the *under* than on the *upper* side of the wing. The explanation by adaptation is therefore untenable; but I will not here pause to confute this view more completely, as I believe I shall be able to show the true cause of the phenomenon.

If seasonal dimorphism does not arise from the *indirect* influence of varying seasons of the year, it may result from the *direct* influence of the varying external conditions of life, which are, without doubt, different in the winter from those of the summer brood.

There are two prominent factors from which such an influence may be expected—temperature

[9] [The fact that moths which, like the Geometræ, rest by day with the wings spread out, are protectively marked on the *upper* side, fully corroborates this statement. R.M.]

and duration of development, i. e., duration of the chrysalis period. The duration of the larval period need not engage our attention, as it is only very little shorter in the winter brood—at least, it was so with the species employed in the experiments.

Starting from these two points of view, I carried on experiments for a number of years, in order to find out whether the dual form of the species in question could be traced back to the direct action of the influences mentioned.

The first experiments were made with *Araschnia Levana*. From the eggs of the winter generation, which had emerged as butterflies in April, I bred caterpillars, and immediately after pupation placed them in a refrigerator, the temperature of the air of which was 8°—10° R. It appeared, however, that the development could not thus be retarded to any desired period by such a small diminution of temperature, for, when the box was taken out of the refrigerator after thirty-four days, all the butterflies, about forty in number, had emerged, many being dead, and others still living. The experiment was so far successful that, instead of the *Prorsa* form which might have been expected under ordinary circumstances, most of the butterflies emerged as the so-called *Porima* (Figs. 3, 4, 7, 8, and 9, Plate I.); that is to say, in a form intermediate between *Prorsa* and *Levana* sometimes found in nature, and possessing more or less the marking of the former, but mixed with much of the yellow of *Levana*.

It should be here mentioned, that similar experiments were made in 1864 by George Dorfmeister, but unfortunately I did not get this information[10] until my own were nearly completed. In these well-conceived, but rather too complicated experiments, the author arrives at the conclusion "that temperature certainly affects the colouring, and through it the marking, of the future butterfly, and chiefly so during pupation." By lowering the temperature of the air during a portion of the pupal period, the author was enabled to produce single specimens of *Porima*, but most of the butterflies retained the *Prorsa* form. Dorfmeister employed a temperature a little higher than I did in my first experiments, viz. 10°—11° R., and did not leave the pupæ long exposed, but after 5½—8 days removed them to a higher temperature. It was therefore evident that he produced transition forms in a few instances only, and that he never succeeded in bringing about a complete transformation of the summer into the winter form.

In my subsequent experiments I always exposed the pupæ to a temperature of 0°—1° R.; they were placed directly in the refrigerator, and

[10] "Über die Einwirkung verschiedener, während der Entwicklungsperioden angewendeter Wärmegrade auf die Färbung und Zeichnung der Schmetterlinge." A communication to the Society of Natural Science of Steiermark, 1864.

taken out at the end of four weeks. I started with the idea that it was perhaps not so much the reduced temperature as the retardation of development which led to the transformation. But the first experiment had shown that the butterflies emerged between 8° and 10° R., and consequently that the development could not be retarded at this temperature.

A very different result was obtained from the experiment made at a lower temperature.[11] Of twenty butterflies, fifteen had become transformed into *Porima*, and of these three appeared very similar to the winter form (*Levana*), differing only in the absence of the narrow blue marginal line, which is seldom absent in the true *Levana*. Five butterflies were uninfluenced by the cold, and remained unchanged, emerging as the ordinary summer form (*Prorsa*). It thus appeared from this experiment, that a large proportion of the butterflies inclined to the *Levana* form by exposure to a temperature of 0°—1° R. for four weeks, while in a few specimens the transformation into this form was nearly perfect.

Should it not be possible to perfect the transformation, so that each individual should take the *Levana* form? If the assumption of the *Prorsa* or *Levana* form depends only on the direct influence of temperature, or on the duration of the period of

[11] See Exp. 9, Appendix I.

development, it should be possible to compel the pupæ to take one or the other form at pleasure, by the application of the necessary external conditions. This has never been accomplished with *Araschnia Prorsa*. As in the experiment already described, and in all subsequent ones, single specimens appeared as the unchanged summer form, others showed an appearance of transition, and but very few had changed so completely as to be possibly taken for the pure *Levana*. In some species of the sub-family *Pierinæ*, however, at least in the case of the summer brood, there was, on the contrary, a complete transformation.

Most of the species of our 'Whites' (*Pierinæ*) exhibit the phenomenon of seasonal dimorphism, the winter and summer forms being remarkably distinct. In *Pieris Napi* (with which species I chiefly experimented) the winter form (Figs. 10 and 11, Plate I.) has a sprinkling of deep black scales at the base of the wings on the upper side, while the tips are more grey, and have in all cases much less black than in the summer form ; on the underside the difference lies mainly in the frequent breadth, and dark greenish-black dusting, of the veins of the hind wings in the winter form, while in the summer form these greenish-black veins are but faintly present.

I placed numerous specimens of the summer brood, immediately after their transformation into chrysalides, in the refrigerator (0°—1° R.), where

I left them for three months, transferring them to a hothouse on September 11th, and there (from September 26th to October 3rd) sixty butterflies emerged, the whole of which, without exception—and most of them in an unusually strong degree—bore the characters of the winter form. I, at least, have never observed in the natural state such a strong yellow on the underside of the hind wings, and such a deep blackish-green veining, as prevailed in these specimens (see, for instance, Figs. 10 and 11). The temperature of the hothouse (12°—24° R.) did not, however, cause the emergence of the whole of the pupæ; a portion hibernated, and produced in the following spring butterflies of the winter form only. I thus succeeded, with this species of *Pieris*, in completely changing every individual of the summer generation into the winter form.

It might be expected that the same result could be more readily obtained with *A. Levana*, and fresh experiments were undertaken, in order that the pupæ might remain in the refrigerator fully two months from the period of their transformation (9-10th July). But the result obtained was the same as before—fifty-seven butterflies emerged in the hothouse[12] from September 19th to October 4th, nearly all of these approaching very near to the winter form, without a single specimen pre-

[12] See Exp. 11, Appendix I.

senting the appearance of a perfect *Levana*, while three were of the pure summer form (*Prorsa*).

Thus with *Levana* it was not possible, by refrigeration and retardation of development, to change the summer completely into the winter form in all specimens. It may, of course, be objected that the period of refrigeration had been too short, and that, instead of leaving the pupæ in the refrigerator for two months, they should have remained there six months, that is, about as long as the winter brood remains under natural conditions in the chrysalis state. The force of this last objection must be recognized, notwithstanding the improbability that the desired effect would be produced by a longer period of cold, since the doubling of this period from four to eight weeks did not produce[13] any decided increase in the strength of the transformation. I should not have omitted to repeat the experiment in this modified form, but unfortunately, in spite of all trouble, I was unable to collect during the summer of 1873 a sufficient number of caterpillars. But the omission thus caused is of quite minor importance from a theoretical point of view.

For let us assume that the omitted experiment had been performed—that pupæ of the summer brood were retarded in their development by cold until the following spring, and that every specimen

[13] See Exps. 4, 9, and 11, Appendix I.,

then emerged in the perfect winter form, *Levana*. Such a result, taken in connexion with the corresponding experiment upon *Pieris Napi*, would warrant the conclusion that the direct action of a certain amount of cold (or of retardation of development) is able to compel all pupæ, from whichever generation derived, to assume the winter form of the species. From this the converse would necessarily follow, viz. that a certain amount of warmth would lead to the production of the summer form, *Prorsa*, it being immaterial from which brood the pupæ thus exposed to warmth might be derived. But the latter conclusion was proved experimentally to be incorrect, and thus the former falls with it, whether the imagined experiment with *Prorsa* had succeeded or not.

I have repeatedly attempted by the application of warmth to change the winter into the summer form, but always with the same negative result. *It is not possible to compel the winter brood to assume the form of the summer generation.*

A. Levana may produce not only two but three broods in the year, and may, therefore, be said to be *polygoneutic*.[14] One winter brood alternates with two summer broods, the first of which appears in July, and the second in August. The latter

[14] It seems to me very necessary to have a word expressing whether a species produces one, two, or more generations in the year, and I have therefore coined the expression *mono-*, *di-*, and *polygoneutic* from γονεύω, I produce.

furnishes a fourth generation of pupæ, which, after hibernation, emerge in April, as the first brood of butterflies in the form *Levana*.

I frequently placed pupæ of this fourth brood in the hothouse immediately after their transformation, and in some cases even during the caterpillar stage, the temperature never falling, even at night, below 12° R., and often rising during the day to 24° R. The result was always the same: all, or nearly all, the pupæ hibernated, and emerged the following year in the winter form as perfectly pure *Levana*, without any trace of transition to the *Prorsa* form. On one occasion only was there a *Porima* among them, a case for which an explanation will, I believe, be found later on. It often happened, on the other hand, that some few of the butterflies emerged in the autumn, about fourteen days after pupation; and these were always *Prorsa* (the summer form), excepting once a *Porima*.

From these experiments it appeared that similar causes (heat) affect different generations of *A. Levana* in different manners. With both summer broods a high temperature always caused the appearance of *Prorsa*, this form arising but seldom from the third brood (and then only in a few individuals), while the greater number retained the *Levana* form unchanged. We may assign as the reason for this behaviour, that the third brood has no further tendency to be accelerated in its development by the action of heat, but that by a longer duration of the pupal stage the *Levana* form must result. On one occasion the chrysalis stage was considerably shortened in this brood by the continued action of a high temperature, many specimens thus having their period of development reduced from six to three months. The supposed explanation above given is, however, in reality no explanation at all, but simply a restatement of the facts. The question still remains, why the third brood in particular has no tendency to be accelerated in its development by the action of heat, as is the case with both the previous broods?

The first answer that can be given to this question is, that the cause of the different action produced by a similar agency can only lie in the *constitution*, i. e., in the *physical nature* of the broods in question, and not in the external influences by which they are acted upon. Now, what is the difference in the physical nature of these respective broods? It is quite evident, as shown by the experiments already described, that cold and warmth cannot be the *immediate* causes of a pupa emerging in the *Prorsa* or *Levana* form, since the last brood always gives rise to the *Levana* form, whether acted on by cold or warmth. The first and second broods only can be made to partly assume, more or less completely, the *Levana* form by the application of cold. In these broods then, a low temperature is the *mediate* cause of the transformation into the *Levana* form.

The following is my explanation of the facts. The form *Levana* is the original type of the species, and *Prorsa* the secondary form arising from the gradual operation of summer climate. When we are able to change many specimens of the summer brood into the winter form by means of cold, this can only depend upon reversion to the original, or ancestral, form, which reversion appears to be most readily produced by cold, that is, by the same external influences as those to which the original form was exposed during a long period of time, and the continuance of which has preserved, in the winter generations, the colour and marking of the original form down to the present time.

I consider the origination of the *Prorsa* from the *Levana* form to have been somewhat as follows :— It is certain that during the diluvial period in Europe there was a so-called 'glacial epoch,' which may have spread a truly polar climate over our temperate zone ; or perhaps a lesser degree of cold may have prevailed with increased atmospheric precipitation. At all events, the summer was then short and comparatively cold, and the existing butterflies could have only produced one generation in the year ; in other words, they were *monogoneutic*. At that time *A. Levana* existed only in the *Levana* form.[15] As the climate gradually be-

[15] [Eng. ed. In the German edition, which appeared in 1874, I was not able to support this hypothesis by geographical data, and could then only ask the question "whether

came warmer, a period must have arrived when *Levana*, in the most northern portion of its area of distribution, appears in two or only in one generation?" This question is now answered by the Swedish Expedition to the Yenisei in 1876. Herr Philipp Trybom, one of the members of this expedition, observed *A. Levana* at the end of June and beginning of July, in the middle of Yenisei, in 60°—63° N. (Dagfjärilar från Yenisei in Översigt ap k. Vertensk. Akad. Förhandlingon, 1877, No. 6.) Trybom found *Levana* at Yenisk on June 23rd, at Worogova (61° 5′) on July 3rd, at Asinova (61° 25′) on July 4th, at Insarowa (62° 5′) on July 7th, and at Alinskaja (63° 25′) on July 9th. The butterflies were especially abundant at the beginning of June, and were all of the typical *Levana* form. Trybom expressly states, "we did not find a single specimen which differed perceptibly from Weismann's Figs. 1 and 2 ('Saison-Dimorphismus' Taf. I.)."

The Swedish expedition soon left the Yenisei, and consequently was not able to decide by observations whether a second generation possessing the *Prorsa* form appeared later in the summer. Nevertheless, it may be stated with great probability that this is not the case. The districts in which *Levana* occurs on the Yenisei have about the same isotherm as Archangel or Haparanda, and therefore the same summer temperature. Dr. Staudinger, whose views I solicited, writes to me:—"In Finnmark (about 67° N.) I observed no species with two generations; even *Polyommatus Phlæas*, which occurs there, and which in Germany has always two, and in the south, perhaps, three generations, in Finnmark has only one generation. A second generation would be impossible, and this would also be the case with *Levana* in the middle of Yenisei. I certainly have *Levana* and *Prorsa* from the middle of Amur, but *Levana* flies there at the end of May, and the summers are very warm." The middle of Amur lies, moreover, in 50° N. lat., and therefore 10°—13° south of the districts of the Yenisei mentioned.

It must thus be certainly admitted that on the Yenisei *A. Levana* occurs only in the *Levana* form, and that consequently this species is at the present time, in the northernmost portion the summer lasted long enough for the interpolation of a second brood. The pupæ of *Levana*, which had hitherto hibernated through the long winter to appear as butterflies in the following summer, were now able to appear on the wing as butterflies during the same summer as that in which they left their eggs as larvæ, and eggs deposited by the last brood produced larvæ which fed up and hibernated as pupæ. A state of things was thus established in which the first brood was developed under very different climatic conditions from the second. So considerable a difference in colour and marking between the two forms as we now witness could not have arisen suddenly, but must have done so gradually. It is evident from the foregoing ex periments that the *Prorsa* form did not originate suddenly. Had this been the case it would simply signify that every individual of this species possessed the faculty of assuming two different forms according as it was acted on by warmth or cold, just in the same manner as litmus-paper becomes red in acids and blue in alkalies. The experiments

of its area of distribution, in the same condition as that in which I conceive it to have been in mid Europe during the glacial period. It would be of the greatest interest to make experiments in breeding with this single-brooded *Levana* from the Yenisei, i. e., to attempt to change its offspring into the *Prorsa* form by the action of a high temperature. If this could not be accomplished it would furnish a confirmation of my hypothesis than which nothing more rigorous could be desired.]

have shown, however, that this is not the case, but rather that the last generation bears an ineradicable tendency to take the *Levana* form, and is not susceptible to the influence of warmth, however long continued; while both summer generations, on the contrary, show a decided tendency to assume the *Prorsa* form, although they certainly can be made to assume the *Levana* form in different degrees by the prolonged action of cold.

The conclusion seems to me inevitable, that the origination of the *Prorsa* form was gradual—that those changes which originated in the chemistry of the pupal stage, and led finally to the *Prorsa* type, occurred very gradually, at first perhaps remaining completely latent throughout a series of generations, then very slight changes of marking appearing, and finally, after a long period of time, the complete *Prorsa* type was produced. It appears to me that the quoted results of the experiments are not only easily explained on the view of the *gradual* action of climate, but that this view is the only one admissible. The action of climate is best comparable with the so-called cumulative effect of certain drugs on the human body; the first small dose produces scarcely any perceptible change, but if often repeated the effect becomes cumulative, and poisoning occurs.

This view of the action of climate is not at all new, most zoologists having thus represented it; only the formal proof of this action is new, and the facts investigated appear to me of special importance as furnishing this proof. I shall again return to this view in considering climatic varieties, and it will then appear that also the nature of the transformation itself confirms the slow operation of climate.

During the transition from the glacial period to the present climate *A. Levana* thus gradually changed from a monogoneutic to a digoneutic species, and at the same time became gradually more distinctly dimorphic, this character originating only through the alteration of the summer brood, the primary colouring and marking of the species being retained unchanged by the winter brood. As the summer became longer a third generation could be interpolated—the species became polygoneutic; and in this manner two summer generations alternated with one winter generation.

We have now to inquire whether facts are in complete accordance with this theory—whether they are never at variance with it—and whether they can all be explained by it. I will at once state in anticipation, that this is the case to the fullest extent.

In the first place, the theory readily explains why the summer but not the winter generations are capable of being transformed; the latter cannot possibly revert to the *Prorsa* form, because this is much the younger. When, however, it happens that out of a hundred cases there occurs one in which a chrysalis of the winter generation, having been forced by warmth, undergoes transformation before the commencement of winter, and emerges in the summer form,[16] this is not in the least inexplicable. It cannot be atavism which determines the direction of the development ; but we see from such a case that the changes in the first two generations have already produced a certain alteration in the third, which manifests itself in single cases under favourable conditions (the influence of warmth) by the assumption of the *Prorsa* form ; or, as it might be otherwise expressed, the *alternating* heredity (of which we shall speak further), which implies the power of assuming the *Prorsa* form, remains latent as a rule in the winter generation, but becomes *continuous* in single individuals.

It is true that we have as yet no kind of insight into the nature of heredity, and this at once shows the defectiveness of the foregoing explanation ; but we nevertheless know many of its external phenomena. We know for certain that one of these consists in the fact that peculiarities of the father do not appear in the son, but in the grandson, or still further on, and that they may be thus transmitted in a latent form. Let us imagine a character so transmitted that it appears in the first, third, and fifth generations, remaining latent

[16] See Exp. 10, Appendix I.

in the intermediate ones; it would not be improbable, according to previous experiences, that the peculiarity should exceptionally, i. e., from a cause unknown to us, appear in single individuals of the second or fourth generation. But this completely agrees with those cases in which "exceptional" individuals of the winter brood took the *Prorsa* form, with the difference only that a cause (warmth) was here apparent which occasioned the development of the latent characters, although we are not in a position to say in what manner heat produces this action. These exceptions to the rule are therefore no objection to the theory. On the contrary, they give us a hint that after one *Prorsa* generation had been produced, the gradual interpolation of a second *Prorsa* generation may have been facilitated by the existence of the first. I do not doubt that even in the natural state single individuals of *Prorsa* sometimes emerge in September or October; and if our summer were lengthened by only one or two months this might give rise to a third summer brood (just as a second is now an accomplished fact), under which circumstances they would not only emerge, but would also have time for copulation and for depositing eggs, the larvæ from which would have time to grow up.

A sharp distinction must be made between the first establishment of a new climatic form and the transference of the latter to newly interpolated generations. The former always takes place very slowly; the latter may occur in a shorter time.

With regard to the duration of time which is necessary to produce a new form by the influence of climate, or to transmit to a succeeding generation a new form already established, great differences occur, according to the physical nature of the species and of the individual. The experiments with *Prorsa* already described show how diverse are individual proclivities in this respect. In Experiment No. 12 it was not possible out of seventy individuals to substitute *Prorsa* for the *Levana* form, even in one solitary case, or, in other words, to change alternating into continuous inheritance; whilst in the corresponding experiments of former years (Experiment 10, for example), out of an equal number of pupæ three emerged as *Prorsa*, and one as *Porima*. We might be inclined to seek for the cause of this different behaviour in external influences, but we should not thus arrive at an explanation of the facts. We might suppose, for instance, that a great deal depended upon the particular period of the pupal stage at which the action of the elevated temperature began—whether on the first, the thirtieth, or the hundredth day after pupation—and this conjecture is correct in so far that in the two last cases warmth can have no further influence than that of somewhat accelerating the emergence of the butterflies, but cannot change the *Levana* into the *Prorsa* form. I have repeatedly exposed a large number of *Levana* pupæ of the third generation to the temperature of an apartment, or even still higher (26° R.), during winter, but no *Prorsa* were obtained.[17]

But it would be erroneous to assume a difference in the action of heat according as it began on the first or third day after transformation; whether during or before pupation. This is best proved by Experiment No. 12, in which caterpillars of the fourth generation were placed in the hothouse several days before they underwent pupation; still, not a single butterfly assumed the *Prorsa* form. I have also frequently made the reverse experiment, and exposed caterpillars of the first summer brood to cold during the act of pupation A regular consequence was the dying off of the caterpillars, which is little to be wondered at, as the sensitiveness of insects during ecdysis is well known, and transformation into the pupal state is attended by much deeper changes.

Dorfmeister thought that he might conclude from his experiments that temperature exerts the greatest influence in the first place during the act

[17] When Dorfmeister remarks that hibernating pupæ which, at an early stage "were taken for development into a room, or not exposed to any cold, gave dwarfed, weakly and crippled," or otherwise damaged butterflies, this is entirely attributable to the fact that this able entomologist had neglected to supply the necessary moisture to the warm air. By keeping pupæ over water I have always obtained very fine butterflies.

of pupation, and in the next place immediately after that period. His experiments were made, however, with such a small number of specimens that scarcely any safe conclusion can be founded on them ; still, this conclusion may be correct, in so far as everything depends on whether, from the beginning, the formative processes in the pupa tended to this or that direction, the final result of which is the *Prorsa* or *Levana* form. If once there is a tendency to one or the other direction, then temperature might exert an accelerating or a retarding influence, but the tendency cannot be further changed.

It is also possible—indeed, probable—that a period may be fixed in which warmth or cold might be able to divert the original direction of development most easily; and this is the next problem to be attacked, the answer to which, now that the main points have been determined, should not be very difficult. I have often contemplated taking the experiments in hand myself, but have abandoned them, because my materials did not appear to me sufficiently extensive, and in all such experiments nothing is to be more avoided than a frittering away of experimental materials by a too complicated form of problem.

There may indeed be a period most favourable for the action of temperature during the first days of the pupal stage; it appears from Experiment No. 12 that individuals tend in different degrees to respond to such influences, and that the disposition to abandon the ordinary course of development is different in different individuals. In no other way can it be explained that, in all the experiments made with the first and second generations of *Prorsa*, only *a portion* of the pupæ were compelled by cold to take the direction of development of *Levana*, and that even from the former only a few individuals completely reverted, the majority remaining intermediate.

If it be asked why in the corresponding experiments with *Pieris Napi* complete reversion always occurred without exception, it may be supposed that in this species the summer form has not been so long in existence, and that it would thus be more easily abandoned; or, that the difference between the two generations has not become so distinct, which further signifies that here again the summer form is of later origin. It might also be finally answered, that the tendency to reversion in different species may vary just as much as in different individuals of the same species. But, in any case, the fact is established that all individuals are impelled by cold to complete reversion, and that in these experiments it does not depend so particularly upon the moment of development when cold is applied, but that differences of individual constitution are much more the cause why cold brings some pupæ to complete, and others to partial, reversion, while yet others are quite uninfluenced. In reference to this, the American *Papilio Ajax* is particularly interesting.

This butterfly, which is somewhat similar to the European *P. Podalirius*, appears, wherever it occurs, in three varieties, designated as var. *Telamonides*, var. *Walshii*, and var. *Marcellus*. The distinguished American entomologist, W. H. Edwards, has proved by breeding experiments, that all three forms belong to the same cycle of development, and in such a manner that the first two appear only in spring, and always come only from hibernating pupæ, while the last form, var. *Marcellus*, appears only in summer, and then in three successive generations. A seasonal dimorphism thus appears which is combined with ordinary dimorphism, winter and summer forms alternating with each other; but the first appears itself in two forms or varieties, vars. *Telamonides* and *Walshii*. If for the present we disregard this complication, and consider these two winter forms as one, we should thus have four generations, of which the first possesses the winter form, and the three succeeding ones have, on the other hand, the summer form, var. *Marcellus*.

The peculiarity of this species consists in the fact that in all three summer generations only a portion of the pupæ emerge after a short period (fourteen days), whilst another and much smaller portion remains in the pupal state during the whole summer and succeeding winter, first emerging in the following spring, and then always in the winter form. Thus, Edwards states that out of fifty chrysalides of the second generation, which had pupated at the end of June, forty-five *Marcellus* butterflies appeared after fourteen days, whilst five pupæ emerged in April of the following year, and then as *Telamonides.*

The explanation of these facts is easily afforded by the foregoing theory. According to this, both the winter forms must be regarded as primary, and the *Marcellus* form as secondary. But this last is not yet so firmly established as *Prorsa,* in which reversion of the summer generations to the *Levana* form only occurs through special external influences; whilst in the case of *Ajax* some individuals are to be found in every generation, the tendency of which to revert is still so strong that even the greatest summer heat is unable to cause them to diverge from their original inherited direction of development, or to accelerate their emergence and compel them to assume the *Marcellus* form. It is here beyond a doubt that it is not different external influences, but internal causes only, which maintain the old hereditary tendency, for all the larvæ and pupæ of many different broods were simultaneously exposed to the same external influences. But, at the same time, it is evident that these facts are not opposed to the present theory; on the contrary, they confirm it, inasmuch as they are readily explained on the basis of the theory, but can scarcely otherwise be understood.

If it be asked what significance attaches to the duplication of the winter form, it may be answered that the species was already dimorphic at the time when it appeared in only one annual generation. Still, this explanation may be objected to, since a dimorphism of this kind is not at present known, though indeed some species exhibit a sexual dimorphism,[18] in which one sex (as, for instance, the case of the female *Papilio Turnus*) appears in two forms of colouring, but not a dimorphism, as is here the case, displayed by both sexes.[19] Another suggestion, therefore, may perhaps be offered.

In *A. Levana* we saw that reversion occurred in very different degrees with different individuals, seldom attaining to the true *Levana* form, and

18 [For other remarkable cases of sexual dimorphism (not *antigeny* in the sense used by Mr. S. H. Scudder, Proc. Amer. Acad., vol. xii. 1877, pp. 150—158) see Wallace "On the Phenomena of Variation and Geographical Distribution, as illustrated by the Papilionidæ of the Malayan Region," Trans. Linn. Soc., vol. xxv. 1865, pp. 5—10. R.M.]

19 [Eng. ed. Dimorphism of this kind has since been made known: the North American *Limenitis Artemis* and *L. Proserpina* are not two species, as was formerly believed, but only one. Edwards bred both forms from eggs of *Proserpina*. Both are single-brooded, and both have males and females. The two forms fly together, but *L. Artemis* is much more widely distributed, and more abundant than *L. Proserpina*. See "Butterflies of North America," vol. ii.]

generally only reaching the intermediate form known as *Porima*. Now it would, at all events, be astonishing if with *P. Ajax* the reversion were always complete, as it is precisely in this case that the tendency to individual reversion is so variable. I might, for this reason, suppose that one of the two winter forms, viz. the var. *Walshii*, is nothing else than an incomplete reversion-form, corresponding to *Porima* in the case of *A. Levana*. Then *Telamonides* only would be the original form of the butterfly, and this would agree with the fact that this variety appears later in the spring than *Walshii*. Experiments ought to be able to decide this.[20] The pupæ of the first

[20] [Eng. ed. Edwards has since proved experimentally that by the application of ice a large proportion of the pupæ do indeed give rise to the var. *Telamonides*. He bred from eggs of *Telamonides* 122 pupæ, which, under natural conditions, would nearly all have given the var. *Marcellus*. After two months' exposure to the low temperature there emerged, from August 24th to October 16th, fifty butterflies, viz. twenty-two *Telamonides*, one intermediate form between *Telamonides* and *Walshii*, eight intermediate forms between *Telamonides* and *Marcellus* more nearly related to the former, six intermediate forms between *Telamonides* and *Marcellus*, but more closely resembling the latter, and thirteen *Marcellus*. Through various mishaps the action of the ice was not complete and equal. See the "Canadian Entomologist," 1875, p. 228. In the newly discovered case of *Phyciodes Tharos* also, Edwards has succeeded in causing the brood from the winter form to revert, by the application of ice to this same form. See Appendix II. for a *résumé* of Edwards' experiments upon both *Papilio Ajax* and *Phyciodes Tharos*. R.M.]

three generations placed upon ice should give, for the greater part, the form *Telamonides*, for the lesser portion *Walshii*, and for only a few, or perhaps no individuals, the form *Marcellus*. This prediction is based on the view that the tendency to revert is on the whole great; that even with the first summer generation, which was the longest exposed to the summer climate, a portion of the pupæ, without artificial means, always emerged as *Telamonides*, and another portion as *Marcellus*. The latter will perhaps now become *Walshii* by the application of cold.

One would expect that the second and third generations would revert more easily, and in a larger percentage, than the first, because this latter first acquired the new *Marcellus* form; but the present experiments furnish no safe conclusion on this point. Thus, of the first summer generation only seven out of sixty-seven pupæ hibernated, and these gave *Telamonides*; while of the second generation forty out of seventy-six, and of the third generation twenty-nine out of forty-two pupæ hibernated. But to establish safer conclusions, a still larger number of experiments is necessary. According to the experience thus far gained, one might perhaps still be inclined to imagine that, with seasonal dimorphism, external influences operating on the individual might directly compel it to assume one or the other form. I long held this view myself, but it is, nevertheless, untenable. That cold does not produce the one kind of marking, and warmth the other, follows from the before-mentioned facts, viz. that in *Papilio Ajax* every generation produces both forms; and, further, in the case of *A. Levana* I have frequently reared the fourth (hibernating) generation entirely in a warm room, and yet I have always obtained the winter form. Still, one might be inclined not to make the temperature *directly* responsible, but rather the retardation or acceleration of development produced through the action of temperature. I confess that I for a long time believed that in this action I had found the true cause of seasonal dimorphism. Both with *A. Levana* and *P. Napi* the difference between the duration of the pupal period in the winter and summer forms is very great, lasting as a rule, in the summer generation of *A. Levana*, from seven to twelve days, and in the winter generation about two hundred days. In this last species the pupal state can certainly be shortened by keeping them at an elevated temperature; but I have, nevertheless, only in one case obtained two or three butterflies at the end of December from caterpillars that had pupated in September, these generally emerging in the course of February and March, and are to be seen on the wing in warm weather during the latter month. The greatest reduction of the pupal period still leaves for this stage more than 100 days.

From this last observation it follows that it is not the duration of development which, in individual cases, determines the form of the butterfly, and which consequently decides whether the winter or summer form shall emerge, but that, on the contrary, the duration of the pupal stage is dependent on the tendency which the forthcoming butterfly had taken in the chrysalis state. This can be well understood when we consider that the winter form must have had a long, and the summer form a short pupal period, during innumerable generations. In the former the habit of slow development must have been just as well established as that of rapid development in the latter; and we cannot be at all surprised if we do not see this habit abandoned by the winter form when the opportunity presents itself. But that it may be occasionally abandoned the more proves that the duration of the pupal development less determines the butterfly form than does the temperature directly, in individual cases.

Thus, for instance, Edwards explicitly states that, whereas the two winter forms of *P. Ajax*, viz. the vars. *Walshii* and *Telamonides*, generally appear only after a pupal period of 150 to 270 days, yet individual cases occur in which the pupal stage is no longer than in the summer form, viz. fourteen days.[21] A similar thing occurs with

[21] Thus from eggs of *Walshii*, laid on April 10th, Edwards obtained, after a pupal period of fourteen days, from the 1st

A. Levana, for, as already explained, not only may the development of the winter form be forced to a certain degree by artificial warmth, but the summer generation frequently produces reversion-forms without protraction of development. The intermediate reversion-form *Porima* was known long before it was thought possible that it could be produced artificially by the action of cold; it appears occasionally, although very rarely, at midsummer in the natural state.

If, then, my explanation of the phenomena is correct, the winter form is primary and the summer the secondary form, and those individuals which, naturally or artificially, assume the winter form must be considered as cases of atavism. The suggestion thus arises whether low temperature alone is competent to bring about this reversion, or whether other external influences are not also effective. Indeed, the latter appears to be the case. Besides purely internal causes, as previously pointed out in *P. Ajax*, warmth and mechanical motion appear to be able to bring about reversion.

That an unusually high temperature may cause reversion, I conclude from the following observation. In the summer of 1869 I bred the first summer brood of *A. Levana*; the caterpillars pupated during the second half of June, and from

to the 6th of June, fifty-eight butterflies of the form *Marcellus*, one of *Walshii*, and one of *Telamonides*.

that time to their emergence, on 28th June—3rd July, great heat prevailed. Now, while the intermediate form *Porima* had hitherto been a great rarity, both in the free state and when bred, having never obtained it myself, for example, out of many hundreds of specimens, there were among the sixty or seventy butterflies that emerged from the above brood, some eight to ten examples of *Porima*. This is certainly not an exact experiment, but there seems to me a certain amount of probability that the high summer temperature in this case brought about reversion.

Neither for the second cause to which I have ascribed the power of producing reversion can I produce any absolute evidence, since the experimental solution of all these collateral questions would demand an endless amount of time. I am in possession of an observation, however, which makes it appear probable to me that continuous mechanical movement acts on the development of the pupæ in a similar manner to cold, that is, retarding them, and at the same time producing reversion. I had, in Freiburg, a large number of pupæ of the first summer brood of *Pieris Napi*, bred from eggs. I changed residence while many caterpillars were in course of transformation and travelled with the pupæ in this state seven hours by rail. Although this brood of *P. Napi*, under ordinary circumstances, always emerges in the summer, generally in July of the same year, as the summer form (var. *Napeæ*), yet out of these numerous pupæ I did not get a single butterfly during the year 1872. In winter I kept them in a warm room, and the first butterflies emerged in January, 1873, the remainder following in February, March, and April, and two females not until June. All appeared, however, as exquisite winter forms. The whole course of development was precisely as though cold had acted on the pupæ; and in fact, I could find no other cause for this quite exceptional deportment than the seven hours' shaking to which the pupæ were exposed by the railway journey, immediately after or during their transformation.

It is obviously a fact of fundamental importance to the theory of seasonal dimorphism, that the summer form can be readily changed into the winter form, whilst the latter cannot be changed into the summer form. I have thus far only made experiments on this subject with *A. Levana*, but the same fact appears to me to obtain for *P. Napi*. I did not, however, operate upon the ordinary winter form of *P. Napi*, but chose for this experiment the variety *Bryoniæ*, well known to all entomologists. This is, to a certain extent, the potential winter form of *P. Napi*; the male (Fig. 14, Plate I.) exactly resembles the ordinary winter form in the most minute detail, but the female is distinguished from *Napi* by a sprinkling of greyish brown scales over the whole of the upper side of the wings (Fig. 15, Plate I.). This type, *Bryoniæ*, occurs in Polar regions as the only form of *Napi*, and is also found in the higher Alps, where it flies in secluded meadows as the only form, but in other localities, less isolated, mixed with the ordinary form of the species. In both regions *Bryoniæ* produces but one generation in the year, and must thus, according to my theory, be regarded as the parent-form of *Pieris Napi*.

If this hypothesis is correct—if the variety *Bryoniæ* is really the original form preserved from the glacial period in certain regions of the earth, whilst *Napi* in its winter form is the first secondary form gradually produced through a warm climate, then it would be impossible ever to breed the ordinary form *Napi* from pupæ of *Bryoniæ* by the action of warmth, since the form of the species now predominant must have come into existence only by a cumulative action exerted on numerous generations, and not *per saltum*.

The experiment was made in the following manner: In the first part of June I caught a female of *Bryoniæ* in a secluded Alpine valley, and placed her in a capacious breeding-cage, where she flew about among the flowers, and laid more than a hundred eggs on the ordinary cabbage. Although the caterpillars in the free state feed upon another plant unknown to me, they readily ate the cabbage, grew rapidly, and pupated at the end of July. I then brought the pupæ into a hothouse in which the temperature fluctuated between 12° and 24° R.; but, in spite of this high temperature, and—what is certainly of more special importance—notwithstanding the want of cooling at night, only one butterfly emerged the same summer, and that a male, which, from certain minute characteristic markings, could be safely identified as var. *Bryoniæ.* The other pupæ hibernated in the heated room, and produced, from the end of January to the beginning of June, 28 butterflies, all of which were exquisite *Bryoniæ.*

Experiment thus confirmed the view that *Bryoniæ* is the parent-form of *Napi*, and the description hitherto given by systematists ought therefore properly to be reversed. *Pieris Bryoniæ* should be elevated to the rank of a species, and the ordinary winter and summer forms should be designated as vars. *Napi* and *Napeæ.* Still I should not like to take it upon myself to increase the endless confusion in the synonomy of butterflies. In a certain sense, it is also quite correct to describe the form *Bryoniæ* as a climatic variety, for it is, in fact, established, if not produced, by climate, by which agency it is likewise preserved; only it is not a secondary, but the primary, climatic variety of *Napi.* In this sense most species might probably be described as climatic varieties, inasmuch as under the influence of another climate they would gradually acquire new characters, whilst, under the influence of the climate now prevailing in their habitats, they have, to a certain extent, acquired and preserved their present form.

The var. *Bryoniæ* is, however, of quite special interest, since it makes clear the relation which exists between climatic variation and seasonal dimorphism, as will be proved in the next section. The correctness of the present theory must first here be submitted to further proof.

It has been shown that the secondary forms of seasonally dimorphic butterflies do not all possess the tendency to revert in the same degree, but that this tendency rather varies with each individual. As the return to the primary form is synonymous with the relinquishing of the secondary, the greater tendency to revert is thus synonymous with the greater tendency to relinquish the secondary form, but this again is equivalent to a lesser stability of the latter; it must consequently be concluded that the individuals of a species are very differently influenced by climatic change, so that with some the new form must become sooner established than with others. From this a variability of the generation concerned must necessarily ensue, i. e., the individuals of the summer generation must differ more in colour and marking than is the case with those of the winter generation. If the theory is correct, the summer generations should be more variable than the winter generations—at least, so long as the greatest possible equalization of individual variations has not occurred through the continued action of warmth, combined with the constant crossing of individuals which have become changed in different degrees. Here also the theory is fully in accord with facts.

In *A. Levana* the *Levana* form is decidedly more constant than the *Prorsa* form. The first is, to a slight extent, sexually dimorphic, the female being light and the male dark-coloured. If we take into consideration this difference between the sexes, which also occurs to a still smaller extent in the *Prorsa* form, the foregoing statement will be found correct, viz. that the *Levana* form varies but little, and in all cases considerably less than the *Prorsa* form, in which the greatest differences occur in the yellow stripes and in the disappearance of the black spots on the white band of the hind wing, these black spots being persistent *Levana* markings. It is, in fact, difficult to find two perfectly similar individuals of the *Prorsa* form. It must, moreover, be considered that the *Levana* marking, being the more complicated, would the more readily show variation. Precisely the same thing occurs in *Pieris Napi*, in which also the var. *Æstiva* is considerably more variable than the var. *Vernalis*. From the behaviour of the var. *Bryoniæ*, on the other hand, which I regard as the parent-form, one might be tempted to raise an objection to the theory; for this form is well known to be extraordinarily variable in colour and marking, both in the Alps and Jura, where it is met with at the greatest altitudes. According to the theory, *Bryoniæ* should be less variable than the winter form of the lowlands, because it is the older, and should therefore be the more constant in its characters. It must not be forgotten, however, that the variability of a species may not only originate in the one familiar manner of unequal response of the individual to the action of varying exciting causes, but also by the crossing of two varieties separately established in adjacent districts and subsequently brought into contact. In the Alps and Jura the ordinary form of *Napi* swarms everywhere from the plains towards the habitats of *Bryoniæ*, so that a crossing of the two forms may occasionally, or even frequently, take place; and it is not astonishing if in some places (Meiringen, for example) a perfect series of intermediate forms between *Napi* and *Bryoniæ* is met with. That crossing is the cause of the great variability of *Bryoniæ* in the Alpine districts, is proved by the fact that in the Polar regions this form "is by no means so variable as in the Alps, but, judging from about forty to fifty Norwegian specimens, is rather constant." My friend, Dr. Staudinger, who has twice spent the summer in Lapland, thus writes in reply to my question. A crossing with *Napi* cannot there take place, as this form is never met with, so that the ancient parent-form *Bryoniæ* has been able to preserve its original constancy. In this case also the facts thus accord with the requirements of the theory.

II.

Seasonal Dimorphism and Climatic Variation.

If, as I have attempted to show, seasonal dimorphism originates through the slow operation of a changed summer climate, then is this phenomenon nothing else than the splitting up of a species into two climatic varieties in the same district, and we may expect to find various connexions between ordinary simple climatic variation and seasonal dimorphism. Cases indeed occur in which seasonal dimorphism and climatic variation pass into each other, and are interwoven in such a manner that the insight into the origin and nature of seasonal dimorphism gained experimentally finds confirmation. Before I go more closely into this subject, however, it is necessary to come to an understanding as to the conception "climatic variation," for this term is often very arbitrarily applied to quite dissimilar phenomena.

According to my view there should be a sharp distinction made between climatic and local varieties. The former should comprehend only such cases as originate through the direct action of climatic influences; while under the general designation of "local forms," should be comprised all variations which have their origin in other causes—such, for example, as in the indirect action of the external conditions of life, or in circumstances which do not owe their present existence to climate and external conditions, but rather to those geological changes which produce isolation. Thus, for instance, ancient species elsewhere long extinct might be preserved in certain parts of the earth by the protecting influence of isolation, whilst others which immigrated in a state of variability might become transformed into local varieties in such regions through the action of 'amixia,'[1] i. e. by not being allowed to cross with their companion forms existing in the other portions of their habitat. In single cases it may be difficult, or for the present impossible, to decide whether we have before us a climatic form, or a local form arising from other causes; but for this very reason we should be cautious in defining climatic variation.

The statement that climatic forms, in the true sense of the word, do exist is well known to me, and has been made unhesitatingly by all zoologists; indeed, a number of authentically observed facts

[1] [The word 'Amixie,' from the Greek *ἀμιξία*, was first adopted by the author to express the idea of the prevention of crossing by isolation in his essay "Über den Einfluss der Isolirung auf die Artbildung," Leipzig, 1872, p. 49. R.M.]

might be produced, which prove that quite constant changes in a species may be brought about by the direct action of changed climatic conditions. With butterflies it is in many cases possible to separate pure climatic varieties from other local forms, inasmuch as we are dealing with only unimportant changes and not with those of biological value, so that natural selection may at the outset be excluded as the cause of the changes in question. Then again the sharply defined geographical distribution climatically governed, often furnishes evidence of transition forms in districts lying between two climatic extremes.

In the following attempt to make clear the relationship between simple climatic variation and seasonal dimorphism, I shall concern myself only with such undoubted climatic varieties. A case of this kind, in which the winter form of a seasonally dimorphic butterfly occurs in other habitats as the only form, i.e., as a climatic variety, has already been adduced in a former paragraph. I allude to the case of *Pieris Napi*, the winter form of which seasonally dimorphic species occurs in the temperate plains of Europe, whilst in Lapland and the Alps it is commonly found as a monomorphic climatic variety which is a higher development of the winter type, viz., the var. *Bryoniæ*.

Very analogous is the case of *Euchloe Belia*, a butterfly likewise belonging to the *Pierinæ*, which extends from the Mediterranean countries to the middle of France, and everywhere manifests a very sharply pronounced seasonal dimorphism. Its summer form was, until quite recently, described as a distinct species, *E. Ausonia*. Staudinger was the first to prove by breeding that the supposed two species were genetically related.[2] This species, in addition to being found in the countries named, occurs also at a little spot in the Alps in the neighbourhood of the Simplon Pass. Owing to the short summer of the Alpine climate the species has in this locality but one annual brood, which bears the characters of the winter form, modified in all cases by the coarser thickly scattered hairs of the body (peculiar to many Alpine butterflies,) and some other slight differences. The var. *Simplonia* is thus in the Alps a simple climatic variety, whilst in the plains of Spain and the South of France it appears as the winter form of a seasonally dimorphic species.

This *Euchloe* var. *Simplonia* obviously corresponds to the var. *Bryoniæ* of *Pieris Napi*, and it is highly probable that this form of *E. Belia* must likewise be regarded as the parent-form of the species surviving from the glacial epoch, although it cannot be asserted, as can be done in the case of *Bryoniæ*, that the type has undergone

[2] [Eng. ed. In 1844, Boisduval maintained this relationship of the two forms. See Speyer's "Geographische Verbreit. d. Schmetterl.," i. p. 455.]

no change since that epoch, for *Bryoniæ* from Lapland is identical with the Alpine form,[3] whilst *E. Simplonia* does not appear to occur in Polar countries.

Very interesting also is the case of *Polyommatus Phlæas*, Linn., one of our commonest *Lycænidæ*, which has a very wide distribution, extending from Lapland to Spain and Sicily.[4] If we compare specimens of this beautiful copper-coloured butterfly from Lapland with those from Germany, no constant difference can be detected; the insect has, however, but one annual generation in Lapland, whilst in Germany it is double-brooded; but the winter and summer generations resemble each other completely, and specimens which had been caught in spring on the Ligurian coast were likewise similarly coloured to those from Sardinia. (Fig. 21, Plate II.). According to these facts we might

[3] According to a written communication from Dr. Staudinger, the female *Bryoniæ* from Lapland are never so dusky as is commonly the case in the Alps, but they often have, on the other hand, a yellow instead of a white ground-colour. In the Alps, yellow specimens are not uncommon, and in the Jura are even the rule.

[4] [According to W. F. Kirby (Syn. Cat. Diurn. Lepidop.), the species is almost cosmopolitan, occurring, as well as throughout Europe, in Northern India (var. *Timeus*), Shanghai (var. *Chinensis*), Abyssinia (var. *Pseudophlæas*), Massachusetts (var. *Americana*), and California (var. *Hypophlæas*). In a long series from Northern India, in my own collection, all the specimens are extremely dark, the males being almost black. R.M.]

believe this species to be extraordinarily indifferent to climatic influence; but the South European summer generation differs to a not inconsiderable extent from the winter generation just mentioned, the brilliant coppery lustre being nearly covered with a thick sprinkling of black scales. (Plate II., Fig. 22.) The species has thus become seasonally dimorphic under the influence of the warm southern climate, although this is not the case in Germany where it also has two generations in the year.[5] No one who is acquainted only with the Sardinian summer form, and not with the winter form of that place, would hesitate to regard the former as a climatic variety of our *P. Phlæas;* or, conversely, the north German form as a climatic variety of the southern summer form—according as he accepts the one or the other as the primary form of the species.

Still more complex are the conditions in another species of *Lycænidæ*, *Plebeius Agestis* (=*Alexis* Scop.), which presents a double seasonal dimorphism. This butterfly appears in three forms; in Germany A and B alternate with each other as winter and summer forms, whilst in Italy B and C succeed each other as winter and summer

[5] [Eng. ed. From a written communication from Dr. Speyer, it appears that also in Germany there is a small difference between the two generations. The German summer brood has likewise more black on the upper side, although seldom so much as the South European summer brood.]

forms. The form B thus occurs in both climates, appearing as the summer form in Germany and as the winter form in Italy. The German winter variety A, is entirely absent in Italy (as I know from numerous specimens which I have caught), whilst the Italian summer form, on the other hand, (var. *Allous*, Gerh.), does not occur in Germany. The distinctions between the three forms are sufficiently striking. The form A (Fig. 18, Plate II.) is blackish brown on the upper side, and has in the most strongly marked specimens only a trace of narrow red spots round the borders; whilst the form B (Fig. 19, Plate II.) is ornamented with vivid red border spots; and C (Fig. 20, Plate II.) is distinguished from B by the strong yellowish-brown of the under side. If we had before us only the German winter and the Italian summer forms, we should, without doubt, regard them as climatic varieties; but they are connected by the form B, interpolated in the course of the development of both, and the two extremes thus maintain the character of mere seasonal forms.

III.

Nature of the Causes producing Climatic Varieties.

It has been shown that the phenomenon of seasonal dimorphism has the same proximate cause as climatic variation, viz. change of climate, and that it must be regarded as identical in nature with climatic variation, being distinguished from ordinary, or, as I have designated it, simple (monomorphic) climatic variation by the fact that, besides the new form produced by change of climate, the old form continues to exist in genetic connexion with it, so that old and new forms alternate with each other according to the season.

Two further questions now present themselves for investigation, viz. (1) by what means does change of climate induce a change in the marking and colouring of a butterfly? and (2) to what extent does the climatic action determine the nature of the change?

With regard to the former question, it must, in the first place, be decided whether the true effect of climatic change lies in the action of a high or low temperature on the organism, or whether it may not perhaps be produced by the accelerated development caused by a high temperature, and the retarded development caused by a low temperature. Other factors belonging to the category of external conditions of life which are included in the term "climate" may be disregarded, as they are of no importance in these cases. The question under consideration is difficult to decide, since, on the one hand, warmth and a short pupal period, and, on the other hand, cold and a long pupal period, are generally inseparably connected with each other; and without great caution one may easily be led into fallacies, by attributing to the influence of causes now acting that which is but the consequence of long inheritance.

When, in the case of *Araschnia Levana*, even in very cold summers, *Prorsa*, but never the *Levana* form, emerges, it would still be erroneous to conclude that it is only the shorter period of development of the winter generation, and not the summer warmth, which occasioned the formation of the *Prorsa* type. This new form of the species did not come suddenly into existence, but (as appears sufficiently from the foregoing experiments) originated in the course of many generations, during which summer warmth and a short development period were generally associated together. From the fact that the winter generation always produces *Levana*, even when the pupæ have not been exposed to cold but kept in a room, it would be equally erroneous to infer that the cold of winter had no influence in determining the type. In this case also the determining causes must have been in operation during innumerable generations. After the winter form of the species has become established throughout such a long period, it remains constant, even when the external influence which produced it (cold) is occasionally withdrawn.

Experiments cannot further assist us here, since we cannot observe throughout long periods of time; but there are certain observations, which to me appear decisive. When, both in Germany and Italy, we see *Polyommatus Phlæas* appearing in two generations, of which both the German ones are alike, whilst in Italy the summer brood is black, we cannot ascribe this fact to the influence of a shorter period of development, because this period is the same both in Germany and Italy (two annual generations), so that it can only be attributed to the higher temperature of summer.

Many similar cases might be adduced, but the one given suffices for proof. I am therefore of opinion that it is not the duration of the period of development which is the cause of change in the formation of climatic varieties of butterflies, but only the temperature to which the species is exposed during its pupal existence. In what manner, then, are we to conceive that warmth acts on the marking and colouring of a butterfly? This is a question which could only be completely answered by gaining an insight into the mysterious chemico-physiological processes by which the butterfly is formed in the chrysalis; and indeed only by such a complete insight into the most minute details, which are far beyond our scrutiny, could we arrive at, or even approximate to, an explanation of the development of any living organism. Nevertheless an important step can be taken towards the solution of this problem, by establishing that the change does not depend essentially upon the action of warmth, but upon the organism itself, as appears from the nature of the change in one and the same species.

If we compare the Italian summer form of *Polyommatus Phlœas* with its winter form, we shall find that the difference between them consists only in the brilliant coppery red colour of the latter being largely suffused in the summer form with black scales. When entomologists speak of a "black dusting" of the upper side of the wings, this statement must not of course be understood literally; the number of scales is the same in both forms, but in the summer variety they are mostly black, a comparatively small number being red. We might thus be inclined to infer that, owing to the high temperature, the chemistry of the material undergoing transformation in *Phlœas* is changed in such a manner that less red and more black pigment is produced. But the case is not so simple, as will appear evident when we consider the fact that the summer forms have not originated suddenly, but only in the course of numerous generations; and when we further compare the two seasonal forms in other species. Thus in *Pieris Napi* the winter is distinguished from the summer form, among other characters, by the strong black dusting of the base of the wings. But we cannot conclude from this that in the present case more black pigment is produced in the winter than in the summer form, for in the latter, although the base of the wings is white, their tips and the black spots on the fore-wings are larger and of a deeper black than in the winter form. The quantity of black pigment produced does not distinguish between the two forms, but the mode of its distribution upon the wings.

Even in the case of species the summer form of which really possesses far more black than the winter form, as, for instance, *Araschnia Levana*, one type cannot be derived from the other simply by the expansion of the black spots present, since on the same place where in *Levana* a black band crosses the wings, *Prorsa*, which otherwise possesses much more black, has a white line. (See Figs. 1—9, Plate I.) The intermediate forms which have been artificially produced by the action of cold on the summer generation present a graduated series, according as reversion is more or less complete; a black spot first appearing in the middle of the white band of *Prorsa*, and then becoming enlarged until, finally, in the perfect *Levana* it unites with another black triangle proceeding from the front of the band, and thus becomes fused into a black bar. The white band of *Prorsa* and the black band of *Levana* by no means correspond in position; in *Prorsa* quite a new pattern appears, which does not originate by a simple colour replacement of the *Levana* marking. In the present case, therefore, there is no doubt that the new form is not produced simply because a certain pigment (black) is formed in larger quantities, but because its mode of distribution is at the same time different, white appearing in some instances where black formerly existed, whilst in other cases the black remains. Whoever compares *Prorsa* with *Levana* will not fail to be struck with the remarkable change of marking produced by the direct action of external conditions.

The numerous intermediate forms which can be produced artificially appear to me to furnish a further proof of the gradual character of the transformation. Ancestral intermediate forms can only occur where they have once had a former existence in the phyletic series. Reversion may only take place completely in some particular characters, whilst in others the new form remains constant—this is in fact the ordinary form of reversion, and in this manner a mixture of characters might appear which never existed as a phyletic stage; but particular characters could certainly never appear unless they were normal to the species at some stage of phyletic development. Were this possible it would directly contradict the idea of reversion, according to which new characters never make their appearance, but only such as have already existed. If, therefore, the ancestral forms of *A. Levana* (which we designate as *Porima*) present a great number of transitional varieties, this leads to the conclusion that the species must have gone through a long series of stages of phyletic development before the summer generation had completely changed into *Prorsa*. The view of the slow cumulative action of climatic influences already submitted, is thus confirmed.

If warmth is thus without doubt the agency which has gradually changed the colour and marking of many of our butterflies, it sufficiently appears from what has just been said concerning the nature of the change that the chief part in the transmutation is not to be attributed to the agency in question, but to the organism which is affected by it. Induced by warmth, there begins a change in the ultimate processes of the matter undergoing transformation, which increases from generation to generation, and which not only consists in the appearance of the colouring matter in one place instead of another, but also in the replacement of yellow, in one place by white and in another by black, or in the transformation of black into white on some portions of the wings, whilst in others black remains. When we consider with what extreme fidelity the most insignificant details of marking are, in constant species of butterflies, transmitted from generation to generation, a total change of the kind under consideration cannot but appear surprising, and we should not explain it by the nature of the agency (warmth), but only by the nature of the species affected. The latter cannot react upon the warmth in the same manner that a solution of an iron salt reacts upon potassium ferrocyanide or upon sulphuretted hydrogen; the colouring matter of the butterfly's wing which was previously black does not become blue or yellow, nor does that which was white become changed into black, but a new marking is developed from the existing one—or, as I may express it in more general terms, the species takes another course of development; the complicated chemico-physical processes in the matter composing the pupa become gradually modified in such a manner that, as the final result, a new marking and colouring of the butterfly is produced.

Further facts can be adduced in support of the view that in these processes it is the constitution of the species, and not the external agency (warmth), which plays the chief part. The latter, as Darwin has strikingly expressed it, rather performs the function of the spark which ignites a combustible substance, whilst the character of the combustion depends upon the nature of the explosive material. Were this not the case, increased warmth would always change a given colour[1] in the same manner in all butterflies, and would therefore always give rise to the production of the same colour. But this does not occur; *Polyommatus Phlæas*, for example, becoming black in the south, whilst the red-brown *Vanessa Urticæ* becomes black in high northern latitudes, and many other cases well known to entomologists might be adduced.[2] It indeed appears that species of similar physical constitution, i.e., nearly allied species, under similar climatic influences, change in an analogous manner. A beautiful example of this is furnished by our *Pierinæ*. Most of the species display seasonal dimorphism; as, for instance, *Pieris Brassicæ*, *Rapæ*, *Napi*, *Krueperi*, and *Daplidice*, *Euchloe Belia* and *Belemia*, and *Leucophasia Sinapis*, in all of which the difference between the winter and the summer forms is of a precisely

[1] [Assuming that in all butterflies similar colours are produced by the same chemical compounds. R.M.]

[2] [Mr. H. W. Bates mentions instances of local variation in colour affecting many distinct species in the same district in his memoir "On the Lepidoptera of the Amazon Valley;" Trans. Linn. Soc., vol. xxiii. Mr. A. R. Wallace also has brought together a large number of cases of variation in colour according to distribution, in his address to the biological section of the British Association at Glasgow in 1876. See "Brit. Assoc. Report," 1876, pp. 100—110. For observations on the change of colour in British Lepidoptera according to distribution see papers by Mr. E. Birchall in "Ent. Mo. Mag.," Nov., 1876, and by Dr. F. Buchanan White, "Ent. Mo. Mag.," Dec., 1876. The colour variations in all these cases are of course not *protective* as in the well-known case of *Gnophos obscurata*, &c. R.M.]

similar nature. The former are characterized by a strong black dusting of the base of the wings, and by a blackish or green sprinkling of scales on the underside of the hind wings, while the latter have intensely black tips to the wings, and frequently also spots on the fore-wings.

Nothing can prove more strikingly, however, that in such cases everything depends upon the physical constitution, than the fact that in the same species the males become changed in a different manner to the females. The parent form of *Pieris Napi* (var. *Bryoniæ*) offers an example. In all the *Pierinæ* secondary sexual differences are found, the males being differently marked to the females; the species are thus sexually dimorphic. Now the male of the Alpine and Polar var. *Bryoniæ*, which I conceive to be the ancestral form, is scarcely to be distinguished, as has already been mentioned, from the male of our German winter form (*P. Napi*, var. *Vernalis*), whilst the female differs considerably.[3] The gradual climatic change which transformed the parent form *Bryoniæ* into *Napi* has therefore exerted a much greater effect on the female than on the male. The external action on the two sexes was exactly the same, but the response of the organism was different, and the cause of the difference can only be sought for in the fine differences of physical constitution which distinguish the male from the female. If we are

[3] See Figs. 10 and 14, 11 and 15, Plate I.

unable to define these differences precisely, we may nevertheless safely conclude from such observations that they exist.

I have given special prominence to this subject because, in my idea, Darwin ascribes too much power to sexual selection when he attributes the formation of secondary sexual characters to the sole action of this agency. The case of *Bryoniæ* teaches us that such characters may arise from purely innate causes; and until experiments have decided how far the influence of sexual selection extends, we are justified in believing that the sexual dimorphism of butterflies is due in great part to the differences of physical constitution between the sexes. It is quite different with such sexual characters as the stridulating organs of male Orthoptera which are of undoubted importance to that sex. These can certainly be attributed with great probability to sexual selection.

It may perhaps not be superfluous to adduce one more similar case, in which, however, the male and not the female is the most affected by climate. In our latitudes, as also in the extreme north, *Polyommatus Phlæas*, already so often mentioned, is perfectly similar in both sexes in colour and marking; and the same holds good for the winter generation of the south. The summer generation of the latter, however, exhibits a slight sexual dimorphism, the red of the fore wings of the female being less completely covered with black than in the male.

IV.

Why all Polygoneutic Species are not Seasonally Dimorphic.

If we may consider it to be established that seasonal dimorphism is nothing else than the splitting up of a species into two climatic varieties in one and the same locality, the further question at once arises why all polygoneutic species (those which produce more than *one* annual generation) are not seasonally dimorphic.

To answer this, it will be necessary to go more deeply into the development of seasonal dimorphism. This evidently depends upon a peculiar kind of periodic, alternating heredity, which we might be tempted to identify with Darwin's "inheritance at corresponding periods of life." It does not, however, in any way completely agree with this principle, although it presents a great analogy to it and must depend ultimately upon the same cause. The Darwinian "inheritance at corresponding periods of life"—or, as it is termed by Haeckel, "homochronic heredity"—is characterized by the fact that new characters always appear in the individuals at the same stage of life as that in which they appeared in their progenitors. The truth of this principle has been firmly established, instances being known in which both the first appearance of a new (especially pathological) character and its transmission through several generations has been observed. Seasonally dimorphic butterflies also furnish a further valuable proof of this principle, since they show that not only variations which arise suddenly (and which are therefore probably due to purely innate causes) follow this mode of inheritance, but also that characters gradually called forth by the influence of external conditions and accumulating from generation to generation, are only inherited at that period of life in which these conditions were or are effective. In all seasonally dimorphic butterflies which I have been able to examine closely, I found the caterpillars of the summer and winter broods to be perfectly identical. The influences which, by acting on the pupæ, split up the imagines into two climatic forms, were thus without effect on the earlier stages of development. I may specially mention that the caterpillars, as well as the pupæ and eggs of *A. Levana*, are perfectly alike both in the summer and winter forms; and the same is the case in the corresponding stages of *P. Napi* and *P. Bryoniæ*.

I shall not here attempt to enter more deeply into the nature of the phenomena of inheritance. It is sufficient to have confirmed the law that influences which act only on certain stages in the development of the individual, even when the action is cumulative and not sudden, only affect those particular stages without having any effect on the earlier or later stages. This law is obviously of the greatest importance to the comprehension of metamorphosis. Lubbock[1] has briefly shown in a very clear manner how the existence of metamorphosis in insects can be explained by the indirect action of varying conditions on the different life-stages of a species. Thus the mandibles of a caterpillar are, by adaptation to another mode of nourishment, exchanged at a later period of life for a suctorial organ. Such adaptation of the various development-stages of a species to the different conditions of life would never give rise to metamorphosis, if the law of homochronic, or periodic, heredity did not cause the characters gradually acquired at a given stage to be transferred to the same stage of the following generation.

The origin of seasonal dimorphism depends upon a very similar law, or rather form, of inheritance, which differs from that above considered only in the fact that, instead of the ontogenetic stages, a whole series of generations is influenced. This form of inheritance may be formulated somewhat as follows:—When dissimilar conditions

[1] "On the Origin and Metamorphoses of Insects," London, 1874.

alternatingly influence a series of generations, a cycle is produced in which the changes are transmitted only to those generations which are acted upon by corresponding conditions, and not to the intermediate ones. Characters which have arisen by the action of a summer climate are inherited by the summer generation only, whilst they remain latent in the winter generation. It is the same as with the mandibles of a caterpillar which are latent in the butterfly, and again make their appearance in the corresponding (larval) stage of the succeeding generation. This is not mere hypothesis, but the legitimate inference from the facts. If it be admitted that my conception of seasonal dimorphism as a double climatic variation is correct, the law of "cyclical heredity,"[2] as I may term it—in contradistinction to "homochronic heredity," which relates only to the ontogenetic stages—immediately follows. All those cases which come under the designation of 'alternation of generation,' can obviously be referred to cyclical heredity, as will be explained further on. In the one case the successive generations deport them-

[2] I at first thought of designating the two forms of cyclical or homochronic heredity as ontogenetic- and phyletic-cyclical heredity. The former would certainly be correct; the latter would be also applicable to alternation of generation (in which actually two or more phyletic stages alternate with each other) but not to all those cases which I attribute to heterogenesis, in which, as with seasonal dimorphism, a series of generations of *the same* phyletic stage constitute the point of departure.

selves exactly in the same manner as do the successive stages of development of the individual in the other; and we may conclude therefrom (as has long been admitted on other grounds) that a generation is, in fact, nothing else than a stage of development in the life of a species. This appears to me to furnish a beautiful confirmation of the theory of descent.

Now if, returning to questions previously solved, the alternating action of cold in winter and warmth in summer leads to the production of a winter and summer form, according to the law of cyclical heredity, the question still remains: why do we not find seasonal dimorphism in all polygoneutic butterflies?

We might at first suppose that all species are not equally sensitive to the influence of temperature: indeed, the various amounts of difference between the winter and summer forms in different species would certainly show the existence of different degrees of sensitiveness to the modifying action of temperature. But even this does not furnish an explanation, since there are butterflies which produce two perfectly similar[3] generations

[3] When Meyer-Dürr, who is otherwise very accurate, states in his "Verzeichniss der Schmetterlinge der Schweiz," (1852, p. 207), that the winter and summer generations of *P. Ægeria* differ to a small extent in the contour of the wings and in marking, he has committed an error. The characters which this author attributes to the summer form are much more appli-

wherever they occur, and which, nevertheless, appear in different climates as climatic varieties. This is the case with *Pararga Ægeria* (Fig. 23, Plate II.), the southern variety of which, *Meione* (Fig. 24, Plate II.), is connected with it by an intermediate form from the Ligurian coast. This species possesses, therefore, a decided power of responding to the influence of temperature, and yet no distinction has taken place between the summer and the winter form. We can thus only attribute this different deportment to a different kind of heredity; and we may therefore plainly state, that changes produced by alternation of climate are not always inherited *alternatingly*, i. e. by the corresponding generations, but sometimes *continuously*, appearing in every generation, and never remaining latent. The causes which determine why, in a particular case, the one or the other form of inheritance prevails, can be only innate, i. e. they lie in the organism itself, and there is as little to be said upon their precise nature as upon that of any other process of heredity. In a similar manner Darwin admits a kind of double inheritance with respect to characters produced by sexual selection; in one form these characters remain limited to the sex which first acquired them, in the other form they are inherited by both sexes, without it

cable to the female sex.' There exists in this species a trifling sexual dimorphism, but no seasonal dimorphism.

being apparent why, in any particular case, the one or the other form of heredity should take place.

The foregoing explanation may obtain in the case of sexual selection, in which it is not inconceivable that certain characters may not be so easily produced, or even not produced at all, in one sex, owing to its differing from the other in physical constitution. In the class of cases under consideration, however, it is not possible that the inherited characters can be prevented from being acquired by one generation owing to its physical constitution, since this constitution was similar in all the successive generations before the appearance of dimorphism. The constitution in question first became dissimilar in the two generations to the extent of producing a change of specific character, through the action of temperature on the alternating broods of each year, combined with cyclical heredity. If the law of cyclical heredity be a general one, it must hold good for all cases, and characters acquired by the summer generation could never have been also transmitted to the winter generation from the very first.

I will not deny the possibility that if alternating heredity should become subsequently entirely suppressed throughout numerous generations, a period may arrive when the preponderating influence of a long series of summer generations may ultimately take effect upon the winter generation. In such a case the summer characters would appear, instead of remaining latent as formerly. In this manner it may be imagined that at first but few, and later more numerous individuals, approximate to the summer form, until finally the dimorphism entirely disappears, the new form thus gaining ascendency and the species becoming once more monomorphic. Such a supposition is indeed capable of being supported by some facts, an observation on *A. Levana* apparently contradicting the theory having been already interpreted in this sense. I refer to the fact that whilst some butterflies of the winter generation emerge in October as *Prorsa*, others hibernate, and appear the following spring in the *Levana* form. The winter form of *Pieris Napi* also no longer preserves, in the female sex, the striking coloration of the ancestral form *Bryoniæ*, a fact which may indicate the influencing of the winter generation by numerous summer generations. The double form of the spring generation of *Papilio Ajax* can be similarly explained by the gradual change of alternating into continuous heredity, as has already been mentioned. All these cases, however, are perhaps capable of another interpretation; at any rate, the correctness of this supposition can only be decided by further facts.

Meanwhile, even if we suppose the above explanation to be correct, it will not apply to the absence of seasonal dimorphism in cases like that of *Pararga Ægeria* and *Meione*, in which only *one* summer generation appears, so that a preponderating inheritance of summer characters cannot be admitted. Another explanation must thus be sought, and I believe that I have found it in the circumstance that the butterflies named do not hibernate as pupæ but as caterpillars, so that the cold of winter does not directly influence those processes of development by which the perfect insect is formed in the chrysalis. It is precisely on this point that the origin of those differences of colour which we designate as the seasonal dimorphism of butterflies appears to depend. Previous experiments give great probability to this statement. From these we know that the eggs, caterpillars, and pupæ of all the seasonally dimorphic species experimented with are perfectly similar in the summer and winter generations, the imago stage only showing any difference. We know further from these experiments, that temperature-influences which affect the caterpillars never entail a change in the butterflies; and finally, that the artificial production of the reversion of the summer to the winter form can only be brought about by operating on the pupæ.

Since many monogoneutic species now hibernate in the caterpillar stage (e. g. *Satyrus Proserpina*, and *Hermione*, *Epinephele Eudora*, *Jurtina*,

Tithonus, *Hyperanthus*, *Ida*, *&c.*), we may admit that during the glacial period such species did not pass the winter as pupæ. As the climate grew warmer, and in consequence thereof a second generation became gradually interpolated in many of these monogoneutic species, there would ensue (though by no means necessarily) a disturbance of the winter generation, of such a kind that the pupæ, instead of the caterpillars as formerly, would then hibernate. It may, indeed, be easily proved *à priori* that whenever a disturbance of the winter generation takes place it only does so retrogressively, that is to say—species which at one time pass the winter as caterpillars subsequently hibernate in the egg, while those which formerly hibernate as pupæ afterwards do so as caterpillars. The interpolation of a summer generation must necessarily delay till further towards the end of summer, the brood about to hibernate; the remainder of the summer, which serves for the development of the eggs and young caterpillars, may possibly under these conditions be insufficient for pupation, and the species which hibernated in the pupal state when it was monogoneutic, may perhaps pass the winter in the larval condition after the introduction of the second brood. A disturbance of this kind is conceivable ; but it is certain that many species suffer no further alteration in their development than that of becoming digoneutic from monogoneutic. This follows from the fact that hibernation takes place in the caterpillar stage in many species of the sub-family *Satyridæ* which are now digoneutic, as well as in the remaining monogoneutic species of the same sub-family. But we cannot expect seasonal dimorphism to appear in all digoneutic butterflies the winter generation of which hibernates in the caterpillar form, since the pupal stage in these species experiences nearly the same influences of temperature in both generations. We are hence led to the conclusion that seasonal dimorphism must arise in butterflies whenever the pupæ of the alternating annual generations are exposed throughout long periods of time to widely different regularly recurring changes of temperature.

The facts agree with this conclusion, inasmuch as most butterflies which exhibit seasonal dimorphism hibernate in the pupa stage. Thus, this is the case with all the *Picrinæ*, with *Papilio Machaon*, *P. Podalirius*, and *P. Ajax*, as well as with *Araschnia Levana*. Nevertheless, it cannot be denied that seasonal dimorphism occurs also in some species which do not hibernate as pupæ but as caterpillars; as, for instance, in the strongly dimorphic *Plebeius Amyntas*. But such cases can be explained in a different manner.

Again, the formation of a climatic variety—and as such must we regard seasonally dimorphic forms—by no means entirely depends on the magnitude of the difference between the temperature which acts on the pupæ of the primary and that which acts on those of the secondary form; it rather depends on the absolute temperature which the pupæ experience. This follows without doubt from the fact that many species, such as our common Swallow-tail (*Papilio Machaon*), and also *P. Podalirius*, in Germany and the rest of temperate Europe, show no perceptible difference of colour between the first generation, the pupæ of which hibernate, and the second generation, the pupal period of which falls in July, whereas the same butterflies in South Spain and Italy are to a small extent seasonally dimorphic. Those butterflies which are developed under the influence of a Sicilian summer heat likewise show climatic variation to a small extent. The following consideration throws further light on these conditions. The mean summer and winter temperatures in Germany differ by about 14.9°R.; this difference being therefore much more pronounced than that between the German and Sicilian summer, which is only about 3.6°R. Nevertheless, the winter and summer generations of *P. Podalirius* are alike in Germany, whilst the Sicilian summer generation has become a climatic variety. The cause of this change must therefore lie in the small difference between the mean summer temperatures of 15.0° R. (Berlin) and 19.4° R. (Palermo). According to this, a given absolute temperature appears to give a tendency to variation in a certain direction, the necessary temperature being different for different species. The latter statement is supported by the facts that, in the first place, in different species there are very different degrees of difference between the summer and winter forms; and secondly, many digoneutic species are still monomorphic in Germany, first becoming seasonally dimorphic in Southern Europe. This is the case with *P. Machaon* and *P. Podalirius*, as already mentioned, and likewise with *Polyommatus Phlœas*. Zeller in 1846-47, during his journey in Italy, recognized as seasonally dimorphic in a small degree a large number of diurnal Lepidoptera which are not so in our climate.[4]

In a similar manner the appearance of seasonal dimorphism in species which, like *Plebeius Amyntas*, do not hibernate as pupæ, but as caterpillars, can be simply explained by supposing that the winter generation was the primary form, and that the increase in the summer temperature since the glacial period was sufficient to cause this particular species to become changed by the gradual interpolation of a second generation. The dimorphism of *P. Amyntas* can, nevertheless, be explained in another manner. Thus, there may

[4] P. C. Zeller, "Bemerkungen über die auf einer Reise nach Italien und Sicilien gesammelten Schmetterlingsarten." Isis, 1847, ii.—xii.

have been a disturbance of the period of development in the manner already indicated, the species which formerly hibernated in the pupal stage becoming subsequently disturbed in its course of development by the interpolation of a summer generation, and hibernating in consequence in the caterpillar state. Under these circumstances we must regard the present winter form (var. *Polysperchon*) as having been established under the influence of a winter climate, this form, since the supposed disturbance in its development, having had no reason to become changed, the spring temperature under which its pupation now takes place not being sufficiently high. The interpolated second generation on the other hand, the pupal period of which falls in the height of summer, may easily have become formed into a summer variety.

This latter explanation agrees precisely with the former, both starting with the assumption that in the present case, as in that of *A. Levana* and the *Pierinæ*, the winter form is the primary one, so that the dimorphism proceeds from the said winter form and does not originate the winter but the summer form, as will be explained. Whether the winter form has been produced by the action of the winter or spring temperature is immaterial in judging single cases, inasmuch as we are not in a position to state what temperature is necessary to cause any particular species to become transformed.

The reverse case is also theoretically conceivable, viz., that in certain species the summer form was the primary one, and by spreading northwards a climate was reached which still permitted the production of two generations, the pupal stage of one generation being exposed to the cold of winter, and thus giving rise to the production of a secondary winter form. In such a case hibernation in the pupal state would certainly give rise to seasonal dimorphism. Whether these conditions actually occur, appears to me extremely doubtful; but it may at least be confidently asserted that the first case is of far more frequent occurrence. The beautiful researches of Ernst Hoffmann[5] furnish strong evidence for believing that the great majority of the European butterflies have immigrated, not from the south, but from Siberia. Of 281 species, 173 have, according to Hoffmann, come from Siberia, 39 from southern Asia, and only 8 from Africa, whilst during the greatest cold of the glacial period, but very few or possibly no species existed north of the Alps. Most of the butterflies now found in Europe have thus, since their immigration, experienced a gradually increasing warmth. Since seasonal dimorphism has been developed in some of these species, the summer form must in all cases have been the secondary one, as the experiments upon

[5] "Isoporien der europäischen Tagfalter." Stuttgart, 1873.

the reversion of *Pieris Napi* and *Araschnia Levana* have also shown.

All the seasonally dimorphic butterflies known to me are found in Hoffmann's list of Siberian immigrants, with the exception of two species, viz., *Euchloe Belemia*, which is cited as an African immigrant, and *Pieris Krueperi*, which may have come through Asia Minor, since at the present time it has not advanced farther west than Greece. No considerable change of climate can be experienced by migrating from east to west, so that the seasonal dimorphism of *Pieris Krueperi* can only depend on a cause similar to that which affected the Siberian immigrants, that is, the gradual increase of temperature in the northern hemisphere since the glacial period. In this species also, the winter form must be the primary one. In the case of *E. Belemia*, on the other hand, the migration northwards from Africa certainly indicates removal to a cooler climate, which may have originated a secondary winter form, even if nothing more certain can be stated. We know nothing of the period of migration into southern Europe; and even migration without climatic change is conceivable, if it kept pace with the gradual increase of warmth in the northern hemisphere since the glacial epoch. Experiments only would in this case be decisive. If the summer generation, var. *Glauce*, were the primary form, it would not be possible by the action of cold on the pupæ of this brood to produce the winter variety *Belemia,* whilst, on the other hand, the pupæ of the winter generation by the influence of warmth would be made to revert more or less completely to the form *Glauce.* It is by no means to be understood that the species would actually comport itself in this manner. On the contrary, I am of opinion that in this case also, the winter form is primary. The northward migration (from Africa to south Spain) would be quite insufficient, and the winter form is now found in Africa as well as in Spain.

V.

On Alternation of Generations.

Seasonal dimorphism has already been designated by Wallace as alternation of generation,[1] a term which cannot be disputed so long as it is confined to a regular alternation of dissimilar generations. But little is gained by this definition, however, unless it can be proved that both phenomena are due to similar causes, and that they are consequently brought about by analogous processes. The causes of alternation of generation have, until the present time, been scarcely investigated, owing to the want of material. Haeckel alone has quite recently subjected these complicated phenomena generally to a searching investigation, and has arrived at the conclusion that the various forms of metagenesis can be arranged in two series. He distinguishes a progressive and a retrogressive series, comprising under the former those species "which, to a certain extent, are still in a transition stage from monogenesis to amphigenesis (asexual to sexual propagation), and the early progenitors

[1] [Trans. Linn. Soc., vol. xxv. 1865, p. 9. R.M.]

of which, therefore, never exclusively propagated themselves sexually" (*Trematoda*, *Hydromedusæ*). Under the other, or retrogressive form of metagenesis, Haeckel includes a "return from amphigenesis to monogenesis," this being the case with all those species which now manifest a regular alternation from amphigenesis to parthenogenesis (*Aphides*, *Rotatoria*, *Daphnidæ*, *Phyllopoda*, &c). Essentially I can but agree entirely with Haeckel. Simply regarding the phenomena of alternation of generation as at present known, it appears to me to be readily admissible that these multiform modes of propagation must have originated in at least two different ways, which can be aptly formulated in the manner suggested by Haeckel.

I will, however, venture to adopt a somewhat different mode of conception, and regard the manner of propagation (whether sexual or asexual) not as the determining, but only as the secondary cause. I will further hazard the separation of the phenomena of alternating generations (in their widest sense) into two main groups according to their origin, designating the cases of one group as true metagenesis and those of the other as heterogenesis.[2] Metagenesis takes its origin from a

[2] It is certainly preferable to make use of the expression "metagenesis" in this special sense instead of introducing a new one. As a general designation, comprehending metagenesis and heterogenesis, there will then remain the expression "alternation of generation," if one does not prefer to say

phyletic series of dissimilar forms, whilst heterogenesis originates from a phyletic series of similar forms—this series, so far as we can at present judge, always consisting of similar sexual generations. The former would thus nearly coincide with Haeckel's progressive, and the latter with his retrogressive metagenesis. Metagenesis may further originate in various ways. In the first place, from metamorphosis, as for example, in the propagation of the celebrated *Cecidomyia* with nursing larvæ. The power which these larvæ possess of propagating themselves asexually has evidently been acquired as a secondary character, as appears from the fact that there are many species of the same genus the larvæ of which do not nurse, these larvæ being themselves undoubted secondary forms produced by the adaptation of this stage of phyletic development to a mode of life widely different from that of the later stages. In the form now possessed by these larvæ they could never have represented the final stage of their ontogeny, neither could they have formerly possessed the power of sexual propagation. The conclusion seems inevitable that metagenesis has here proceeded from metamorphosis; that is to say, one stage of the ontogeny, by acquiring asexual propagation, has changed the originally existing metamorphosis into metagenesis.

"cyclical propagation." The latter may be well used in contradistinction to "metamorphosis."

Lubbock[3] is undoubtedly correct when, for cases like that just mentioned, he attempts to derive alternation of generations from metamorphosis. But if we exclude heterogenesis there still remain a large number of cases of true metagenesis which cannot be explained from this point of view.

It must be admitted, with Haeckel, that the alternation of generations in the Hydromedusæ and Trematoda does not depend, as in the case of *Cecidomyia*, upon the larvæ having acquired the power of nursing, but that the inferior stages of these species always possessed this power which they now only preserve. The nursing Trematode larvæ now existing may possibly have been formerly able to propagate themselves also sexually, this mode of propagation having at the present time been transferred to a later phyletic stage. In this case, therefore, metagenesis was not properly produced by metamorphosis, but arose therefrom in the course of the phyletic development, the earlier phyletic stages abandoning the power of sexual reproduction, and preserving the asexual mode of propagation. A third way in which metagenesis might originate is through polymorphosis. When the latter is combined with asexual reproduction, as is especially the case with the Hydrozoa, metagenesis may be derived therefrom. The successive stages of transformation of one and the same physiological in-

[3] *Loc. cit.* chap. iv.

dividual do not in these cases serve as the point of departure for alternation of generation, but the different contemporary forms living gregariously into which the species has become divided through functional differentiation of the various individuals of the same stock. Individuals are here produced which alone acquire the power of sexual reproduction, and metagenesis is thus brought about, these individuals detaching themselves from the stock on which they originated, while the rest of the individuals remain in combination, and retain the asexual mode of propagation. No sharp distinction can be otherwise drawn between this and the cases previously considered.[4] The difference consists only in the whole cycle of reproduction being performed by one stock; both classes have the common character that the different phyletic stages never appear in the same individual (metamorphosis), but in the course of further phyletic development metagenesis at the same time arises, i.e. the division of these stages among a succession of individuals. We are therefore able to distinguish this *primary metagenesis* from the *secondary metagenesis* arising from metamorphosis.

[4] The idea that alternation of generation is derived from polymorphism (not the reverse, as usually happens; i. e. polymorphism from alternation of generation) is not new, as I find whilst correcting the final proof. Semper has already expressed it at the conclusion of his interesting memoir, "Über Generationswechsel bei Steinkorallen," &c. See "Zeitschrift f. wiss. Zool." vol. xxii. 1872.

It is not here my intention to enter into the ultimate causes of metagenesis ; in this subject we should only be able to advance by making vague hypotheses. The phenomenon of seasonal dimorphism, with which this work has mainly to deal, is evidently far removed from metagenesis, and it was to make this clear that the foregoing observations were brought forward. The characters common in the origin of metagenesis are to be found, according to the views previously set forth, in the facts that here the faculty of asexual and of sexual reproduction is always distributed among *several* phyletic stages of development which succeed each other in an ascending series (progressive metagenesis of Haeckel), whereas I find differences only in the fact that the power of asexual propagation may (in metagenesis) be either newly acquired (larva of *Cecidomyia*) or preserved from previous ages (*Hydroida*). It seems that in this process sexual reproduction is without exception lost by the earlier, and remains confined solely to the most recent stages.

From the investigations on seasonal dimorphism it appears that a cycle of generations can arise in an entirely different way. In this case a series of generations originally alike are made dissimilar by external influences. This appears to me of the greatest importance, since seasonal dimorphism is without doubt closely related to that mode of reproduction which has hitherto been exclusively designated as heterogenesis, and a knowledge of its mode of origination must therefore throw light on the nature and origin of heterogenesis in general.

In seasonal dimorphism, as I have attempted to show, it is the direct action of climate, and indeed chiefly that of temperature, which brings about the change in some of the generations. Since these generations have been exposed to the alternating influence of the summer and winter temperature a periodical dimorphism has been developed—a regular cycle of dissimilar generations. It has already been asserted that the consecutive generations of a species comport themselves with respect to heredity in a manner precisely similar to that of the ontogenetic stages, and at the same time such succeeding generations point out the parallelism between metamorphosis and heterogenesis. If influences capable of directly or indirectly producing changes operate on any particular stage of development, these changes are always transmitted to the same stage. Upon this metamorphosis depends. In a precisely similar manner changes which operated periodically on certain generations (1, 3, 5, for instance) are transmitted to these generations only, and not to the intermediate ones. Upon this depends heterogenesis. We have just been led to the comprehension of heterogenesis by cyclical heredity, by the fact that a cycle is produced whenever a series of generations exists under regularly alternating influences.

In this cycle newly-acquired changes, however minute in character at first, are only transmitted to a later, and not to the succeeding generation, appearing only in the one corresponding, i.e. in that generation which exists under similar transforming influences. Nothing can more clearly show the extreme importance which the conditions of life must have upon the formation and further development of species than this fact. At the same time nothing shows better that the action of these conditions is not suddenly and violently exerted, but that it rather takes place by small and slow operations. In these cases the long-continued accumulation of imperceptibly small variations proves to be the magic means by which the forms of the organic world are so powerfully moulded. By the application of even the greatest warmth nobody would be able to change the winter form of *A. Levana* into the summer form ; nevertheless, the summer warmth, acting regularly on the second and third generations of the year, has, in the course of a lengthened period, stamped these two generations with a new form without the first generation being thereby changed. In the same region two different climatic varieties have been produced (just as in the majority of cases climatic varieties occur only in separate regions) which alternate with each other, and thus give rise to a cycle of which each generation propagates itself sexually.

But even if seasonal dimorphism is to be ascribed to heterogenesis, it must by no means be asserted that those cases of cyclical propagation hitherto designated as heterogenesis are completely identical with seasonal dimorphism. Their identity extends only to their origin and manner of development, but not to the mode of operation of the causes which bring about their transformation. Both phenomena have a common mode of origination, arising from similar (monomorphic) sexual generations and course of development, a cycle of generations with gradually diverging characters coming into existence by the action of alternating influences. On the other hand, the nature of the changes by which the secondary differs from the primary generation may be referred to another mode of action of the exciting causes. In seasonal dimorphism the differences between the two generations are much less than in other cases of heterogenesis. These differences are both quantitatively less, and are likewise qualitative, affecting only characters of biological insignificance.[5] The variations in question are mostly restricted to the marking and colouring of the wings and body, occasionally affecting also the form of the wing, and in a few cases the size of the body (*Plebeius Amyntas*), whilst the bodily structure—so far at

[5] See my essay "Über den Einfluss der Isolirung auf die Artbildung." Leipzig, 1872.

least as my investigations extend—appears to be the same in both generations.[6]

The state of affairs is quite different in the remaining cases of heterogenesis; here the entire structure of the body appears to be more or less changed, and its size is often very different, nearly all the internal organs differing in the two generations. According to Claus,[7] "we can scarcely find any other explanation of the mode of origination of heterogenesis than the gradual and slow advantageous adaptation of the organization to important varying conditions of life"—a judgment in which this author is certainly correct. In all such cases the change does not affect unimportant characters, as it does in butterflies, but parts of biological or physiological value; and we cannot, therefore, consider such changes to have originated through the *direct* action of altered conditions of life, but *indirectly* through natural selection or adaptation.

Thus, the difference between seasonal dimorphism and the other known cases of heterogenesis

[6] [In the case of monogoneutic species which, by artificial 'forcing,' have been made to give two generations in the year, it has generally been found that the reproductive system has been imperfectly developed in the second brood. A minute anatomical investigation of the sexual organs in the two broods of seasonally-dimorphic insects would be of great interest, and might lead to important results. R.M.]

[7] "Grundzüge der Zoologie." 2nd ed. Leipzig, 1872. Introduction.

consists in the secondary form in which the species appears in the former originating through the direct action of external conditions, whilst in the latter this form most probably originates through the indirect action of such influences. The first half of the foregoing proposition is alone capable of provisional proof, but it is in the highest degree probable that the latter half is also correct. Naturally we cannot say to what extent the direct action of external conditions plays also a part in true heterogenesis, as there have been as yet no experiments made on its origin. That direct action, working to a certain extent co-operatively, plays only a secondary part, while the chief cause of the change is to be found in adaptation, no one can doubt who keeps in view, for instance, the mode of propagation discovered by Leuckart in *Ascaris nigrovenosa.* In this worm, the one generation lives free in the water, and the other generation inhabits the lungs of frogs, the two generations differing from one another in size of body and structure of internal organs to an extent only possible with the true Nematoda.

To prevent possible misunderstanding, let it be finally noted—even if superfluous—that the changes causing the diversity of the two generations in seasonal dimorphism and heterogenesis are not of such a nature that the value of different "specific characters" can be attached to them.

Distinctly defined specific characters are well known not to occur generally, and it would therefore be erroneous to attach but little value to the differences in seasonal dimorphism because these chiefly consist in the colouring and marking of the wings. The question here under consideration is not whether two animal forms have the value of species or of mere varieties—a question which can never be decided, since the reply always depends upon individual opinion of the value of the distinctions in question, and the idea of both species and varieties is moreover purely conventional. The question is, rather, whether the distinguishing characters possess an equal constancy—that is, whether they are transmitted with the same force and accuracy to all individuals; and whether they occur, therefore, in such a manner that they can be practically employed as specific characters. With respect to this, it cannot be doubtful for a moment that the colouring and marking of a butterfly possess exactly the same value as the constant characters in any other group of animals, such as the palate-folds in mice, the structure of the teeth in mammals, the number and form of the wing and tail feathers in birds, &c. We have but to remember with what wonderful constancy often the most minute details of marking are transmitted in butterflies. The systematist frequently distinguishes between two nearly allied species, as for instance in the *Lycænidæ*, chiefly by the position of certain insignificant black spots on the under side of the wing (*P. Alexis* female, and *P. Agestis*) ; and this diagnosis proves sufficient, since *P. Alexis*, which has the spots in a straight row, has a different caterpillar to *P. Agestis*, in which the central spot is nearer the base of the hind wing!

For the reasons just given, I maintain that it is neither justifiable nor useful to designate the di- and polymorphism of butterflies as di- and polychroism, and thereby to attribute but little importance to these phenomena.[8] This designation would be only justifiable if the differences of colour were due to other causes than the differences of form, using this last word in a narrow sense. But it has been shown that the same direct action of climate which originates new colours, produces also in some species differences of form (contour of wing, size, &c.) ; whilst, on the other hand, it has long been known that many protective colours can only be explained by the indirect action of external conditions.

When I raise a distinction in the nature of the changes between seasonal dimorphism and the remaining known cases of heterogenesis, this must be taken as referring only to the biological or physiological result of the change in the transformed organism itself. In seasonal dimorphism only

[8] With reference to this subject, see the discussion by the Belgian Entomological Society, Brussels, 1873.

insignificant characters become prominently changed, characters which are without importance for the welfare of the species; while in true heterogenesis we are compelled to admit that useful changes, or adaptations, have occurred.

Heterogenesis may thus be defined either in accordance with my proposal or in the manner hitherto adopted, since it may be regarded as more morphological than the cyclical succession of differently formed sexual generations; or, with Claus, as the succession of different sexual generations, "living under different conditions of existence"—a definition which applies in all cases to seasonal dimorphism. Varying conditions of existence, in their widest sense, are the result of the action of different climates; and a case has been made known recently in which it is extremely probable that the climatic differences of the seasons have produced a cycle of generations by influencing the processes of nutrition. This case is quite analogous to that which we have observed in the seasonal dimorphism of butterflies, but with the distinction that the difference between the winter and summer generations does not, at least entirely, consist in the form of the reproductive adult, but almost entirely in its ontogeny—in the mode of its development. A comparison of this case with the analogous phenomenon in butterflies, may be of interest. In the remarkable fresh-water Daphnid, *Leptodora hyalina* Lillejeborg, it was proved some years ago by P. E. Müller,[9] who studied the ontogeny, that this last was direct, since the embryo, before leaving the egg, already possesses the form, members, and internal organs of the adult. This was, at least, the case with the summer eggs. It was subsequently shown by Sars[10] that this mode of development only holds good for the summer brood, the winter eggs producing an embryo in the spring which possesses only the three first pairs of limbs, and, instead of compound eyes, only a single frontal eye, thus exhibiting briefly, at first, the structure of a *Nauplius*, and gradually acquiring that of *Leptodora*. The mature form derived from the winter eggs is not distinguishable from the later generations, except by the presence of the simple larval eye, which appears as a small black spot. The generations when fully developed are thus distinguished only by this minute marking, but the summer generation undergoes direct development, whilst the winter generation, on the contrary, is only developed by metamorphosis, beginning with the simplest Crustacean type, and thus fairly representing the phyletic development of the species. We therefore see, in this case, the combination of a metamorphic and a direct development taking place to a certain

[9] P. E. Müller, "Bidrag til Cladocerners Fortplantingshistorie," 1868.

[10] Sars, in "Förhandlinger i Videnskabs Selskabet i Christiania," 1873, part i.

extent under our eyes. It cannot be proved with certainty what the cause of this phenomenon may be, but the conjecture is almost unavoidable that it is closely related to the origin of the seasonal dimorphism of butterflies, since both depend on the alternating climatic influences of summer and winter: it is most probable that these influences have directly [11] brought about a shortening of the period of development in summer. Thus we have here a case of heterogenesis nearly related to the seasonal dimorphism of butterflies in a twofold manner—first, because the cycle of generations is also in this case brought about by the direct action of the external conditions of life; and secondly, the winter form is here also the primary, and the summer form the secondary one.

In accordance with the idea first introduced into science by Rudolph Leuckart, we have hitherto understood heterogenesis to be only the alternation of dissimilar sexual generations. From this point of view the reproduction of *Leptodora* can be as little ascribed to heterogenesis as can that of *Aphis* or *Daphnia*, although the apparent agamic reproduction of the winter and a portion of the summer generation is undoubtedly partheno-

[11] [Eng. ed. Recent researches on alternation of generation in the Daphnoidea have convinced me that *direct* action of external conditions does not in these cases come into consideration, but only *indirect* action.]

genesis and not propagation by nursing.[12] As has already been said, however, I would attribute no fundamental importance to the criterion of agamic reproduction—the more especially because we are ignorant of the physiological significance of the two modes of propagation; and further, because this principle of classification is entirely external, and only valuable in so far as no better one can be substituted for it. A separation of the modes of cyclical propagation according to their genesis appears to me—especially if practicable—not alone to be of greater value, but the only correct one, and for this the knowledge of the origin of seasonal dimorphism seems to me to furnish a possible method.

If, as was indicated above, we designate as metagenesis (in the narrow sense) all those cases in which it must be admitted that a series of differently aged phyletic stages have furnished the points of departure, and as heterogenesis those cases in which similar phyletic stages have been compelled to produce a cycle of generations by the periodic action of external influences, it is clear that the scope of heterogenesis is by this means considerably extended, and at the same time sharply and precisely defined.

Under heterogenesis then is comprised, not only

[12] See my memoir, "Über Bau und Lebenserscheinungen der *Leptodora hyalina*," Zeitschrift f. wiss. Zool., vol. xxiv. part 3, 1874.

as heretofore the reproduction of *Ascaris nigrovenosa*, of *Leptodora appendiculata*, and of the cattle-lice, but also that of the *Aphides*, *Coccidæ*, *Daphnidæ*, *Rotatoria*, and *Phyllopoda*, and, in short, all those cases in which we can determine the former identity of the two kinds of generations from their form, anatomical structure, and mode of reproduction. This conclusion is essentially supported by a comparison of the most closely allied species. Thus, for instance, when we see the genus *Aphis* and its allies related on all sides to insects which propagate sexually in all generations, and when we further observe the great similarity of the whole external and internal structure in the two kinds of generations of *Aphis*, we are forced to the conjecture that the apparent asexual reproduction of the *Aphidæ* is in reality parthenogenesis, i. e., that it has been developed from sexual reproduction. Neither can it be any longer disputed that in this case, as well as in that of *Leptodora* and other *Daphnidæ*, the same female alternately propagates parthenogenetically, and produces eggs requiring fertilization. This was established by Von Heyden[13] some years ago, in the case of *Lachnus Querci*, and has been since confirmed by Balbiani.[14]

There can be no doubt that in all these cases the cycle of generations has been developed from

[1] Stettin. entom. Zeit., vol. xviii p. 83, 1857.

[14] Compt. Rend., vol. lxxvii. p. 1164, 1873.

phyletically similar generations. But instances are certainly conceivable which present themselves with less clearness and simplicity. In the first place, we do not know whether parthenogenesis may not finally settle down into complete asexual reproduction. Should this be the case, it might be possible that from heterogenesis a mode of propagation would ultimately arise, which was apparently indistinguishable from pure metagenesis. Such a state of affairs might result, if the generations settling into asexual reproduction (as, for instance, the plant-lice), at the same time by adaptation to varying conditions of life, underwent considerable change of structure, and entered upon a metamorphosis to some extent retrogressive. We should then be inclined to regard these generations as an earlier phyletic stage, whilst, in fact, they would be a later one, and the idea of metagenesis would thus have been formed after the manner of heterogenesis.

On the other hand, it is equally conceivable that heterogenesis may have been developed from true metagenesis in the case of larvæ which, having acquired the faculty of asexual propagation, are similar in function to sexually mature insects. This possibility is not at first sight apparent. If the nursing-larvæ of the *Cecidomyiæ* were as much like the sexual insects as are the young Orthoptera to the sexually mature forms, we should not know whether to regard them as degraded sexual insects, or as true larvæ which had attained the power of asexual propagation. Their propagation would be considered to be parthenogenesis; and as it could not be denied that heterogenesis was here manifest, the mode of development of their particular kind of propagation might be proved, i. e., it might be demonstrated, that the generations now parthenogenetic were formerly mere reproductive larval stages.

I have only offered these last observations in order to show on what uncertain ground we are still standing with regard to this subject whenever we deal with the meaning of any particular case, and how much still remains to be done. It appears certain that the two forms of cyclical propagation, heterogenesis and metagenesis, originate in entirely distinct ways, so that it must be admitted that, under these circumstances, the idea of the existing conditions respecting the true genesis may possibly be erroneous. To indicate the manner in which the cyclical mode of propagation has arisen in any single case, would only be possible by a searching proof and complete knowledge of existing facts in addition to experiments.

VI.

General Conclusions.

I shall not here give a repetition and summary of the results arrived at with respect to seasonal dimorphism, but rather the general conclusions derived from these results; and, at the same time, I may take the opportunity of raising certain questions which have not hitherto found expression, or have been but briefly and casually stated.

It must, in the first place, be admitted that differences of specific value can originate through the direct action of external conditions of life only. Of the truth of this proposition there can be no doubt, after what has been above stated concerning the difference between the two forms of any seasonally dimorphic species. The best proof is furnished by the older systematists, to whom the genetic relationship of the two forms was unknown, and who, with unprejudiced taxonomy, in many cases indicated their distinctness by separate specific names. This was the case with *Araschnia Levana* and *Prorsa*, *Euchloe Belia* and *Ausonia*, *E. Belemia* and *Glauce*, *Plebeius Polysperchon* and *Amyntas*. In the presence of these facts it can scarcely be doubted that new species can be formed in the manner indicated; and I believe that this was and is still the case, with butterflies at least, to a considerable extent; the more so with these insects, because the striking colours and markings of the wings and body, being in most cases without biological significance, are useless for the preservation of the individual or the species, and cannot, therefore, be objects of natural selection.

Darwin must have obtained a clear insight into this, when he attempted to attribute the markings of butterflies to sexual and not to natural selection. According to this view, every new colour or marking first appears in one sex accidentally,[1] and is there fixed by being preferred by the other sex to the older coloration. When the new ornamentation becomes constant (in the male for example), Darwin supposes that it becomes transferred to the female by inheritance, either partially or completely, or not at all; so that the species, therefore, remains more or less sexually dimorphic, or (by complete transference) becomes again sexually monomorphic.

The admissibility of such different, and, to a certain extent, arbitrarily limited inheritance, has already been acknowledged. The question here concerned is, whether Darwin is correct when he

[1] ["Accidental" in the sense of our being in ignorance of the laws of variation, as so frequently insisted upon by Darwin. R.M.]

in this manner attributes the entire coloration of butterflies to sexual selection. The origin of seasonal dimorphism appears to me to be against this view, howsoever seductive and grand the latter may seem. If differences as important as those which exist between the summer and winter forms of many butterflies can be called forth by the direct action of a changed climate, it would be extremely hazardous to attribute great importance to sexual selection in this particular case.

The principle of sexual selection appears to me to be incontestible, and I will not deny that it is also effective in the case of butterflies; but I believe that as a final explanation of colour this agency can be dispensed with, inasmuch as we see that considerable changes of colour can occur without the influence of sexual selection.[2]

[2] [Eng. ed. Since this was written I have studied the ornamental colours of the *Daphnidæ;* and, as a result, I no longer doubt that sexual selection plays a very important part in the marking and colouring of butterflies. I by no means exclude both transforming factors, however; it is quite conceivable, on the contrary, that a change produced directly by climate may be still further increased by sexual selection. The above given case of *Polyommatus Phlæas* may perhaps be explained in this manner. That sexual selection plays a part in butterflies, is proved above all by the odoriferous scales and tufts of the males discovered by Fritz Müller.] [For remarks on the odours emitted by butterflies and moths, see Fritz Müller in "Jena. Zeit. f. Naturwissen.," vol. xi. p. 99; also "Notes on Brazilian Entomology," Trans. Ent. Soc. 1878, p. 211. The odoriferous organs of the female *Heliconinæ* are fully described in a paper in "Zeit. f. Wissen. Zool.," vol. xxx. p. 167. The position of

The question now arises, how far does the transforming influence of climate extend? When a species has become transformed by climatic change to such an extent that its new form possesses the systematic value of a new species, does it return to its older form by removal to the old climatic conditions? or would it under these circumstances become again transformed in a new manner? This question is not without importance, inasmuch as in the first case climatic influences would be of little value in the formation of species, and there would result at most only a fluctuation between two extremes. In the same manner as in seasonally dimorphic species the summer and winter forms now alternate with each other every year, so would the forms produced by warmth and cold then alternate in the greater periods of the earth's history. Other groups of animals are certainly changed by the action of different climatic influences; but in but-

the scent-tufts in the sphinx-moths is shown in Proc. Entom. Soc. 1878, p. ii. Many British moths, such as *Phlogophora meticulosa*, *Cosmia trapezina*, &c. &c., have tufts in a similar position. The fans on the feet of *Acidalia bisetata*, *Herminia barbalis*, *H. tarsipennalis*, &c., are also probably scent organs. A large moth from Jamaica, well known to possess a powerful odour when alive (*Erebus odorus* Linn.), has great scent-tufts on the hind legs. For the application of the theory of sexual selection to butterflies, see, in addition, to Darwin's "Descent of Man," Fritz Müller in "Kosmos," vol. ii. p. 42; also for January, 1879, p. 285; and Darwin in "Nature," vol. xxi. January 8th, 1880, p. 237. R.M.]

terflies, as I believe I have proved, temperature plays the chief part, and as this only oscillates between rather narrow limits, it admits of no great differences of coloration.

The question thus suggests itself, whether species of butterflies only oscillate between two forms, or whether climatic change, when sufficiently great to produce variation, does not again originate a new form. Inasmuch as the reversion experiments with seasonally dimorphic butterflies appear to correspond with the latter view, I believe that this must be admitted. I am of opinion that an old form never again arises through change of climate, but always a new one; so that a periodically recurring change of climate is alone sufficient, in the course of a long period of time, to admit of new species arising from one another. This, at least, may be the case with butterflies.

My views rest essentially upon theoretical considerations. It has already been insisted upon, as results immediately from the experiments, that temperature does not act on the physical constitution of the individual in the same manner as acid or alkali upon litmus paper, i. e., that one and the same individual does not produce this or that coloration and marking according as it is exposed to warmth or cold; but rather that climate, when it influences in a similar manner many succeeding generations, gradually produces such a change in the physical constitution of the species that this manifests itself by other colours and markings. Now when this newly acquired physical constitution, established, as we may admit, throughout a long series of generations, is again submitted to a constant change of climate, this influence, even if precisely similar to that which obtained during the period of the first form of the species, cannot possibly reproduce this first form. The nature of the external conditions may be the same, but not so the physical constitution of the species. Just in the same manner as a *Pieris* (as has been already shown), a *Lycæna*, or a *Satyrus*, produces quite different varieties under the transforming influence of the same climate, so must the variation originating from the transformed species of our present case after the beginning of the primary climate be different from that primary form of the species, although perhaps in a less degree. In other words, if only two different climates alternated with each other during the earth's geological periods, every species of butterfly submitted to these changes of climate would give rise to an endless series of different specific forms. The difference of climate would in reality be greater than supposed, and for any given species the climatic variation would not only occur through the periodic shifting of the ecliptic, but also through geological changes and the migrations of the species itself, so that a continuous change of species must have gone on from this sole cause of alternation of climate. When we consider that many species elsewhere extinct have become locally preserved, and when, further, to these we add those local forms which have arisen by the prevention of crossing (amixia), and finally take into consideration the important effects of sexual selection, we can no longer be astonished at the vast numbers of species of butterflies which we now meet with on the earth.

Should any one be inclined to conclude, from my reversion experiments with seasonally dimorphic butterflies, that the secondary species when exposed to the same climate as that which produced it must revert to the primary, he forgets that this reversion to the winter form is nothing but a reversion—i. e., a sudden return to a primary form through peculiar laws of inheritance—and by no means a gradual re-acquisition of this primary form under the gradual influence of the primary climate. Reversion to the winter form occurs also through other influences, as, for instance, by high temperature. Reversions of this kind, depending on laws of heredity, certainly happen with those cases of transmutation which do not alternate with the primary form, as in seasonal dimorphism, but which occur continuously. They would, however probably be more quickly suppressed in such cases than in seasonal dimorphism, where the constant alternation of the primary and secondary forms must always maintain the tendency of the latter to produce the former.

That the above conclusion is correct—that a secondary species, when exposed to the external conditions under the influence of which the primary form originated, does not again revert to the latter—is proved by experience with plants. Botanists[3] assure us "that cultivated races which become wild, and are thus brought back to their former conditions of life, do not become changed into the original wild form, but into some new one."

A second point which appears to me to be elucidated by seasonal dimorphism, is the origin of variability. It has already been prominently shown that secondary forms are for the most part considerably more variable than primary forms. From this it follows that similar external influences either induce different changes in the different individuals of a species, or else change all individuals in the same manner, variability arising only from the unequal time in which the individuals are exposed to the external influence. The latter is undoubtedly the case, as appears from the differences which are shown by the various individuals of a secondary form. These

[3] Nägeli, "Entstehung und Begriff der naturhistorischen Art," Munich, 1865, p. 25. The author interprets the facts above quoted in a quite opposite sense, but this is obviously erroneous.

are always only differences of degree and not of kind, as is perhaps most distinctly shown by the very variable *A. Prorsa* (summer form), in which all the occurring variations differ only by the *Levana* marking being more or less absent, and, at the same time, by approximating more or less to the pure *Prorsa* marking; but changes in a totally different direction never occur. It is likewise further evident, as has been mentioned above, that allied species and genera, and even entire families (*Pieridæ*), are changed by similar external inducing causes in the same manner—or, better, in the same direction.

In accordance with these facts the law may be stated, that, in butterflies at least, all the individuals of a species respond to the same external influences by similar changes, and that, consequently, the changes brought about by climatic influences take a fixed direction, determined by the physical constitution of the species. When, however, new climatic forms of butterflies, in which natural selection is completely excluded, and the nature of the species itself definitely determines the direction of the changes, nevertheless show variability from the very beginning, we may venture to conclude that every transformation of a species generally begins with a fluctuation of its characters. But when we find the primary forms of butterflies always far more constant, this shows that the continued crossing of the individuals of a species to a certain extent balances the fluctuations of form. Both facts taken together confirm the law formerly enunciated by me,[4] that in every species a period of variability alternates with one of (relative) constancy—the latter indicating the culmination, and the former the beginning or end, of its development. I here call to mind this law, because the facts which I advanced at that time, viz., Hilgendorf's history of the phyletic development of the Steinheim fossil shells, having since become somewhat doubtful, one might easily be inclined to go too far in mistrusting them and refuse to give them any weight at all.[5]

In the essay just indicated I traced the origin of a certain class of local forms to local isolation. I attempted to show that when a species finds itself in an isolated district in a condition (period) of variability, it must there necessarily acquire somewhat deviating characters by being prevented from crossing with the individuals of other regions, or, what comes to the same thing, a local form must originate. This production of local forms

[4] See my essay, "Über den Einfluss der Isolirung auf die Artbildung." Leipzig, 1872.

[5] [Eng. ed. In the summer of 1877, Dr. Hilgendorf again investigated the Steinheim fossil shells, and found his former statements to be completely confirmed. At the meeting of the German Naturalists and Physicists at Munich, in 1877, he exhibited numerous preparations, which left no doubt that the chief results of his first research were correct, and that there have been deposited a series of successively derived species together with their connecting intermediate forms.]

results because the different variations which, for the time being, constitute the variability of the species, would always be in a different numerical proportion in the isolated district as compared with other regions; and further, because constancy is produced by the crossing of these (isolated) varieties among themselves; so that the resultant of the various components is (local) variation. If the components are dissimilar the resultant would also be different, and thus, from a theoretical point of view, there seems to me no obstacle in the way of the production of such local forms by the process of 'amixia.' I believe that I have further shown that numerous local forms can be conceived to have arisen through this process of preventive crossing, whilst they cannot be explained by the action of climatic influences.

That I do not deny the existence of true climatic forms in admitting this principle of 'amixia,' as has been frequently imagined, appears sufficiently from the treatise in question. The question arises, however, whether climatic influences may not also originate forms by 'amixia' by making a species variable. It would be difficult at present to decide finally upon this subject. If, however, in all cases a variation in a certain fixed direction occurred through climatic influences, a form could not arise by 'amixia' from such a variability, since the components could then produce resultants different only in degree and not in kind. But we are not yet able to extend our researches to such fine distinctions.

As a final, and not unimportant result of these investigations, I may once more insist that dissimilar influences, when they alternatingly affect a long series of originally similar generations in regularly recurring change, only modify the generations concerned, and not intermediate ones. Or, more briefly, cyclically acting causes of change produce cyclically recurring changes : under their influence series of monomorphic generations become formed into a cycle of di- or polymorphic generations.

There is no occasion to return here to the immediate evidence and proof of the foregoing law. In the latter, however, is comprised the question—is not the cycle of generations produced by cyclical heredity ultimately equivalent to Darwin and Haeckel's homochronic heredity which forms the ontogenetic stages into a cycle ? It is possible that from this point, in the future, the nature of the processes of heredity, which are still so obscure, may be penetrated into, and both phenomena traced to the same cause, as can now be only surmised but not clearly perceived.

Finally, the most general, and in so far chief result of these investigations, appears to me to lie in the conclusion, which may be thus formulated :—A species is only caused to change through the influence of changing external conditions of life, this change being in a fixed direction which entirely depends on the physical nature of the varying organism, and is different in different species, or even in the two sexes of the same species.

I am so little disposed to speak in favour of an unknown transforming power that I may here again insist that the transformation of a species only partly depends upon external influences, and partly on the specific constitution of the particular form. I designate this constitution 'specific,' inasmuch as it responds to the same inciting cause in a manner different to the constitution of another species. We can generally form a clear conception why this should be the case; for not only is there in another species a different kind of latent vital activity, but each species has also a different developmental history. It must be admitted that, from the earliest period of the formation of an organism, and throughout all its intermediate stages, properties which have become established, such as growth, nutrition, or tendency to development, have been transferred to the species now existing, each of which bears these tendencies in itself to a certain extent. It is these innate tendencies which determine the external and internal appearance of the species at every period of its life, and which, by their reaction to external factors, represent the life of the individual as well as that of the species. Since the sum of these inherited tendencies must vary more or less in every species, not only is the different external appearance of species as well as their physiological and biological diversity thus explained, but it necessarily follows therefrom, that different species must respond differently to those external causes which tend to produce a change in their form.

Now, this last conclusion is equivalent to the statement that every species, through its physical constitution, (in the sense defined) is impressed with certain fixed powers of variation, which are evidently extraordinarily numerous in the case of each species, but are not unlimited; they permit of a wide range for the action of natural selection, but they also limit its functions, since they certainly restrain the course of development, however wide the latter may be. I have elsewhere previously insisted[6] that too little is ascribed to the part played by the physical constitution of species in the history of their transformation, when the course of this transformation is attributed entirely to external conditions. Darwin certainly admits the importance of this factor, but only so far as it concerns the individual variation, the nature of which appears to him to depend on the physical constitution of the species. I believe, however, that in this directive influence lies the precise reason why, under the most favourable external circumstances, a bird can never become transformed into a mam-

[6] See my essay, "Über die Berechtigung der Darwin'schen Theorie." Leipzig, 1868.

mal—or, to express myself generally, why, from a given starting-point, the development of a particular species cannot now attain, even under the most favourable external conditions, any desired goal; and why, from this starting-point, given courses of development, even when of considerable latitude, must be restricted, just as a ball rolling down a hill is diverted by a fixed obstacle in a direction determined by the position of the latter, and depending on the direction of motion and the velocity at the moment of being diverted.

In this sense I agree with Askenasy's "fixed" direction of variation; but not if another new physical force directing variation itself is thereby intended.[7] The explanation of the phenomena does not appear to me to require such an admission, and, if unnecessary, it is certainly not legitimate. According to my view, transmutation by purely internal causes is not to be entertained. If we could absolutely suspend the changes of the external conditions of life, existing species would remain stationary. The action of external inciting causes, in the widest sense of the word, is alone able to produce modifications; and even the never-failing "individual variations," together with the inherited dissimilarity of constitution, appear to

[7] I expressly insist upon this here, because the notice of Askenasy's thoughtful essay which I gave in the "Archiv für Anthropologie" (1873) has frequently been misunderstood.

me to depend upon unlike external influences, the inherited constitution itself being dissimilar because the individuals have been at all times exposed to somewhat varying external influences.

A change arising from purely internal causes seems to me above all quite untenable, because I cannot imagine how the same material substratum of physical constitution of a species can be transferred to the succeeding generation as two opposing tendencies. Yet this must be the case if the direction of development transferred by heredity is to be regarded as the ultimate ground both of the similarity and dissimilarity to the ancestors. All changes, from the least to the greatest, appear to me to depend ultimately only on external influences; they are the response of the organism to external inciting causes. It is evident that this response must be different when a physical constitution of a different nature is affected by the same inciting cause, and upon this, according to my view, depends the great importance of these constitutional differences.

If, under "heredity," we comprise the totality of inheritance—that is to say, the physical constitution of a species at any time, and therefore the restricted and, in the foregoing sense, pre-determined power of variation, whilst under "adaptation" we comprehend the direct and indirect response of this physical constitution to the changes in the conditions of life, I can agree with Haeckel's mode of expression, and with him trace the transformation of species to the two factors of heredity and adaptation.

APPENDIX I.

EXPERIMENTS.

EXPERIMENTS WITH ARASCHNIA LEVANA.

1. BRED from eggs laid by a female of the winter form on 12th—15th May, 1868, in a breeding-cage. The caterpillars emerged on 20th—22nd May, and pupated on 7th—9th June. The pupæ, kept at the ordinary temperature, produced :—

On the	19th	of June	4	butterflies.
"	20th	"	5	"
"	21st	"	10	"
"	22nd	"	9	"
"	23rd	"	7	"
"	25th	"	13	"
	Total	. .	48	"

All these butterflies were of the *Prorsa* type, 3 females having a considerable amount of yellow, but none with so much as figs. 3, 4, 7, 8, or 9. Pl. I.

2. August 12th, 1868, found larvæ of the third generation, which pupated at the beginning of September, and were kept in a room not warmed. In September three butterflies emerged in the *Prorsa* form, the remainder hibernating and producing, after being placed in a heated room at the end of February, from the 1st to the 17th of March, 1869, more butterflies, all of the *Levana* form.

3. Larvæ found on the 17th June, 1869, were sorted according to colour; the yellow ones, with light brown spines, produced, at the ordinary temperature, on 8th—12th July, 13 butterflies, 12 of which showed the ordinary *Prorsa* type, and one, a male, possessing more yellow than fig. 3, Pl. I., must be considered as a *Porima* type.

4. From caterpillars of the second generation, found at the same time as those of Exp. 3, 30 pupæ were placed in the refrigerator (temperature 8°—10° R.) on June 25th. When the box was opened on August 3rd, almost all had emerged, many being dead, and all, without exception, were of the intermediate form (*Porima*), although nearer the *Prorsa* than the *Levana* type.

5. A large number of caterpillars of the second generation, found at the same time, pupated, and were kept at a high summer temperature. After a pupal period of about 19 days, some 70 butterflies emerged from 28th June to 5th July, all of the *Prorsa* form, with the exception of 5, which were strongly marked with yellow (*Porima*).

6. The 70 butterflies of the foregoing experiment were placed in an enclosure 6 feet high, and 8 feet long, in which, during warm weather, they freely swarmed on flowers. Copulation was only once observed, and but one female laid eggs on nettle on July 4th. At the high summer temperature prevailing at the time, these eggs produced butterflies after 30—31 days (third generation). All were *Prorsa*, with more or less yellow; among 18 none were completely *Porima*.

7. Young larvæ of the fourth generation, found on the 8th of August, were reared in a hothouse (17°—20° R.). They pupated on 21st—23rd August. Of these:—

A. 56 pupæ were placed on ice (0°—1° R.) for five weeks, and then allowed to hibernate in a room not warmed. In April, 1870, they all gave the *Levana* form, with the exception of a single *Porima*.

B. About an equal number of pupæ were placed in the hothouse, but without any result; for, notwithstanding a temperature of 12°—24° R., not a single butterfly emerged in the course of October and November. The pupæ were then allowed to hibernate in an unheated room, and in April and May gave nothing but *Levana*.

8. Caterpillars of the second generation, found at the beginning of June, 1870, pupated on 13th—15th June, and gave, at the ordinary temperature, on June 29th—30th. 7 butterflies of the *Prorsa* form.

9. Pupæ of the same (second) generation were placed immediately after pupation on June 18th, 1870, in a refrigerator (0°—1° R.), and after remaining there four weeks (till July 18th) gave, at the ordinary summer temperature:—

On the 22nd of July,	2	*Prorsa*.
" 23rd "	3	"
" 24th "	6	*Porima*, 4 of which were very similar to *Levana*.
" 25th "	1	*Levana*, without the blue marginal line.
" 26th "	2	*Levana*, also without the blue marginal line.
" 2nd August,	6	*Porima*.
Total . .	20	

Of these 20 butterflies only 5 were of the pure *Prorsa* form.

10. Full grown larvæ of the fourth generation, found on August 20th, 1870, pupated on August 26th to September 5th. The pupæ were divided into three portions:—

A. Placed in the hothouse (12°—25° R.), immediately after pupation and left there till October 20th. Of about 40 pupæ only 4 emerged, 3 of which were *Prorsa* and 1 *Porima*. The remaining pupæ hibernated and all changed into *Levana* the following spring.

B. Kept in a room heated to 6°—15° R. from November. Not a single specimen emerged the same year. This lot of pupæ were added to C from November.

C. Placed on ice for a month immediately after pupation; then, from September 28th to October 19th in the hothouse, where no more butterflies emerged. The pupæ hibernated, together with those from lot B, in a room heated by water to 6°—15° R., and gave :—

On the	6th of February,		1 female *Levana*.
"	22nd	"	1 male *Levana*.
"	23rd	"	1 male *Levana*.
"	24th	"	1 female *Levana*.
"	25th	"	1 male and 1 female *Levana*.
"	28th	"	1 male and 1 female *Levana*.
"	1st of March,		1 male *Levana*.
"	13th	"	1 female *Levana*.
"	15th	"	1 female *Levana*.
"	19th	"	1 male *Levana*.
"	2nd of April,		2 male and 1 female *Levana*.
"	7th	"	1 female *Levana*.
"	21st	"	1 female *Levana*.
"	2nd of May		1 female *Levana*.
	Total . .		18 *Levana*, 10 of which were females.

The exact record of the time of emergence is interesting, because it is thereby rendered apparent that different individuals respond more in different degrees to a higher than to the ordinary temperature. Whilst with many an acceleration of development of 1—2 months occurred, others emerged in April and May, i. e. at the time of their appearance in the natural state.

11. Reared the second generation from eggs of the first generation. Emerged from the eggs on June 6th, 1872, pupated on July 9th. The pupæ were placed on ice (0°—1° R.) from July 11th till September 11th, and then transferred to a hothouse, where all emerged :—

On the 19th of September,		3 male *Prorsa*, 1 male *Porima*.
,,	21st ,,	13 *Porima* (12 males, 1 female) 2 female *Levana*.
,,	22nd ,,	14 *Porima* (12 males, 2 females) and 1 female *Levana*.
,,	23rd ,,	10 female *Levana*, 3 male *Porima*.
,,	24th ,,	5 female *Levana*.
,,	25th ,,	1 female *Levana*.
,,	27th ,,	3 female *Levana*.
,,	4th of October,	1 male *Porima*.

Total . . 57 butterflies (32 males and 25 females), only 3 of which were *Prorsa*, 32 *Porima*, and 22 *Levana*.

It must be pointed out, however, that among those specimens marked as "*Levana*" there were none which entirely corresponded with the natural *Levana*, or which indeed approximated so nearly to this form as did some of the specimens in Exp. 9. All were larger than the natural *Levana*, and possessed, notwithstanding the large amount of yellow, more black than any true *Levana*. In all artificially bred *Levana* the black band of the basal half of the hind wings is always interrupted with yellow, which is seldom the case with true *Levana*. The whole appearance of the artificial *Levana* is also coarser, and the contour of the wings somewhat different, the fore-wings being broader and less pointed. (See figs. 7 to 9, Pl. I.).

12. Larvæ of the fourth generation, found on September 22nd, 1872, were divided into two portions :—

A. Placed for pupation in an orchid-house at 12°—25° R., and allowed to remain there till December. In spite of the high temperature not a single butterfly emerged during this time, whilst pupæ of *Vanessa C-album* and *Pyrameis Atalanta*, found at the same time, and placed in the same hothouse, emerged in the middle of October. From the middle of December the pupæ were kept in an unheated room, and they emerged very late in the spring of 1873, all as *Levana* :—

On the 6th	of June,	7	*Levana*.
„ 8th	„	2	„
„ 11th	„	2	„
„ 12th	„	1	„
„ 15th	„	6	„
„ 16th	„	1	„
„ 19th	„	2	„
Total . .		21	„

B. Kept in an unheated room during the winter. The butterflies emerged from the 28th of May, all as *Levana*.

Experiments with Pierinæ.

13. Females of *Pieris Rapæ*, captured in April, laid eggs on *Sisymbrium Alliaria*. From these caterpillars were obtained, which pupated on 1st—3rd June. The pupæ were placed on ice from June 3rd till September 11th (0°—1° R.), and from September 11th till October 3rd in the hothouse (12°—24° R.), where there emerged :—

On the 23rd	of October,	1 female.
„ 24th	„	1 female.
„ 25th	„	2 males, 1 female.
„ 26th	„	1 female.
„ 28th	„	1 male, 1 female.
Total . .		3 males, 5 females.

All these were sharply impressed with the characters of the winter form, the females all strongly yellow on the upper side, the males pure white ; on the under side a strong black dusting on the hind wings, particularly on the discoidal cell. One pupa did not emerge in the hothouse, but hibernated, and gave in a heated room on January 20th, 1873, a female, also of the winter form.

14. Females of *Pieris Napi*, captured on 27th—28th April, 1872, laid eggs on *Sisymbrium Alliaria*. The larvæ bred from these pupated on May 28th to June 7th. The pupæ, shortly after transformation, were placed on ice, where they remained till Sept. 11th (three months). Transferred to the hothouse on October 3rd, they produced, up to October 20th, 60 butterflies, all with the sharply-defined characters of the winter form. The remaining pupæ hibernated in a room, and produced :—

On the	28th of April,	3 males, 6 females.
„	4th of May,	1 female.
„	12th „	4 males.
„	15th „	1 male, 1 female.
„	16th „	1 male.
„	18th „	1 male, 1 female.
„	19th „	1 female.
„	20th „	2 males, 1 female.
„	23rd „	2 males.
„	26th „	1 male.
„	29th „	1 female.
„	3rd of June,	3 females.
„	6th „	1 female.
„	9th „	1 female.
„	21st „	1 female.
„	2nd of July,	1 female.
	Total . .	15 males, 19 females.

15. Several butterflies from Exp. 14, which emerged in May, 1873, were placed in a capacious breeding-

house, where they copulated and laid eggs on rape. The caterpillars fed on the living plants in the breeding-house, and after pupation were divided into two portions :—

A. Several pupæ, kept at the ordinary summer temperature, gave butterflies on July 2nd, having the characters of the summer form.

B. The remainder of the pupæ were placed on ice immediately after transformation, and remained over three months in the refrigerator (from July 1st till October 10th). Unfortunately most of them perished through the penetration of moisture into the box. Only 8 survived, 3 of which emerged on the 20th of October as the winter form; the others hibernated in an unheated room, and emerged at the beginning of June, 1874. All 5 were females, and all exhibited the characters of the winter form. Notwithstanding a pupal period of eleven months, they did not possess these characters to a greater extent than usual, and did not, therefore, approximate to the parent form *Bryoniæ.*

16. On June 12th, 1871, specimens of *Pieris Napi,* var. *Bryoniæ,* were captured on a mountain in the neighbourhood of Oberstorf (Allgäuer Alpen), and placed in a breeding-house, where they flew freely about the flowers; but although copulation did not take place, several females laid eggs on the ordinary garden cabbage. From these caterpillars were hatched, which at all stages of growth were exactly like those of the ordinary form of *Napi.* They throve well until shortly before pupation, when a fungoid epidemic decimated them, so that from 300 caterpillars only about 40 living pupæ were obtained. These also completely resembled the ordinary form of *Napi,* and showed the same polymorphism, some being beautifully green, others (the majority) straw yellow, and others yellowish grey. Only one butterfly emerged the same summer, a male, which, by the black

dusting of the veins on the margin of the wings (upper side), could be with certainty recognized as var. *Bryoniæ.* The remaining pupæ hibernated in a heated room, and gave, from the end of January to the beginning of June, 10 males and 5 females, all with the characters of the var. *Bryoniæ.* They emerged :—

On the	22nd of January,	1 male.
„	26th „	1 male.
„	3rd of February,	1 male.
„	4th „	1 male.
„	5th „	1 male.
„	7th „	1 female.
„	9th „	1 male.
„	24th „	1 male.
„	4th of March,	1 female.
„	11th „	1 male, 1 female.
„	6th of April,	1 female.
„	17th „	1 male.
„	11th of May,	1 female.
„	3rd of June,	1 male.

We here perceive that the tendency to accelerate development through the action of warmth is, in this case, also very different in different individuals. Of the 16 butterflies only 1 kept to the normal period of development from July 27th to June 3rd, fully ten months ; all the others had this period abbreviated, 1 male to eleven days, 8 specimens to six months, 4 to seven months, 2 to eight months, and 1 to nine months.

APPENDIX II.

Experiments with Papilio Ajax.[1]

From eggs of var. *Telamonides* laid on the last of May larvæ were obtained, which gave on June 22nd—26th, 122 pupæ. These, as fast as formed, were placed on ice in the refrigerator in small tin boxes, and when all the larvæ had become transformed the pupæ were transferred to a cylindrical tin box (4 in. diam. and 6 in. high), and packed in layers between fine shavings. The tin box was set in a small wooden one, which was put directly on the ice and kept there till July 20th. From that date, by an unfortunate accident, the box, instead of being kept on the surface of the ice in an ice-house, as was intended, was placed on straw near the ice, so that the action of the cold was modified, the outside pupæ certainly experiencing its full effects, but the inside ones were probably at a somewhat higher temperature. The ice failed on August 20th, so that the pupæ had been subjected to an equable low temperature in the refrigerator for three to four weeks, and to a lesser degree of cold in the ice-house for five weeks, the tem-

[1] The experiments upon *Papilio Ajax* and *Phyciodes Tharos*, described in this Appendix, were made by Mr. W. H. Edwards (see his "Butterflies of North America;" also the "Canadian Entomologist," vol. vii. p. 228—240, and vol. ix. p. 1—10, 51—5, and 203—6); and I have added them, together with some hitherto unpublished results, to Dr. Weismann's Essay, in order to complete the history of the subject of seasonal dimorphism up to the present time.—R.M.

perature of the last place rising daily, as the ice had all thawed by August 20th. On opening the box it was found (probably owing to the cold not having been sufficiently severe) that the butterflies had commenced to emerge. Twenty-seven dead and crippled specimens were removed, together with several dead pupæ. One butterfly that had just emerged was taken out and placed in a box, and when its wings had fully expanded it was found to be a "*Telamodides* of the most pronounced type." The experimenter then states:—"Early in the morning I made search for the dead and rejected butterflies, and recovered a few. It was not possible to examine them very closely from the wet and decayed condition they were in, but I was able to discover the broad crimson band which lies above the inner angle of the hind wings, and which is usually lined on its anterior side with white, and is characteristic of either *Walshii* or *Telamonides*, but is not found in *Marcellus*. And the tip only of the tail being white in *Walshii*, while both tip and sides are white in *Telamonides*, enabled me to identify the form as between these two. There certainly were no *Walshii*, but there seemed to be a single *Marcellus*, and excepting that all were *Telamonides*."

The remaining pupæ were kept in a light room where 3 *Telamonides* emerged the following day, and by September 4th 14 specimens of the same variety had emerged, but no *Marcellus* or intermediate forms. From the 4th to the 20th of September a few more *Telamonides* appeared, but between the 4th and 15th of the month 12 out of 26 butterflies that had emerged were intermediate between *Telamonides* and *Marcellus*, some approximating to one form and some to the other form. The first pure *Marcellus* appeared on September 4th, and was followed by one specimen on the 6th, 8th, 13th and 15th respectively. From this last date to

October 3rd, 6 out of 10 were *Marcellus* and 3 intermediate. On September 3rd, a specimen intermediate between *Telamonides* and *Walshii* emerged, "in which the tails were white tipped as in *Walshii*, but in size and other characters it was *Telamonides*, though the crimson band might have belonged to either form." Butterflies continued to emerge daily up to September 20th, after which date single specimens appeared at intervals of from four to six days, the last emergence being on October 16th. Thus, from the time the box was removed from the ice-house, the total period of emerging was fifty-seven days, some specimens having emerged before the removal of the box. With specimens of *P. Ajax* which appear on the wing the first season the natural pupal period is about fourteen days, individuals rarely emerging after a period of four to six weeks.

Between August 20th and October 16th, the 50 following butterflies emerged :—

On the	20th	of August,	1 male *Telamonides*.
"	21st	"	1 male and 2 female *Telamonides*.
"	22nd	"	1 female *Telamonides*.
"	24th	"	1 female *Telamonides*.
"	29th	"	1 male *Telamonides*.
"	31st	"	1 female *Telamonides*.
"	1st	September	1 female *Telamonides*.
"	2nd	"	1 female *Telamonides*.
"	3rd	"	1 female intermediate between *Telamonides* and *Walshii*.
"	"	"	1 male *Telamonides*.
"	4th	"	4 males and 1 female *Telamonides*.
"	"	"	2 males, medium, nearest *Telamonides*.

On the	4th	of September,	2 males, medium, nearest *Marcellus.*
„	„	„	2 males, *Marcellus.*
„	5th	„	1 male and 1 female *Telamonides.*
„	„	„	1 male medium, nearest *Telamonides.*
„	6th	„	1 male *Marcellus.*
„	7th	„	1 male *Telamonides.*
„	8th	„	1 male *Marcellus* and 1 female *Telamonides.*
„	9th	„	1 male *Marcellus* and 1 female medium, nearest *Marcellus.*
„	13th	„	1 male medium, nearest *Marcellus.*
„	„	„	1 male medium, nearest *Telamonides.*
„	„	„	1 male *Marcellus*
„	14th	„	1 male *Marcellus* and 1 female medium, nearest *Marcellus.*
„	„	„	1 male medium, nearest *Telamonides.*
„	15th	„	1 male *Marcellus.*
„	16th	„	1 female *Marcellus* and 1 male *Telamonides.*
„	18th	„	1 male medium, nearest *Marcellus.*
„	19th	„	1 female *Marcellus.*
„	20th	„	1 male *Telamonides.*
„	24th	„	1 male *Marcellus.*
„	30th	„	1 female *Marcellus.*
„	2nd	of October,	1 female *Marcellus.*
„	3rd	„	1 female medium, nearest *Telamonides.*

On the 8th of October, 1 female medium, nearest *Telamonides.*
„ 16th „ 1 female medium, nearest *Telamonides.*

Total.

Telamonides	22	12 males,	10 females
Telamonides partly *Walshii* .	1		1 female.
Medium, nearest *Telamonides*	8	5 males,	3 females
Medium, nearest *Marcellus* .	6	4 males,	2 females
Marcellus	13	9 males,	4 females.
	50	30 males,	20 females.

All these butterflies were very uniform in size, being about that of the ordinary *Telamonides.* The specimens of *Telamonides* especially were "strongly marked, the crimson band in a large proportion of them being as conspicuous as is usual in *Walshii,* and the blue lunules near the tail were remarkably large and bright coloured. Of the *Marcellus,* in addition to the somewhat reduced size, the tails were almost invariably shorter than usual and narrower, and instead of the characteristic single crimson spot, nearly all had two spots, often large. In all these particulars they approach *Telamonides.*"

Adding to the *Telamonides* which emerged after August 20th most of those specimens which were found dead in the box at that date, the total number of this form is thus brought up to nearly 50. Of the 122 pupæ with which Mr. Edwards started, 28 remained in a state fit for hibernation, several having died without emerging. Previous experiments had shown that 28 out of 122 pupæ is not an unreasonalble number to hibernate, so that the author concludes that the butterflies which emerged the same season would have done so naturally, and the effect of the artificial cold was not "to precipitate the emerging of any which would have slept" till the following spring. Now under ordinary circumstances all the butterflies which emerged the same season would have been of the *Marcellus* form, so that the cold changed a large part of these into the form *Telamonides*, some (probably from those pupæ which experienced the lowest temperature) being completely changed, and others (from those pupæ which were only imperfectly subjected to the cold) being intermediate, *i. e.*, only partly changed. It appears also that several pupæ experienced sufficient cold to retard their emergence and stunt their growth, but not enough to change their form, these being the 13 recorded specimens of *Marcellus*. Had the degree of cold been equal and constant, the reversion would probably have been more complete. The application of cold produced great confusion in the duration of the pupal period, the emergence, instead of taking place fourteen days after the withdrawal of the cold, as might have been expected from Dr. Weismann's corresponding experiment with *Pieris Napi* (Appendix I. Exps. 13 and 14), having been extended over more than two months.

From the results of this experiment it must be concluded that *Telamonides* is the primary form of the species.

Additional Experiments with Papilio Ajax.

[*Communicated by* Mr. W. H. Edwards, *November* 18*th*, 1879.]

Exp. 1.—In 1877 chrysalides of *P. Ajax* and *Grapta Interrogationis* (the eggs laid by females of the form *Fabricii)* were experimented upon; but the results were not satisfactory, for the reason that the author having been absent from home most of the time while the pupæ were in the ice-box, on his return found the temperature above 5°—6° R. And so far as could be told, the ice had been put in irregularly, and there might have been intervals during which no ice at all was in the box. Six chrysalides of the *Grapta* so exposed produced unchanged *Umbrosa*, the co-form with *Fabricii*. But all chrysalides from the same lot of eggs, and not exposed to cold, also produced *Umbrosa*. Nothing was learnt, therefore, respecting this species.

But chrysalides of *Ajax*, exposed at same time, did give changed butterflies to some extent. From a lot of 8, placed in the box when under twelve hours from pupation, and left for twenty-four days, there came 5 males and 3 females. Of these was 1 *Telamonides* in markings and coloration, and all the rest were between *Marcellus* and *Telamonides*. Two other chrysalides on ice for twenty-three days gave *Telamonides*, but 3 more exposed twenty-six days, and all one hour old when put on ice, were unchanged, producing *Marcellus*.

During the same season 6 other *Ajax* chrysalides were placed in the box, and kept at about 0°—1° R. One was one hour old, and remained for five days; 1 was one hour old, and remained for two days and three-quarters; 3 at three hours old for eight days; and 1 (age omitted), six days. All these gave unchanged butterflies of the form *Marcellus*.

Exp. 2.—In May, 1878, many chrysalides were placed in the ice-box, being from eggs laid by *Ajax*, var. *Walshii*. The youngest were but ten to fifteen minutes from pupation, and were soft; others at intervals up to twenty-four hours (the chrysalis is hard at about twelve hours); after that, each day up to eight days after pupation. All were removed from the box on the same day, 28th May. The exposure was from nineteen to five days, those chrysalides which were put on ice latest having the shortest exposure. The author wished to determine if possible whether, in order to effect any change, it was necessary that cold should be applied immediately after pupation or if one or several days might intervene between pupation and refrigeration. Inasmuch as no colour begins to show itself in the pupæ till a few hours, or at most a day or two, before the butterfly emerges, it was thought possible that cold applied shortly before that time would be quite as effective as if applied earlier and especially very soon after pupation. The result was, that more than half of the chrysalides exposed before they had hardened died: 1 exposed at ten minutes, 2 at one hour, 1 at two hours, 2 at three hours after pupation. On the other hand 1 at fifteen minutes produced a butterfly, 1 at two hours, another at twelve hours. The temperature was from 0°—1° R. most of the time, but varied somewhat each day as the ice melted. The normal chrysalis period is from eleven to fourteen days, in case the butterfly emerges the same season, but very rarely an individual will emerge several weeks after pupation.

On the 14th day after taking the pupæ from the ice, 1 *Telamonides* emerged from a chrysalis which had been placed in the ice-box three days after pupation, and was on ice sixteen days.

On 19th day, 1 *Telamonides* emerged from a pupa put on the ice twelve hours after pupation, and kept there eleven days.

On 19th day, 1 *Walshii* emerged from a pupa two hours old, and on ice eleven days.

All the rest emerged *Marcellus*, unchanged, but at periods prolonged in a surprising way.

1 on	43rd day	exposed	15 minutes	after pupation.
„	46th	„	2 hours	„
„	53rd	„	24 hours	„
„	62nd	„	6 days	„
„	63rd	„	4 days	„
„	66th	„	7 days	„
„	77th	„	4 days	„
„	81st	„	12 hours	„
„	91st	„	5 days	„
„	96th	„	19 hours	„

Five chrysalides lived through the winter, and all gave *Telamonides* in the spring of 1879.

It appeared, therefore, that the only effect produced by cold in all chrysalides exposed more than three days after pupation was to retard the emergence of the butterfly. But even in some of these earliest exposed, and kept on the ice for full nineteen days, the only effect seemed to be to retard the butterfly.

EXP. 3.—In June, 1879, eggs of the form *Marcellus* were obtained, and in due time gave 104 chrysalides. Of these one-third were placed in the ice-box at from twelve to twenty-four hours after pupation, and were divided into 3 lots.

1st, 9 pupæ, kept on ice			14 days.
2nd, 12	"	"	20 days.
3rd, 11	"	"	25 days.

Temperature 0°—1° R. most of the time, but varying somewhat as the ice melted. (Both in 1878 and 1879 Mr. Edwards watched the box himself, and endeavoured to keep a low temperature.)

Of the 69 chrysalides not exposed to cold, 34 gave butterflies at from eleven to fourteen days after pupation, and 1 additional male emerged 11th August, or twenty-two days at least past the regular period of the species.

Of the iced chrysalides, from lot No. 1 emerged 4 females at eight days and a half to nine days and a half after removal from the ice, and 5 are now alive (Nov. 18) and will go over the winter.

From lot No. 2 emerged 1 male and 5 females at eight to nine days; another male came out at forty days; and 5 will hibernate.

From lot No. 3 emerged 4 females at nine to twelve days; another male came out at fifty-four days; and 6 were found to be dead.

In this experiment the author wished to see as exactly as possible—First, in what points changes would occur.

Second, if there would be any change in the shape of the wings, as well as in markings or coloration—that is, whether the shape might remain as that of *Marcellus*, while the markings might be of *Telamonides* or *Walshii;* a summer form with winter markings. Third, to ascertain more closely than had yet been done what length of exposure was required to bring about a decided change, and what would be the effect of prolonging this period. After the experiments with *Phyciodes Tharos*, which had resulted in a suffusion of colour, the author hoped that some similar cases might be seen in *Ajax.* The decided changes in 1878 had been produced by eleven and sixteen days' cold. In 1877, an exposure of two days and three-quarters to eight days had failed to produce an effect.

From these chrysalides 11 perfect butterflies were obtained, 1 male and 10 females. Some emerged crippled, and these were rejected, as it was not possible to make out the markings satisfactorily.

From lot No. 1, fourteen days came:—

1 female between *Marcellus* and *Telamonides.*

2 females, *Marcellus.*

These 2 *Marcellus* were pale coloured, the light parts a dirty white; the submarginal lunules on hind wings were only two in number and small; at the anal angle was one large and one small red spot; the frontal hairs were very short. The first, or intermediate female, was also pale black, but the light parts were more green and less sordid; there were 3 large lunules; the anal red spot was double and connected, as in *Telamonides;* the frontal hairs short, as in *Marcellus.* These are the most salient points for comparing the several forms of *Ajax.* In nature, there is much difference in shape between *Marcellus* and *Telamonides*, still more between *Marcellus* and *Walshii*; and the latter may be distinguished from the other winter forms by the white tips of the tails. It is also smaller, and the anal spot is larger, with a broad white edging.

From lot No. 2, twenty days, came :—

1 female *Marcellus,* with single red spot.

1 female between *Marcellus* and *Telamonides ;* general coloration pale ; the lunules all obsolescent ; 2 large red anal spots not connected ; frontal hairs medium length, as in *Telamonides.*

1 female between *Marcellus* and *Telamonides ;* colour bright and clear ; 3 lunules ; 2 large red spots ; frontal hairs short.

1 female *Telamonides ;* colours black and green ; 4 lunules ; a large double and connected red spot ; frontal hairs medium.

2 female *Telamonides ;* colours like last ; 3 and 4 lunules ; 2 large red spots ; frontal hairs medium.

From lot No. 3, twenty-five days, came:—

1 male *Telamonides ;* clear colours ; 4 large lunules ; 1 large, 1 small red spot ; frontal hairs long.

1 female *Telamonides ;* medium colours ; 4 lunules ; large double connected red spot ; frontal hairs long.

In general shape all were *Marcellus,* the wings produced, the tails long.

From this it appeared that those exposed twenty-five days were fully changed ; of those exposed twenty days, 3 were fully, 2 partly, 1 not at all ; and of those exposed fourteen days, 1 partly, 2 not at all.

The butterflies from this lot of 104 chrysalides, but which had not been iced, were put in papers. Taking 6 males and 6 females from the papers just as they came to hand, Mr. Edwards set them, and compared them with the iced examples.

Of the 6 males, 4 had 1 red anal spot only, 2 had 1 large 1 small ; 4 had 2 green lunules on the hind wings, 2 had

3, and in these last there was a 4th obsolescent, at outer angle; all had short frontal hairs.

Of the 6 females, 5 had but 1 red spot, 1 had 1 large 1 small spot; 5 had 2 lunules only, 1 had 3; all had short frontal hairs.

Comparing 6 of the females from the iced chrysalides, being those in which a change had more or less occurred, with the 6 females not iced:

1. All the former had the colours more intense, the black deeper, the light, green.
2. In 5 of the former the green lunules on hind wings were decidedly larger; 3 of the 6 had 4 distinct lunules, 1 had 3, 1 had 3, and a 4th obsolescent. Of the 6 females not iced none had 4, 2 had 2, and a 3rd, the lowest of the row, obsolescent; 3 had 3, the lowest being very small; one had 3, and a 4th, at outer angle, obsolescent.
3. In all the former the subapical spot on fore wing and the stripe on same wing which crosses the cell inside the common black band, were distinct and green; in all the latter these marks were either obscure or obsolescent.
4. In 4 of the former there was a large double connected red spot, and in one of the 4 it was edged with white on its upper side; 2 had 1 large and 1 small red spot. Of the latter 5 had 1 spot only, and the 6th had 1 spot and a red dot.
5. The former had all the black portions of the wing of deeper colour but less diffused, the bands being narrower; on the other hand, the green bands were wider as well as deeper coloured. Measuring the width of the outermost common green band along the middle of the upper medium interspace on fore wing in tenths of a millimetre, it was found to be as follows:

On the iced pupæ . . . 81, 66, 76, 76, 66, 66.
On the not iced . . . 56, 56, 51, 51, 46, 51.

Measuring the common black discal band across the middle of the lower medium interspace on fore wing:

On the iced pupæ . . . 51, 66, 51, 51, 56, 61.
On the not iced . . . 76, 71, 66, 63, 71, 76.

In other words the natural examples were more melanic than the others.

No difference was found in the length of the tails or in the length and breadth of wings. In other words, the cold had not altered the shape of the wings.

Comparing 1 male iced with 6 males not iced:

1, The former had a large double connected red anal spot, edged with white scales at top. Of the 6 not iced, 3 had but 1 red spot, 2 had 1 large 1 small, 1 had 1 large and a red dot.
2. The former had 4 green lunules; of the latter 3 had 3, 3 had only 2.
3. The former had the subapical spot and stripe in the cells clear green; of the latter 1 had the same, 5 had these obscure or obsolescent.
4. The colours of the iced male were bright; of the others, 2 were the same, 4 had the black pale, the light sordid white or greenish-white.

Looking over all, male and female, of both lots, the large size of the green submarginal lunules on the fore wings in the iced examples was found to be conspicuous as compared with all those not iced, though this feature is included in the general widening of the green bands spoken of.

In all the experiments with *Ajax*, if any change at all has been produced by cold, it is seen in the enlarging or doubling of the red anal spot, and in the increased number of clear green lunules on the hind wings. Almost always the frontal hairs are lengthened and the colour of the wings deepened, and the extent of the black area is also diminished. All these changes are in the direction of *Telamonides*, or the winter form.

That the effect of cold is not simply to precipitate the appearance of the winter form, causing the butterfly to emerge from the chrysalis in the summer in which it began its larval existence instead of the succeeding year, is evident from the fact that the butterflies come forth with the shape of *Marcellus*, although the markings may be of *Telamonides* or *Walshii*. And almost always some of the chrysalides, after having been iced, go over the winter, and then produce *Telamonides*, as do the hibernating pupæ in their natural state. The cold appears to have no effect on these individual chrysalides.[2]

With every experiment, however similar the conditions seem to be, and are intended to be, there is a difference in results ; and further experiments—perhaps many—will be required before the cause of this is understood. For example, in 1878, the first butterfly emerged on the fourteenth day after removal from ice, the period being exactly what it is (at its longest) in the species in nature. Others emerged at 19—96 days. In 1879, the emergence began on the ninth day, and by the twelfth day all had come out, except three belated individuals, which came out at twenty, forty, and fifty-four days. In the last experiment, either the cold had not fully suspended the changes which the insect undergoes in the chrysalis, or its action was to hasten them after the chrysalides were taken from the ice. In the first experiment, apparently the changes were absolutely suspended as long as the cold remained.

It might be expected that the application of heat to the hibernating chrysalides would precipitate the appearance of the summer form, or change the mark-

[2] This is a striking illustration of the diversity of individual constitution so frequently insisted on by Dr. Weismann in the foregoing portion of this work.

ings of the butterfly into the summer form, even if the shape of the wings was not altered; that is, to produce individuals having the winter shape but the summer markings. But this was not found to occur. Mr. Edwards has been in the habit for several years of placing the chrysalides in a warm room, or in the greenhouse, early in the winter, thus causing the butterflies to emerge in February, instead of in March and April, as otherwise they would do. The heat in the house is 19° R. by day, and not less than 3·5°R. by night. But the winter form of the butterfly invariably emerged, usually *Telamonides*, occasionally *Walshii*.

EXPERIMENTS WITH PHYCIODES THAROS.

EXP. 1.—In July, 1875, eggs of *P. Tharos* were obtained on *Aster Nova-Angliæ* in the Catskill Mountains, and the young larvæ, when hatched, taken to Coalburgh, West Virginia. On the journey the larvæ were fed on various species of *Aster*, which they ate readily. By the 4th of September they had ceased feeding (after having twice moulted), and slept. Two weeks later part of them were again active, and fed for a day or two, when they gathered in clusters and moulted for the third time, then becoming lethargic, each one where it moulted with the cast skin by its side. The larvæ were then placed in a cellar, where they remained till February 7th, when those that were alive were transferred to the leaves of an *Aster* which had been forced in a greenhouse, and some commenced to feed the same day. In due time they moulted twice more, making, in some cases, a total of five moults. On May 5th the first larva pupated, and its butterfly emerged after thirteen days. Another emerged on the 30th, after eight days pupal period, this stage being shortened as the weather became warmer. There emerged altogether 8 butterflies, 5 males and 3 females, all of the form *Marcia*, and all of the variety designated C, except 1 female, which was var. B.[3]

Exp. 2.—On May 18th the first specimens (3 male *Marcia*) were seen on the wing at Coalburgh; 1 female was taken on the 19th, 2 on the 23rd, and 2 on the 24th, these being all that were seen up to that date, but shortly after both sexes became common. On the 26th, 7 females were captured and tied up in separate bags on branches of *Aster*. The next day 6 out of the 7 had laid eggs in clusters containing from 50 to 225 eggs in each. Hundreds of caterpillars were obtained, each brood being kept separate, and the butterflies began to emerge on June 29th, the several stages being:—egg six days, larva twenty-two, chrysalis five. Some of the butterflies did not emerge till the 15th of July. Just after this date one brood was taken to the Catskills, where they pupated, and in this state were sent back to Coalburgh. There was no difference in the length of the different stages of this brood and the others which had been left at Coalburgh, and none of either lot became lethargic. The butterflies from these eggs of May were all *Tharos*, with the exception of 1 female *Marcia*, var. C. Thus the first generation of *Marcia* from the hibernating larvæ furnishes a second generation of *Tharos*.

Exp. 3.—On July 16th, at Coalburgh, eggs were obtained from several females, all *Tharos*, as no other form was flying. In four days the eggs hatched; the larval stage was twenty-two, and the pupal stage seven days; but, as before, many larvæ lingered. The first butterfly emerged on August 18th. All were *Tharos*, and none

[3] The reader who wishes to acquire a detailed knowledge of the different varieties of this butterfly, of which a very large number are known, must consult the plates and descriptions in Edwards' "Butterflies of North America," vol. ii..

of the larvæ had been lethargic. This was the third generation from the second laying of eggs.

EXP. 4.—On August 15th, at Coalburgh, eggs were obtained from a female *Tharos*, and then taken directly to the Catskill Mountains, where they hatched on the 20th. This was the fourth generation from the third laying of eggs. In Virginia, and during the journey, the weather had been exceedingly warm, but on reaching the mountains it was cool, and at night decidedly cold. September was wet and cold, and the larvæ were protected in a warm room at night and much of the time by day, as they will not feed when the temperature is less than about 8° R. The first pupa was formed September 15th, twenty-six days from the hatching of the larvæ, and others at different dates up to September 26th, or thirty-seven days from the egg. Fifty-two larvæ out of 127 became lethargic after the second moult on September 16th, and on September 26th fully one half of these lethargic larvæ commenced to feed again, and moulted for the third time, after which they became again lethargic and remained in this state. The pupæ from this batch were divided into three portions:—

A. This lot was brought back to Coalburgh on October 15th, the weather during the journey having been cold with several frosty nights, so that for a period of thirty days the pupæ had at no time been exposed to warmth. The butterflies began to emerge on the day of arrival, and before the end of a week all that were living had come forth, viz., 9 males and 10 females. "Of these 9 males 4 were changed to *Marcia* var. C, 3 were var. D, and 2 were not changed at all. Of the 10 females 7 were changed, 5 of them to var. B, 3 to var. C. The other 2 females were not different from many *Tharos* of the summer brood, having large discal patches on under side of hind wing, besides the markings common to the summer brood."

B. This lot, consisting of 10 pupæ, was sent from the Catskills to Albany, New York, where they were kept in a cool place. Between October 21st and Nov. 2nd, 6 butterflies emerged, all females, and all of the var. B. Of the remaining pupæ 1 died, and 3 were alive on December 27th. According to Mr. Edwards this species never hibernates in the pupal state in nature. The butterflies of this lot were more completely changed than were those from the pupæ of lot A.

C. On September 20th 18 of the pupæ were placed in a tin box directly on the surface of the ice, the temperature of the house being 3°—4° R. Some were placed in the box within three hours after transformation and before they had hardened; others within six hours, and others within nine hours. They were all allowed to remain on the ice for seven days, that being the longest summer period of the chrysalis. On being removed they all appeared dead, being still soft, and when they had become hard they had a shrivelled surface. On being brought to Coalburgh they showed no signs of life till October 21st, when the weather became hot (24°—25° R.), and in two days 15 butterflies emerged, "every one *Marcia*, not a doubtful form among them in either sex." Of these butterflies 10 were males and 5 females; of the former 5 were var. C, 4 var. D, and 1 var. B, and of the latter 1 was var. C, and 4 var. D. The other 3 pupæ died. All the butterflies of this brood were diminutive, starved by the cold, but those from the ice were sensibly smaller than the others. All the examples of var. B were more intense in the colouring of the under surface than any ever seen by Mr. Edwards in nature, and the single male was as deeply coloured as the females, this also never occurring in nature.

Mr. Edwards next proceeds to compare the behaviour

of the Coalburgh broods with those of the same species in the Catskills :—

EXP. 5.—On arriving at the Catskills, on June 18th, a few male *Marcia*, var. D, were seen flying, but no females. This was exactly one month later than the first males had been seen at Coalburgh. The first female was taken on June 26th, another on June 27th, and a third on the 28th, all *Marcia*, var. C. Thus the first female was thirty-eight days later than the first at Coalburgh. No more females were seen, and no *Tharos*. The three specimens captured were placed on *Aster*, where two immediately deposited eggs [4] which were forwarded to Coalburgh, where they hatched on July 3rd. The first chrysalis was formed on the 20th, its butterfly emerging on the 29th, so that the periods were : egg six, larva seventeen, pupa nine days. Five per cent. of the larvæ became lethargic after the second moult. This was, therefore, the second generation of the butterfly from the first laying of eggs. All the butterflies which emerged were *Tharos*, the dark portions of the wings being intensely black as compared with the Coalburgh examples, and other differences of colour existed, but the general peculiarities of the *Tharos* form were retained. This second generation was just one month behind the second at Coalburgh, and since, in 1875, eggs were obtained by Mr. Mead on July 27th and following days, the larvæ from which all hibernated, this would be the second laying of eggs, and the resulting butterflies the first generation of the following season.

Thus in the Catskills the species is digoneutic, the first generation being *Marcia* (the winter form), and the second the summer form. A certain proportion of the

[4] Mr. Edwards has shown also that *Argynnis Myrina* can lay fertile eggs when but a few hours out of the chrysalis. Canad. Ent., September, 1876, vol. viii. No. 9.

larvæ from the first generation hibernate, and apparently all those from the second.

Discussion of Results.—There are four generations of this butterfly at Coalburgh, the first being *Marcia* and the second and third *Tharos*. None of the larvæ from these were found to hibernate. The fourth generation under the exceptional conditions above recorded (Exp. 4) produced some *Tharos* and more *Marcia* the same season, a large proportion of the larvæ also hibernating. Had the larvæ of this brood been kept at Coalburgh, where the temperature remained high for several weeks after they had left the egg, the resulting butterflies would have been all *Tharos*, and the larvæ from their eggs would have hibernated.

The altitude of the Catskills, where Mr. Edwards was, is from 1650 to 2000 feet above high water, and the altitude of Coalburgh is 600 feet. The transference of the larvæ from the Catskills to Virginia (about 48° lat.) and *vice-versa*, besides the difference of altitude, had no perceptible influence on the butterflies of the several broods except the last one, in which the climatic change exerted a direct influence on part of them both as to form and size. The stages of the June Catskill brood may have been accelerated to a small extent by transference to Virginia, but the butterflies reserved their peculiarities of colour. (See Exp. 5.) So also was the habit of lethargy retained.[5] The May brood, on the other hand,

[5] Mr. Edwards remarks that the habit of becoming lethargie is of great service to a digoneutic species in a mountain region where it is exposed to sharp changes of temperature. "If the fate of the species depended on the last larval brood of the year, and especially if the larvæ must reach a certain stage of growth before they were fitted to enter upon their hibernation, it might well happen that now and then an early frost or a tempestuous season would destroy all the larvæ of the district."

taken from Virginia to the Catskills, suffered no retardation of development. (See Exp. 2.) It might have been expected that all the larvæ of this last brood taken to the mountains would have become lethargic, but the majority resisted this change of habit. From all these facts it may be concluded "that it takes time to naturalize a stranger, and that habits and tendencies, even in a butterfly, are not to be changed suddenly." [6]

It has been shown that *Tharos* is digoneutic in the Catskills and polygoneutic in West Virginia, so that at a great altitude, or in a high latitude, we might expect to find the species monogoneutic and probably restricted to the winter form *Marcia*. These conditions are fulfilled in the Island of Anticosti, and on the opposite coast of Labrador (about lat. 50°), the summer temperature of these districts being about the same. Mr. Edwards states, on the authority of Mr. Cooper, who collected in the Island, that *Tharos* is a rare species there, but has a wide distribution. No specimens were seen later than July 29, after which date the weather became cold, and very few butterflies of any sort were to be seen. It seems probable that none of the butterflies of Anticosti or Labrador produce a second brood. All the specimens examined from these districts were of the winter form.

In explanation of the present case Dr. Weismann wrote to Mr. Edwards :—"*Marcia* is the old primary form of the species, in the glacial period the only one. *Tharos* is the secondary form, having arisen in the course of many generations through the gradually working influence of summer heat. In your experiments cold has caused the summer generation to revert to the primary form. The reversion which occurred was complete in the females, but not in all the males. This

[6] Compare this with Weismann's remarks, pp. 19—22, and 53.

proves, as it appears to me, that the males are changed or affected more strongly by the heat of summer than the females. The secondary form has a stronger constitution in the males than in the females. As I read your letter, it at once occurred to me whether in the spring there would not appear some males which were not pure *Marcia*, but were of the summer form, or nearly resembling it; and when I reached the conclusion of the letter I found that you especially mentioned that this was so! And I was reminded that the same thing is observable in *A. Levana*, though in a less striking degree. If we treated the summer brood of *Levana* with ice, many more females than males would revert to the winter form. This sex is more conservative than the male—slower to change."

The extreme variability of *P. Tharos* was alluded to more than a century ago by Drury, who stated :—" In short, Nature forms such a variety of this species that it is difficult to set bounds, or to know all that belongs to it." Of the different named varieties, according to Mr. Edwards, " A appears to be an offset of B, in the direction most remote from the summer form, just as in *Papilio Ajax*, the var. *Walshii* is on the further side of *Telamonides*, remote from the summer form *Marcellus*." Var. C leads from B through D directly to the summer form, whilst A is more unlike this last variety than are several distinct species of the genus, and would not be suspected to possess any close relationship were it not for the intermediate forms. The var. B is regarded as nearest to the primitive type for the following reasons:— In the first place it is the commonest form, predominating over all the other varieties in W. Virginia, and moreover appears constantly in the butterflies from pupæ submitted to refrigeration. Its distinctive peculiarity of colour occurs in the allied species *P. Phaon* (Gulf States) and *P. Vesta* (Texas), both of which are seasonally dimorphic, and both apparently restricted in their winter broods to the form corresponding to B of *Tharos.* In their summer generation both these species closely resemble the summer form of *Tharos*, and it is remarkable that these two species, which are the only ones (with the exception of the doubtful *Batesii*) closely allied to *Tharos*, should alone be seasonally dimorphic out of the large number of species in the genus.

Mr. Edwards thus explains the case under consideration:—"When *Phaon*, *Vesta*, and *Tharos* were as yet only varieties of one species, the sole coloration was that now common to the three. As they gradually became permanent, or in other words, as these varieties became species, *Tharos* was giving rise to several sub-varieties, some of them in time to become distinct and well marked, while the other two, *Phaon* and *Vesta*, remained constant. As the climate moderated and the summer became longer, each species came to have a summer generation; and in these the resemblance of blood-relationship is still manifest. As the winter generations of each species had been much alike, so the summer generations which sprung from them were much alike. And if we consider the metropolis of the species *Tharos*, or perhaps of its parent species, at the time when it had but one annual generation, to have been somewhere between 37° and 40° on the Atlantic slope, and within which limits all the varieties and sub-varieties of both winter and summer forms of *Tharos* are now found in amazing luxuriance, we can see how it is possible, as the glacial cold receded, that only part of the varieties of the winter form might spread to the northward, and but one of them at last reach the sub-boreal regions and hold possession to this day as the sole representative of the species. And at a very early period the primary form, together with *Phaon* and *Vesta*, had made its way southward, where all three are found now."

Experiments with Grapta Interrogationis.[7]

[*Communicated by* Mr. W. H Edwards, *November* 15*th*, 1879.]

The experiments with this species were made in June, 1879, on pupæ from eggs laid by the summer form *Umbrosa* of the second brood of the year, and obtained by confining a female in a bag on a stem of hop. As the pupæ formed, and at intervals of from six to twenty-four hours after pupation (by which time all the older ones had fully hardened), they were placed in the ice-box. In making this experiment Mr. Edwards had three objects in view. 1st. To see whether it was essential that the exposure should take place immediately after pupation, in order to effect any change. 2ndly. To see how short a period would suffice to bring about any change. 3rdly. Whether exposing the summer pupæ would bring about a change in the form of the resulting butterfly. Inasmuch as breeding from the egg of *Umbrosa*, in June, in a former year,[7] gave both *Umbrosa* (11) and *Fabricii* (6), the butterflies from the eggs obtained, if left to nature, might be expected to be of both forms. The last or fourth brood of the year having been found up to the present time to be *Fabricii*, and the 1st brood of the spring, raised from eggs of *Fabricii* (laid in confinement), having been found to be wholly *Umbrosa*, the latter is probably the summer and *Fabricii* the winter form. The two intervening broods, *i. e.* the 2nd and 3rd, have yielded both forms. This species hibernates in the imago state.

After the pupæ had been in the ice-box fourteen days they were all removed but 5, which were left in six days longer. Several were dead at the end of fourteen days. The temperature most of the time was 1°—2° R. ; but for

[7] See Canad. Ent., vol. ix. p. 69.

a few hours each day rose as the ice melted, and was found to be 3°—6° R.

From the fourteen-day lot 7 butterflies were obtained, 3 males and 4 females. From the twenty-day lot 4 males and 1 female; every one *Umbrosa*. All had changed in one striking particular. In the normal *Umbrosa* of both sexes,[8] the fore wings have on the upper side on the costal margin next inside the hind marginal border, and separated from it by a considerable fulvous space, a dark patch which ends a little below the discoidal nervule; inside the same border at the inner angle is another dark patch lying on the submedian interspace. Between these two patches, across all the median interspaces, the ground colour is fulvous, very slightly clouded with dark.

In all the 4 females exposed to cold for fourteen days a broad black band appeared crossing the whole wing, continuous, of uniform shade, covering the two patches, and almost confluent from end to end with the marginal border, only a streak of obscure fulvous anywhere separating the two. In the case of the females from pupæ exposed for twenty days, the band was present, but while broad, and covering the space between the patches, it was not so dark as in the other females, and included against the border a series of obscure fulvous lunules. This is just like many normal females, and this butterfly was essentially unchanged.

In all the males the patches were diffuse, that at the apex almost coalescing with the border. In the 3 from chrysalides exposed fourteen days these patches were connected by a narrow dark band (very different from the broad band of the females), occupying the same position as the clouding of the normal male, but blackened and somewhat diffused. In the 4 examples

[8] Figures of the different forms of this species are given in vol. i. of Edward's "Butterflies of North America."

from the twenty-day pupæ, this connecting band was scarcely as deeply coloured and continuous as in the other 3. Beyond this change on the submarginal area, whereby a band is created where naturally would be only the two patches, and a slight clouding of the intervening fulvous surfaces, there was no difference of the upper surface apparent between these examples of both sexes, and a long series of natural ones placed beside them.

On the under side all the males were of one type, the colours being very intense. There was considerably more red, both dark and pale, over the whole surface, than in a series of natural examples in which shades of brown and a bluish hue predominate. No charge was observed in the females on the under side.

It appears that fourteen days were as effective in producing changes as a longer period. In fact, the most decided changes were found in the females exposed the shorter period. It also appears that with this species cold will produce change if applied after the chrysalis has hardened. The same experiments were attempted in 1878 with pupæ of *Grapta Comma.* They were put on ice at from ten minutes to six hours after forming, and subjected to a temperature of about 0°—1° R. for eighteen to twenty days, but every pupa was killed. Chrysalides of *Papilio Ajax* in the same box, and partly exposed very soon after pupation, were not injured. It was for this reason that none of the *Interrogationis* pupæ were placed in the box till six hours had passed.

It appears further that cold may change the markings on one part of the wing only, and in cases where it does change dark or dusky markings melanises them;or it may deepen the colours of the under surface (as in the females of the present experiment). The females in the above experiment were apparently most susceptible to the cold, the most decided changes having been effected in them.

The resulting butterflies were all of one form, although both might have been expected to appear under natural circumstances.

Dr. Weismann's remarks on the foregoing experiments. —The author of the present work has, at my request, been good enough to furnish the following remarks upon Mr. Edward's experiments with *G. Interrogationis*:—

The interesting experiments of Mr. Edwards are here principally introduced because they show how many weighty questions in connexion with seasonal dimorphism still remain to be solved. The present experiments do not offer a *direct* but, at most, only an *indirect* proof of the truth of my theory, since they show that the explanation opposed to mine is also in this case inadmissible. Thus we have here, as with *Papilio Ajax*, two out of the four annual generations mixed, *i.e.*, consisting of summer and winter forms, and the conclusion is inevitable that these forms were not produced by the *gradual* action of heat or cold. When, from pupæ of the same generation which are developed under precisely the same external conditions, both forms of the butterfly are produced, the cause of their diversity cannot lie in these conditions. It must rather depend on causes innate in the organism itself, *i. e.*, on inherited duplicating tendencies which meet in the same generation, and to a certain extent contend with each other for precedence. The two forms must have had their origin in earlier generations, and there is nothing against the view that they have arisen through the gradual augmentation of the influences of temperature.

In another sense, however, one might perceive, in the facts discovered by Edwards, an objection to my theory.

By the action of cold the form *Umbrosa*, which flies in June, was produced. Now we should be inclined to regard the var. *Umbrosa* as the summer form, and the var. *Fabricii*, which emerges in the autumn, hibernates in the imago state, and lays eggs in the spring, as the winter form. It would then be incomprehensible why the var. *Umbrosa* (*i.e.*, the summer form) should be produced by cold.

But it is quite as possible that the var. *Umbrosa* as that the var. *Fabricii* is the winter form. We must not forget that, in this species, *not one of the four annual generations is exposed to the cold of winter in the pupal state.* When, therefore, we have in such cases seasonal dimorphism, to which complete certainty can only be given by continued observations of this butterfly, which does not occur very commonly in Virginia, this must depend on the fact that the species formerly hibernated in the pupal stage. This question now arises, which of the existing generations was formerly the hibernating one—the first or the last?

Either may have done so *à priori*, according as the summer was formerly shorter or longer than now for this species. If the former were the case, the var. *Fabricii* is the older winter form; were the latter the case, the var. *Umbrosa* is the original winter form, as shall now be more closely established.

Should the experiments which Mr. Edwards has performed in the course of his interesting investigations be repeated in future with always the same results, I should be inclined to explain the case as follows:—

It is not the var. *Fabricii*, but *Umbrosa*, which is the winter generation. By the northward migration of the species and the relative shortening of the summer, this winter generation would be pushed forward into the summer, and would thereby lose only a portion of the winter characters which it had till that time possessed. The last of the four generations which occurs in Virginia is extremely rare, so that it must be regarded either as one of the generations now supposed to be originating, or as one now supposed to be disappearing. The latter may be admitted. Somewhat further north this generation would be entirely suppressed, and the third brood would hibernate, either in the imago state or as pupæ or caterpillars. In Virginia this third generation consists of both forms. We may expect that further north, at least, where it hibernates as pupæ, it will consist entirely, or almost entirely, of the var. *Umbrosa.* Still further north in the Catskill Mountains, as Edwards states from his own observations, the species has only two generations, and one might expect that the var. *Umbrosa* would there occur as the winter generation.

Should the foregoing be correct, then the fact that the second generation assumes the *Umbrosa* form by the action of cold, as was the case in Edward's experiments, becomes explicable. The new marking peculiar to this form produced by this means must be regarded as a complete reversion to the true winter form, the characters of which are becoming partly lost as this generation is no longer exposed to the influence of winter, but has become advanced to the beginning of summer.

The foregoing explanation is, of course, purely hypothetical; it cannot at present be asserted that it is the correct one. Many investigations based on a sufficiently large number of facts are still necessary to be able to attempt to explain this complicated case with any certainty. Neither should I have ventured to offer any opinion on the present case, did I not believe that even such a premature and entirely uncertain explanation may always be of use in serving the inventive principle, *i. e.*, in pointing out the way in which the truth must be sought.

As far as I know, no attempt has yet been made to prove from a general point of view the interpolation of new generations, or the omission of single generations from the annual cycle, with respect to causes and effects. An investigation of this kind would be of the greatest importance, not only for seasonal dimorphism, but also for the elucidation of questions of a much more general nature, and would be a most satisfactory problem for the scientific entomologist. I may here be permitted to develope in a purely theoretical manner the principles in accordance with which such an investigation should be made :—

On the change in the number of generations of the annual cycle.—A change in the number of generations which a species produces annually must be sought chiefly in changes of climate, and therefore in a lengthening or shortening of the period of warmth, or in an increase or diminution of warmth within this period ; or, finally, in both changes conjointly. The last case would be of the most frequent occurrence, since a lengthening of the period of warmth is, as a rule, correlated with an elevation of the mean temperature of this period, and *vice versâ.* Of other complications I can here perceive the following :—

Climatic changes may be *active* or *passive, i. e.,* they occur by a change of climate or by a migration and extension of the species over new districts having another climate.

By a lengthening of the summer, as I shall designate the shorter portion of the whole annual period of warmth, the last generation of the year would be advanced further in its development than before ; if, for instance, it formerly hibernated in the pupal state, it would now pass the winter in the imago stage. Should a further lengthening of the summer occur, the butterflies might emerge soon enough to lay eggs in the autumn, and by a still greater lengthening the eggs also might hatch, the larvæ grow up and hibernate as pupæ. In this manner we should have a new generation interpolated, owing to the generation which formerly hibernated being made to recede step by step towards the autumn and the summer. *By a lengthening of the summer there occurs therefore a retrogressive interruption of generations.*

The exact opposite occurs if the summer should become shortened. In this case the last generation would no longer be so far developed as formerly; for instance, it might not reach the pupal stage, as formerly, at the beginning of winter, and would thus hibernate in a younger stage, either as egg or larvæ. Finally, by a continual shortening of the summer it would no longer appear at the end of this period but in the following spring; in other words, it would be eliminated. *By a shortening of the summer accordingly the interruption of generations occurs by advancement.*

The following considerations, which submit themselves with reference to seasonal dimorphism, are readily conceivable, at least, in so far as they can be arrived at by purely theoretical methods. Were the summer to become shorter the generation which formerly hibernated in the pupal stage would be advanced further into the spring. It would not thereby necessarily immediately lose the winter characters which it formerly possessed. Whether this would happen, and to what extent, would rather depend upon the intensity of the action of the summer climate on the generation in question, and on the number of generations which have been submitted to this action. Hitherto no attempts have been made to expose a monomorphic species to an elevated temperature throughout several generations, so as to obtain an approximate measure of the rapidity with which such climatic influences can bring about changes. For this reason we must for the present refrain from all hypothesis relating to this subject.

The disturbance of generations by the shortening of summer might also occur to a species in such a manner that the generation which formerly hibernated advances into the spring, the last of the summer generations at the same time reaching the beginning of winter. The latter would then hibernate in the pupal state, and would sooner or later also assume the winter form through the action of the cold of winter. We should, in this case, have two generations possessing more or less completely the winter form, the ancient winter generation now gradually losing the winter characters, and the new winter generation gradually acquiring these characters.

In the reverse case, *i. e.*, by a lengthening of the summer, we should have the same possibilities only with the difference that the disturbance of generations would occur in a reverse direction. In this case it might happen that the former winter generation would become the autumnal brood, and more or less preserve its characters for a long period. Here also a new winter generation would be produced as soon as the former spring brood had so far retrograded that its pupæ hibernated.

I am only too conscious how entirely theoretical are these conjectures. It is very possible that observation of nature will render numerous corrections necessary. For instance, I have assumed that every species is able, when necessary, to adapt any one of its developmental stages to hibernation. Whether this is actually the case must be learnt from further researches ; at present we only know that many species hibernate in the egg stage, others in the larval state, others as pupæ, and yet others in the perfect state. We know also that many species hibernate in several stages at the same time, but we do not know whether each stage of every species has an equal power of accommodation to cold. Should this not be the case the above conjectures would have to be considerably modified. To take up this subject, so as to completely master all the facts connected therewith, naturalists would have to devote their whole time and energy to the order Lepidoptera, which I have been unable to do.

From the considerations offered, it thus appears that the phenomena of seasonal dimorphism may depend on extremely complex processes, so that one need not be surprised if only a few cases now admit of certain analysis. We must also admit, however, that it is more advantageous to science to be able in the first place to analyze the simplest cases by means of breeding experiments, than to concern oneself in guessing at cases which are so complicated as to make it impossible at present to procure all the materials necessary for their solution.

Plate I.

Aug Weismann pinx. Lith. J.A. Hofmann, Würzburg.

Plate II.

Aug. Weismann pinx.

Lith. J. A. Hofmann, Würzburg.

EXPLANATION OF THE PLATES.

PLATE I.

Fig. 1. Male *Araschnia Levana*, winter form.

Fig. 2. Female *A. Levana*, winter form.

Fig. 3. Male *A. Levana*, artificially bred intermediate form (so-called *Porima*).

Fig. 4. Female *A. Levana*, intermediate form (*Porima*), artificially bred from the summer generation, agreeing perfectly in marking with the winter form, and only to be distinguished from it by the somewhat darker ground colour.

Fig. 5. Male *A. Levana*, summer form (*Prorsa*).

Fig. 6. Female *A. Levana*, summer form (*Prorsa*).

Figs. 7 to 9. Intermediate forms (*Porima*), artificially bred from the first summer generation.

Figs. 10 and 11. Male and female *Pieris Napi*, winter form, artificially bred from the summer generation ; the yellow ground colour of the underside of the hind wings brighter than in the natural winter form.

Figs. 12 and 13. Male and female *Pieris Napi*, summer form.

Figs. 14 and 15. *Pieris Napi*, var. *Bryoniæ*, male and female reared from eggs.

PLATE II.

Fig. 16. *Papilio Ajax*, var. *Telamonides*, winter form.

Fig. 17. *P. Ajax*, var. *Marcellus*, summer form.

Fig. 18. *Plebeius Agestis* (*Alexis*, Scop.), German winter form.

Fig. 19. *P. Agestis* (*Alexis*, Scop.), German summer form.

Fig. 20. *P. Agestis* (*Alexis*, Scop.), Italian summer form. (The chief difference between figs. 19 and 20 lies on the under-side, which could not be here represented.)

Fig. 21. *Polyommatus Phlæas*, winter form, from Sardinia; the German winter and summer generations are perfectly similar.

Fig. 22. *P. Phlæas*, summer form, from Genoa.

Fig. 23. *Pararga Ægeria*, from Freiburg, Baden.

Fig. 24. *P. Meione*, southern climatic form of *Ægeria* from Sardinia.

END OF PART I.

Part II.

THE ORIGIN OF THE MARKINGS OF CATERPILLARS,

AND

PHYLETIC PARALLELISM IN METAMORPHIC SPECIES.

STUDIES IN THE THEORY OF DESCENT.

Part II.

ON THE FINAL CAUSES OF TRANSFORMATION.

I.

THE ORIGIN OF THE MARKINGS OF CATERPILLARS.

INTRODUCTION.

THE general idea which has instigated the researches described in the present essay has already been expressed in the Preface, where it has also been explained why the markings of caterpillars, and especially those of the Sphinx-larvæ, were chosen for testing this idea.

The task presented itself in the following form :—In order to test the idea referred to, it must be investigated whether all the forms of marking which occur in the Sphinx-larvæ can or cannot be traced to known transforming factors.

That natural selection produces a large number of characters can be as little doubted as that many varying external influences can bring about changes in an organism by direct action. That these two transforming factors, together with their correlatively induced changes, are competent to produce *all* characters, howsoever insignificant, has indeed been truly asserted, but has never yet been proved. The solution of the problem, however, appeared to me to depend particularly on this point. We are now no longer concerned in proving that a changing environment reacts upon the organism—this has already been shown—but we have to deal with the question whether *every change* is the result of the action of the environment upon the organism. Were it possible to trace all the forms of markings which occur, to one of the known factors of species transformation, it could be thus shown that here at least an "innate power of development" was of no effect; were this not possible, *i. e.* did there remain residual markings which could not be explained, then the notion of an "innate principle of development" could not be at once entirely discountenanced.

The attempt to solve this problem should commence by the acquisition of a morphological groundwork, so that the phyletic development of the markings might by this means be represented as far as possible. It cannot be stated with certainty, *primâ facie*, whether some form of development conformable to law is here to be found, but it soon becomes manifest that such is certainly the case in a great measure. In all species the young caterpillars are differently marked to the adults, and in many the markings change with each of the five stages of growth indicated by the four ecdyses, this gradational transformation of the markings being a "development" in the true sense of the word, *i. e.*, an origination of the complex from the simple, the development of characters from those previously in existence, and never an inconstant, unconnected series of *per saltum* changes. This development of the markings in individuals very well reveals their phyletic development, since there can be no doubt but that we have here preserved to us in the ontogeny, as I shall establish more fully further on, a very slightly altered picture of the phyletic development. The latter can have been but slightly "falsified" in these cases, although it is indeed considerably abbreviated, and that in very different degrees; to the greatest extent in those species which are most advanced in their phyletic development, and to the least extent in those which are less advanced. From this the value of being able to compare a large number of species with respect to their ontogeny will appear. Unfortunately, however, this has only been possible to a very limited extent.

The youngest larval stages are those which are of the most importance for revealing the phyletic development, because they make us acquainted with the markings of the progenitors of the existing species. For these investigations it is therefore in the first place necessary to obtain fertile eggs. Female *Sphingidæ*, however, do not generally lay eggs in confinement,[1] or at most only a very small number. In the case of many species (*Deilephila Galii, D. Lineata, D. Vespertilio, D. Hippophaës*) I have for this reason unfortunately been unable to observe the entire development, and such observations would in all probability have given especially valuable information.

I was certainly successful in finding the young larvæ of some of the above as well as of other species on their food-plants, but even in the most favourable instances only individuals of the second stage and generally older. When, however, notwithstanding this imperfection of the materials, and in spite of the important gaps thus inevitably caused in these series of observations, it has nevertheless been possible to form a picture, on the whole tolerably complete, of the phyletic development of the Sphinx-markings, this well indi-

[1] Only the species of *Smerinthus* can be made to lay eggs regularly in confinement; *Macroglossa Stellatarum* laid a number in a large gauze-covered breeding-cage; the species of *Deilephila* could not be induced to lay more than single ones in such a cage. From species of *Chærocampa* also I never obtained but a few eggs, and from *Sphinx* and *Acherontia* never more than single ones.

cates what a fertile field is offered by the investigation of this subject, and will, I trust, furnish an inducement to others, not only to fill up the various gaps in the small family of the *Sphingidæ*, but also to treat other Lepidopterous families in a similar manner. Such an investigation of the *Papilionidæ* appears to me to be especially desirable; not only of the few European but also of the American and Indian species. We know practically nothing, of the youngest stages of the *Papilio* larvæ from this point of view. No entomological work gives any description of the form and marking of the newly hatched larvæ, even in the case of our commonest species (*Papilio Machaon* and *P. Podalirius*), and I believe that I do not go too far when I assert that up to the present time nobody has observed them at this early stage.[2] When, however, we consider that

[2] [Eng. ed. Since the appearance of the German edition of this work, numerous descriptions of the young stages of caterpillars have been given, but in all cases without representing the relationship of the forms.] [In the excellent figures of larvæ at various stages of growth, given in some of the more recent works on Lepidoptera, there will be found much material which may be regarded as a contribution to the field of research entered on by the author in the present essay, *i. e.* the ontogeny and comparative morphology of larval markings, although it is much to be regretted that the figures and descriptions have not been given from this point of view. In his "Butterflies of North America," for example, W. H. Edwards figures the young as well as the adult larvæ of species of *Apatura*, *Argynnis*, *Libythea*, *Phyciodes*, *Limenitis*, *Colias*, *Papilio*, &c.

in these young caterpillars we have preserved to us the parent-form, extinct for centuries, of the

Burmeister, in his recently published "Lépidoptères de la République Argentine," figures the young stages of species of *Caligo*, *Opsiphanes*, *Callidryas*, *Philampelus*, &c. Messrs. Hellins and Buckler have figured and described the early stages of large numbers of the caterpillars of British Lepidoptera, but their figures remain unpublished. The larvæ of many of our native species belonging to the genera *Liparis*, *Tæniocampa*, *Epunda*, *Cymatophora*, *Calocampa*, &c., are dull when young, but become brightly coloured at the last moult. Such changes of colour are probably associated with some change, either in the habits or in the environment; and a careful study of the ontogenetic development of such species in connection with their life-history would furnish results of great value to the present inquiry. The same remarks apply to those *Noctuæ* larvæ which are brightly coloured in their young stages, and become dull when adult.

Among other papers which may be considered as contributions to the present subject, I may mention the following:— In 1864 Capt. Hutton published a paper, "On the Reversion and Restoration of the Silkworm, Part II." (Trans. Ent. Soc. 1864, p. 295), in which he describes the various stages of development of several species of *Bombycidæ*. In 1867 G. Semper published accounts of the early stages of several Sphinx-larvæ ("Beiträge zur Entwicklungsgeschichte einiger ostasiatischer Schmetterlinge," Verhandl. k.k. Zoolog.-botan. Gesell. in Wien, vol. xvii.). The question as to the number of claspers in young *Noctuæ* larvæ has been raised in notes by Dr. F. Buchanan White ("Ent. Mo. Mag.," vol. v. p. 204) and B. Lockyer ("Entomologist," 1871, p. 438). A valuable paper, "On the Embryonic Larvæ of Butterflies," was published in 1871 by S. H. Scudder ("Ent. Mo. Mag.," vol. viii. p. 122). For remarks on the development of the larva of *Papilio Merope*, see J. P. Mansel Weale in Trans. Ent. Soc., 1874, p. 131, and Pl. I.; also this author on the young stages of the larva of *Gynanisa Isis*, Trans. Ent. Soc., 1878, p. 184. For an existing species of *Papilio*, it must assuredly be of the greatest interest to become accurately acquainted with them, to compare them with the earliest stages of allied species, and to follow the gradual divergence of the succeeding stages in different directions, thus forming a picture of the phyletic development of an evolving group. In the course of such observations numerous collateral results would doubtless come out. Investigations of this kind, whether conducted on this or on any other group, would, above all, show the true systematic affinities of the forms, *i. e.*, their *genealogical* affinities, and that in a better way than could be shown by the morphology of the perfect insects or the adult caterpillars alone. If I am diffident in founding these conclusions upon the development of the Sphinx-markings treated

account of the development of the larvæ of certain North American species of *Satyrus*, see W. H. Edwards in the "Cănadian Entom.," vol. xii. p. 21. Mr. P. H. Gosse's recent description of the newly hatched caterpillar of *Papilio Homerus* (Proc. Ent. Soc. 1879, p. lv), furnishes a good illustration of the value of studying the ontogeny. The natural affinities of the *Papilionidæ* were at one time much disputed, some systematists placing this family at the head of the Lepidoptera, and others regarding them as being more closely allied to the moths. Mr. Gosse's observation tends to confirm the latter view, now generally received by Lepidopterists, since he states that the larva in question "suggests one of the great *Saturniadæ*, such as *Samia Cecropia*." Mr. Scudder, in the paper above referred to, adopts an analogous argument to show the close relationship between the *Papilionidæ* and *Hesperidæ*. R.M.]

of in the present essay, this arises entirely from a knowledge of the imperfections in the basis of facts. If however, through the united labours of many investigators, the individual development of all the species of *Sphingidæ* now existing should at some future period be clearly laid before us, we should then not only have arrived at a knowledge of the relative ages of the different species, genera and families, but we should also arrive at an explanation of the nature of their affinities.

It is erroneous to assert that Classification has only to take form-relationship into consideration; that it should and can be nothing else than the expression of form-relationship. The latter is certainly our only measure of blood-relationship, but those who maintain the assertion that form- and blood-relationship are by no means always synonymous, are undoubtedly correct. I shall in a future essay adduce facts which leave no doubt on this point, and which prove at the same time that modern systematists—especially in the order Lepidoptera—have always endeavoured—although quite unconsciously—to make the blood-relationship the basis of their classification. For this reason alone, larvæ and pupæ would have an important bearing upon the establishment of systematic groups, although certainly in a manner frequently irregular.

It must be admitted that so long as we are able to compare the species of one group with those of another *in one form only*, we are often unable to ascertain the blood-relationship.[3] In such cases we can only determine the latter from the form-relationship, and as these are not always parallel, any conclusion based on a single form must be very unsound. If, for instance, butterflies emerged from the egg directly, without passing through any larval stage, a comparison of their resemblances of form would alone be of systematic value; we should unite them into groups on the ground of these resemblances only, and the formation of these groups would then much depend upon the weight assigned to this or that character. We might thus fall into error, not only through a different valuation of characters but still more because two species of near blood-relationship frequently differ from one another in form to a greater extent than from other species. We should have no warrant that our conception of

[3] [Mr. A. G. Butler has recently furnished a good illustration of the danger of classifying Lepidoptera according to the affinities of the perfect insects only, in his paper, "On the Natural Affinities of the Lepidoptera hitherto referred to the Genus *Acronycta* of authors," Trans. Ent. Soc. 1879, p. 313. If the author's views are ultimately accepted, the species at present grouped under this genus will be distributed among the *Arctiidæ*, *Liparidæ*, *Notodontidæ*, and *Noctuæ*. Mr. Butler's determination of the affinities of the species supposed to belong to the genus mentioned, is based chiefly upon a comparative examination of the larvæ, and this is far more likely to show the true blood-relationship of the species than a comparison of the perfect insects only. A study of the comparative ontogeny can alone give a final answer to this question. R.M.]

the form-relationship expressed the genealogical connection of the species. But it would be quite different if every species presented itself in two or three different forms. If in two species or genera the butterflies as well as the larvæ and pupæ exhibited the same degree of form-relationship, the probability that this expressed also the blood-relationship would then be exceedingly great. Now this agreement certainly does not always occur, and when these different stages are related in form in unequal degrees, the problem then is to decide which of these relationships expresses the genealogy. This decision may be difficult to arrive at in single cases, since the caterpillar may diverge in form from the next blood-related species to a greater extent than the butterfly, or, conversely, the butterfly may diverge more widely from its nearest blood-related species than the caterpillar.

For such cases there remains the developmental history of the caterpillar, which will almost always furnish us to a certain extent with information respecting the true genealogical relationship of the forms, because it always reveals a portion of the phyletic (ancestral) development of the species. If we see two species of butterflies quite dissimilar in form of wing and other characters, we should be inclined, in spite of many points of agreement between them, to place them in entirely different genera. But should we then find that not only did their adult larvæ agree in every detail of marking, but also that the entire phyletic development of these markings, as revealed by the ontogeny of the larvæ, had taken precisely the same course in both species, we should certainly conclude that they possessed a near blood-relationship, and should place them close together in the same genus. Such an instance is afforded by the two Hawk-moths, *Chærocampa Elpenor* and *C. Porcellus*, as will appear in the course of these investigations. These two species were placed by Walker in different genera, the form relationship of the imagines being thus correctly represented, since *Porcellus* (imago), is indeed more nearly related in form to the species of the genus *Pergesa*, Walker, than to those of the genus *Chærocampa*.[4] Nevertheless, these species must remain in the same genus, as no other arrangement expresses their degree of blood-relationship.

An intimate knowledge of the development-stages of caterpillars thus offers, even from a systematic point of view, an invaluable means of judging the degree of blood-relationship, and from this standpoint we must regard the study of the caterpillar as of more importance than that of the perfect insect. Certainly all groups would not be

[4] [In his recent revision of the *Sphingidæ*, Mr. A. G. Butler (Trans. Zoo. Soc., vol. ix. part x.) retains Walker's arrangement. R.M.]

so rich in information as the *Sphingidæ*, or, as I am inclined to believe, the *Papilionidæ*, since all families of caterpillars do not possess such a marked and diversified pattern, nor do they present such a varied and characteristic bodily form. The representation of the true, *i. e.*, the blood-relationship, and through this the formation of natural groups with any completeness, can certainly only be looked for when we are intimately acquainted with the different stages of development of the larvæ of numerous species in every group, from their emergence from the egg to their period of pupation. The genealogical relationship of many forms at present of doubtful systematic position would then be made clear. This investigation, however, could not be the work of a single individual ; not only because the materials for observation are too great, but, above all, because they are spread over too wide a field. It is not sufficient to study the European types only—we should endeavour to learn as much as possible of the Lepidoptera of the whole world. But such observations can only be made on the spot. Why should it not be possible to trace the development from the egg, even under a tropical sky, and to devote to breeding and observing, a portion of that time which is generally spent in mere collecting? I may perhaps be able to convince some cf the many excellent and careful observers among entomologists, that beyond the necessary and valuable search for new forms, there is another field which may be successfully worked, viz., the precise investigation of the development of known species.

The first portion of the present essay consists of the determination of this development in those species of *Sphingidæ* which have been accessible to me. Seven genera are successively treated of, some completely, and others only in some of their stages ; and thus I have sought to present a picture of the course of development of the markings in each genus, by comparing the species with each other, and with allied forms in cases where the young stages were unknown. In this portion, as far as possible, the facts only have been given, the working up of the latter into general conclusions upon the development of marking being reserved for the second portion. A complete separation of facts from generalizations could not, however, be carried out ; it appeared convenient to close the account of each genus with a summary of the results obtained from the various species.

After having established that the markings of the Sphinx-caterpillars had undergone an extremely gradual phyletic development, conformable to law, in certain fixed directions, it appeared desirable to investigate the causes of the first appearance of these markings, as well as of their subsequent development. The question as to the biological significance of marking here presented itself in the first place for solution, and the third section is devoted to this subject. If it is maintained that marking is of no importance to the life of the insect, or that it is so only exceptionally, and that it is in reality, as it appears to be, a character of purely morphological, *i. e.*, physiological, insignificance, then its striking phylogenetic development conformable to law cannot be explained by any of the known factors of species transformation, and we should have to assume the action of an innate transforming power. In the present investigations, this subject in particular has been extensively treated of, and not only the markings of Sphinx-caterpillars, but also those of caterpillars in general, have been taken into consideration. The results arrived at are indeed quite opposed to this assumption—marking is shown to be a character of extreme importance to the life of the species, and the admission of a phyletic vital force must, at least from the present point of view, be excluded. This leads to the fifth section, in which I have attempted to test certain objections to the admission of a "phyletic vital force." The sixth section finally gives a summary of the results obtained.

I may now add a few explanations which are necessary for understanding the subsequent descriptions. It was found impossible to avoid the introduction of some new technicalities for describing the various elements of larval markings, especially as the latter had to be treated of scientifically. I have therefore chosen the simplest and most obvious designations, all of which have already been employed by various authors, but not in any rigorously defined sense. I understand by the "dorsal line" that which runs down the middle of the back; the lines above and below the spiracles will be respectively distinguished as the "supra-" and "infra-spiracular" lines, and the line between the dorsal and spiracular as the "subdorsal line." The distinction between "ring-spots" and "eye-spots" will be made manifest in the course of the investigation. A glance at any of the existing descriptions of larvæ will show how necessary it was to introduce a precise terminology. Even when the latter is exact as far as it goes, the want of precise expressions not only makes the descriptions unnecessarily long, but it also considerably increases the difficulty of comparing one species with another, since we can never be sure whether the same designation applies to the same homologous character. For instance, when the larva of *Chærocampa Elpenor* is said to have "a light longitudinal line on the sides of the thoracic segments," this statement is indeed correct; but it is not apparent whether the line is above or below, and consequently it does not appear whether it is the equivalent of the longitudinal line "on the sides" of the segments in other species. If, however, it is said that this line is "*subdorsal* on the thoracic segments, and on the eleventh abdominal segment," it is thereby indicated that we have here a residue of the same marking which is found completely developed in many other Sphinx-larvæ, and indeed in the young stages of this same species. The mode of describing caterpillars hitherto in vogue is in fact unscientific ; the descriptions have not been made with a view to determining the *morphology* of the larvæ, but simply to meet the practical want of being able to readily identify any species that may be found : even for this purpose, however, it would have been better to have employed a more precise mode of description.

I.

Ontogeny and Morphology of Sphinx Markings.

THE GENUS CHÆROCAMPA, DUPONCHEL.

Although by no means in favour of the excessive subdivision of genera, I am of opinion that Ochsenheimer's genus *Deilephila* has been correctly separated by Duponchel into the two genera *Chærocampa* and *Deilephila, sensû strictiori.* Such a division may appear but little necessary if we examine the perfect insects only; but the developmental history of the caterpillars shows that there is a wide division between the two groups of species, these groups however being branches of one stem.

Chærocampa Elpenor, Linn.

Some captured females laid single eggs sparsely on grass, wood, and especially on the tarlatan with which the breeding-cage was covered. The eggs are nearly spherical, but somewhat compressed, of a grass-green colour, a little lighter, and somewhat larger (1.2 millim.) than those of *Deilephila Euphorbiæ.* During the development of the embryo the eggs first became yellowish-green, and finally yellowish.

First Stage.

The young caterpillars are four millimeters in length, and immediately after hatching are not green, but of a yellowish-white opalescent colour, the large and somewhat curved caudal horn being black. The caterpillars were so transparent that under a low magnifying power the nervous, tracheal, and alimentary systems could be beautifully seen. As soon as the larvæ began to feed (on *Epilobium parviflorum*) they became green in consequence of the food appearing through the skin, but the latter also gradually acquired a dark green colour (Pl. IV., Fig. 17). All the specimens (some twenty in number) were exactly alike, and showed *no trace of marking*.

Second Stage.

The first ecdysis occurred after 5—6 days, the length of the caterpillars being from nine to ten millimeters. After this first moult they appeared of a shining green, the horn, which was black during the first stage, becoming a little red at the base, while a fine white subdorsal line extended from the horn to the head (Fig. 18). The head and legs were green; the divisions between the segments appeared as fine light rings, and the entire upper surface of the segments was also crossed by fine transverse rings, as was also the case in the first stage.

At the beginning of the present stage no trace of the eye-spots could be detected ; but a few days after the first moult it was observed that the white subdorsal line was no longer straight on the fourth and fifth segments, but had become curved upwards into two small crescents. The latter soon stood out more strongly, owing to the filling up of their concavities with darker green. These are the first rudiments of the eye-spots (Figs. 19 and 30). A very fine white line now connected the spiracles (infra-spiracular line), and could be traced from the last segment to the head. This line takes no further part in the subsequent development of the markings, but disappears in the following stage. The blood-red colour of the base of the black caudal horn is retained till the fifth stage, and then also disappears.

Before the second moult, which occurs after another period of 5—6 days, the caterpillars, which were about 1.3 centimeters in length, had assumed their characteristic tapering, slug-like form. I did not notice that the larvæ at this stage possessed the power of withdrawing the three foremost segments into the two succeeding ones, as is so frequently to be observed in the adults ; neither were these two segments so strikingly enlarged as they are at an earlier period.

Third Stage.

After the second ecdysis the marking and colouring only undergo change with respect to the eye-spots. The concavities of the crescent-shaped portions of the subdorsal line become black,[1] the remainder of this line at the same time losing much of its whiteness, and thus becoming less distinct, whilst the crescents assume the appearance of small eye-spots (Fig 20). During this stage the curved, crescent-formed portions become prepared for complete separation from the remainder of the subdorsal line; and just before the third moult the eye-spots become sharply defined both in front and behind, whilst the black ground-colour curves upwards, and the white spots gradually become lenticular and commence to enlarge (Fig. 21).

Fourth Stage.

The third moult takes place after another interval of 5—6 days, the eye-spots then becoming very prominent. The white nucleus of the front spot is kidney-shaped, and that of the hind spot egg-shaped; whilst the black ground-colour extends as a slender border upwards along the sides of the spots, but does not completely surround them till towards the end of the present stage (Fig. 21). The

[1] The deposition of black pigment may commence immediately before ecdysis.

central portion of the white spots at the same time becomes of a peculiar violet-brown colour inclining to yellow above, the peripheral region alone remaining pure white.

Of the subdorsal line only traces are now to be recognized, and these are retained, with almost unchanged intensity, sometimes into the last stage, remaining with the greatest persistence on the three front and on the penultimate segments, whilst on those containing the eye-spots, *i. e.*, the fourth and fifth, not a trace remains. At the present stage the peculiar mingling of colours becomes apparent over the whole of the upper surface; the green is no longer uniform, but a mixture of short and gently sinuous, dark-green striations on a lighter ground now appear. On the sides of the caterpillar these stripes, which are at first indistinct, but become more strongly pronounced in the next stage, are arranged obliquely on the spiracles, with the lower portions directed forwards.

Fifth Stage.

The fourth moult occurs 7—8 days after the third, the caterpillar being 4—5 centimeters in length. Whilst all the specimens hitherto observed were with one exception light green, they now mostly changed their colour and became dark brown. In one case only did the brown colour appear in the previous (fourth) stage. The striations previously mentioned appear as dull and interrupted dirty yellow streaks, the same dirty yellow colour showing itself continuously on the sides of the four front segments. Of the subdorsal line only a distinct trace is now to be seen on the eleventh and on the three front segments, whilst on the third segment the formation of another eye-spot commences to be plainly perceptible by a local deposition of black (Fig. 23). This third spot does not, however, become completely developed, either in this or in the last stage, but the subdorsal line remains continuous on the three front segments. Among other changes at this stage, there occurs a considerable shortening of the caudal horn, which at the same time loses its beautiful black and red colours and becomes brownish.

The two large eye-spots have now nearly attained complete development. The kidney-shaped white spot has become entirely surrounded by black; and on the brown, red, and yellow tints present in this spot during the last stage, a nearly black spot has been developed—the pupil of the eye (Fig. 33). In order to establish a definite terminology for the different portions of the eye-spot, I shall designate the pupil as the "nucleus," the light ground on which the pupil stands as the "mirror," and the black ground which surrounds the mirror as the "ground-area."

In this fifth stage the larva attains a length

of six centimeters, after which the fifth moult takes place, the caterpillar becoming ready for pupation in the sixth stage. No striking changes of colouring or marking occur after the present stage, but only certain unimportant alterations, which are, however, of the greatest theoretical interest.

Sixth Stage.

In this stage the eye-like appearance of the spots on the front segments becomes still more distinct than in the fifth stage; at the same time these spots repeat themselves on all the other segments from the fifth to the eleventh, although certainly without pupils, and appearing only as diffused, deep black spots, of the morphological significance of which, however, there cannot be the least doubt. They are situated in precisely the same positions on the 5—11 segments as those on the third and fourth—near the front, and above and below the subdorsal line. A feeble indication of the latter can often be recognized (Fig. 23).

In all dark-brown specimens the repeated spots can only be detected in a favourable light, and after acquiring an intimate knowledge of the caterpillar; but in light-brown and green specimens they appear very sharply defined.

There is one other new character which I have never observed at an earlier period than the sixth stage, viz. the small dots which appear in pairs near the posterior edge of segments 5—11. These dots cannot have been developed from the subdorsal line, as they are situated higher than the latter. Their colour varies according to the ground-colour of the caterpillar, but it is always lighter, being light-green in green specimens, dull yellow in those that are light brown, and grey in the blackish-brown caterpillars. These "dorsal spots," as I shall term them, are chiefly of interest because they are present in *Chærocampa Porcellus,* in which species they appear one stage earlier than in *C. Elpenor.*

CHÆROCAMPA PORCELLUS, LINN.

Females captured on the wing, laid in the breeding-cage single eggs of a light green colour, spheroidal in form, and very similar to those of *C. Elpenor.*

First Stage.

The caterpillars on first hatching measure 3.5 millimeters in length, and are of a uniform light green colour, with a fine white transverse line on the posterior edge of each segment, precisely similar to that which appears in the second stage of *C. Elpenor.* They resemble the latter species still further in showing a fine white subdorsal line, which can easily be recognized by the naked eye (Fig. 24). Although the adult larva is distinguished from all the other known species of *Chærocampa* by the absence of a caudal horn, a distinct but very small one is nevertheless present at this first stage, and is indeed retained throughout the entire course of development, but does not increase further in size, and thus gradually becomes so small in proportion to the size of the caterpillar that it may be entirely overlooked.

The first moult takes places after 4—5 days.

Second Stage.

The blue-green coloration remains unchanged; but a somewhat darker green dorsal line becomes apparent down the middle of the back (the dorsal vessel?), and the subdorsal line now becomes very broad and pure white, being much more conspicuous than in any stage of *C. Elpenor* (Fig. 25). The tapering of the three front segments occurs at this stage, and oblique, dark-green striations on a lighter ground stand out distinctly on the spiracles. As with *C. Elpenor*, the first traces of the future eye-spots appear during the second stage; not in the present case as a curvature of the subdorsal line, but as a spot-like widening of the latter, of a brighter white than the somewhat greenish colour of the remainder of the line.

Third Stage.

After the second moult, the formation of the dark "ground area" of the eye-spots commences by the appearance of a little brown on the under edge of the foremost of the white spots, this coloration gradually increasing in extent and in depth. At the same time both spots become more sharply distinguishable from the subdorsal line, which becomes constantly greener (Fig. 27). The brown colour soon grows round the white of the front eye-spot, which becomes so far perfected; whilst the completion of the hind spot is effected slowly afterwards. The formation of the eye-spots does not therefore proceed any more rapidly in this species than in *C. Elpenor.*

At the end of the present stage the length of the caterpillar is about four centimeters; the ground colour is still sea-green; the subdorsal line is much diminished, completely fading away at its lower edge, but remaining sharply defined above, against the green ground-colour (Fig. 26).

Fourth Stage.

After the third moult all the caterpillars (5) became brown, this change occurring therefore one stage earlier than is generally the case with *C. Elpenor.* In single instances the brown colour appeared in the third stage. The subdorsal line had disappeared from all the segments but the three first and the last. The eye-spots now rapidly attained complete development; they contained a black pupil, and gave the insect a truly repulsive appearance when, on being threatened by danger, it drew in the front segments, and expanded the fourth (Fig. 28). The eye-spots of the fifth segment are much less developed than in *C. Elpenor*; they remain small, and are not readily detected. On the other hand, there now appear on all the segments with the exception of the last, just as in the sixth stage of *C. Elpenor*, distinct rudiments of eye-spots, which present the appearance of irregular, roundish, black spots on the front borders of the segments, at the height of the former subdorsal line. In this latter region the black pigment is disposed as a longitudinal streak, and to this a median line is added, the whole forming a marking which perhaps makes the caterpillar appear still more alarming to its foes. This marking is, however, only to be distinctly recognized on the three first segments. The "dorsal spots" mentioned in the case of *C. Elpenor* then appear very distinctly on segments 5—11.

The caterpillars continued to feed for eleven days after the third moult, at the end of which period the fourth moult took place, but without the occurrence of any change of marking. The larvæ then buried themselves, the complete development having taken 28—29 days.

The development of the *Porcellus* caterpillar was twice followed; in 1869 in twelve, and in 1874 in five specimens. In no case did I obtain caterpillars which remained green throughout the entire course of development, although this colour is stated in the books to occur occasionally in these larvæ; neither have I been able to find any figure of an adult green specimen, so that it must in the meantime be admitted that such specimens, if they occur at all, are exceptional instances.[2] The theoretical bearing of this admission will appear later on.

Results of the development of Chærocampa Elpenor and C. Porcellus; comparison of these with the other known species of Chærocampa.

The first stage of *Elpenor* shows that the most remote ancestor of the genus possessed no kind of marking, but was uniformly green. At a later period, the white longitudinal stripe which I have designated the "subdorsal line" made its appearance, and at a still later period this line vanished, with the exception of a few more or less distinct remnants, whilst, at the same time, from certain

[2] [Mr. Herbert Goss states (Proc. Ent. Soc. 1878, p. v.) that according to his experience, the green and brown varieties of *C. Porcellus* (erroneously printed as *Elpenor* in the passage referred to) are about equally common, the former colour not being in any way confined to young larvæ. Mr. Owen Wilson in his recent work, "The Larvæ of British Lepidoptera and their food-plants," figures (Pl. VIII., Figs. 3 and 3a) the two forms, both apparently in the adult state. During the years 1878-79, my friend, Mr. J. Evershed, jun., took five of these full-grown larvæ in Surrey, one of these being the green variety. In order to get more statistics on this subject, I applied this year (1880) to Messrs. Davis of Dartford, who informed me that among 18—20 adult caterpillars of *Porcellus* in their possession, there was only one green specimen. R.M.]

portions of it, the eye-spots of the fourth and fifth segments became developed. After the perfecting of the eye-spots, weak repetitions of the latter appeared as black spots on all the segments except the last.

In *Porcellus* the caterpillar emerges from the egg with the subdorsal line, the first stage of *Elpenor* being omitted, From this fact we may venture to conclude that *Porcellus* is the younger species, or, what comes to the same thing, that it has further advanced in development. The whole subsequent history of *Porcellus* agrees with this view, its course of development being essentially but a repetition of the phenomena displayed by *Elpenor*, and differing only in one point, viz. that all new characters make their appearance one stage earlier than in the latter species. This is the case with the transformation of the green into a brown ground-colour; with the repetition of the eye-spots on the remaining segments in the form of suffused black spots; and with the appearance of the light "dorsal spots." Only the eye-spots themselves appear, and the snout-like tapering of the front segments occurs in the same stage as in *Elpenor*, *i. e.* the second.

From these data alone, we may venture to infer the occurrence of four chief stages in the phyletic development of the genus. The first stage was simply green, without any marking; the second showed a subdorsal line; the third, eye-spots on the third and fourth segments; and the fourth stage showed a repetition of the eye-spots, although but rudimentary, on all the remaining segments with the exception of the twelfth.

Now if we compare the other known species of *Chærocampa* larvæ with the above, we shall arrive at the interesting conclusion that all these species can be arranged in three groups, which correspond exactly with the three last phyletic stages as just deduced from the ontogeny of *C. Elpenor* and *Porcellus.*

Of the genus *Chærocampa*,[3] over fifty species have been described,[4] of which the larvæ of only

[3] I unite the genera *Pergesa* and *Darapsa* of Walk. with *Chærocampa*, Dup.; the first appears to me to be quite untenable, since it is impossible that two species, of which the caterpillars agree so completely as those of *C. Elpenor* and *Porcellus*, can be located in different genera. *Porcellus* indeed was referred to the genus *Pergesa* because of its different contour of wings, an instance which distinctly shows how dangerous it is to attempt to found Lepidopterous genera without considering the caterpillars. The genus *Darapsa* also appears to me to be of very doubtful value, and in any case requires further confirmation with respect to the larval forms.

[4] [Mr. A. G. Butler (Trans. Zoo. Soc., vol. ix., part. x., 1876) gives a list of about eighty-four species of *Chærocampa*, and sixteen of *Pergesa*, besides numerous other species belonging to several genera placed between *Chærocampa* and *Pergesa*. Of *Darapsa*, he states "that this genus was founded upon most heterogeneous material, the first three species being referable to Hübner's genus *Otus*, the fifth to Walker's genus *Diodosida*, the sixth and eighth to the genus *Daphnis* of Hübner, the seventh, ninth, and tenth to *Chærocampa* of Duponchel; there therefore remains only the fourth species, allied to *Chærocampa*, but

fifteen are known in the form which they possess at the last ontogenetic stage.

Group i.—I can furnish but little information with respect to this group. The first species with which I became acquainted was *Chærocampa Syriaca*,[5] of which I saw two blown caterpillars in Staudinger's collection, and which I have figured in Pl. IV., Fig. 29. The larva is green, and has the short oblique stripes over the legs common to so many species of *Chærocampa*, the only marking besides these being a simple white subdorsal line, without any trace of eye-spots. This species exactly corresponds therefore with the second ontogenetic stage of *C. Elpenor* and *Porcellus*. The account of the species, both in the larval and perfect state, is unfortunately so imperfect, that we cannot with certainty infer the age of the two caterpillars from their size. If the moth were of the same size as *Elpenor*, then the caterpillar figured, having a length of 5.3 centimeters, would not be in the last but in the penultimate stage, and it remains doubtful whether it may not acquire eye-spots in the last stage.

That species exist, however, which in their last stage correspond to the second stage of *Elpenor*, is shown by two of the forms belonging to Walker's genus *Darapsa*, which was founded on

apparently sufficiently distinct." The species still retained in the genus *Darapsa* is *D. rhodocera*, Wlk., from Haiti. R.M.]

[5] [*Otus Syriacus* of Butler's revision. R.M.]

the characters of the imagines only. Ten species of this genus are given in Gray's catalogue, the adult larva of two of these being known through the excellent figures of Abbot and Smith.[6] These two caterpillars possess the characteristic tapering form in a very marked degree; one is figured in the attitude so often assumed by our species of *Chærocampa* on the approach of danger, the three front segments being withdrawn into the fourth. (Fig. 34, Pl. IV., is copied from this Plate). There are no eye-spots either in *D. Myron* or *D. Chœrilus,*[7] but only a broad white subdorsal line; underneath which, and to a certain extent proceeding from it, there are oblique white stripes, precisely similar to those which meet the subdorsal line in the third stage of *C. Porcellus.*[8]

[6] Abbot and Smith. "The Natural History of the rarer Lepidopterous Insects of Georgia, collected from the observations of John Abbot, with the plants on which they feed." London, 1797, 2 vols. fol.

[7] [*Otus Chœrilus* and *O. Myron* of Butler's revision. R.M.]

[8] [To this group may also be added *Ampelophaga Rubiginosa*, Ménétriés, from China and Japan, the caterpillar of which, having the distinct subdorsal line without any trace of eye-spots, is figured by Butler (*loc. cit.*, Pl. XCI., Fig. 4). This author also gives a figure of another species belonging to the subfamily *Chærocampinæ* (Pl. XC., Fig. 11), viz. *Acosmeryx Anceus*, Cram., from Amboina, Java, Silhet, and S. India; the caterpillar is green, with seven oblique yellow stripes along the sides, and a very conspicuous white subdorsal line with a red border above. As there are no eye-spots, this species may be referred to the present group provisionally, although its general marking is very distinct from that of the *Chærocampa* group. R.M.]

GROUP 2.—This group contains numerous species which, like our native *C. Elpenor* and *Porcellus*, show eye-spots on the fourth and fifth segments, whilst these markings are absent, or at most only present in traces, on the remainder. To this section there belong, besides the two species mentioned, five others, viz. in Europe, *C. Celerio* and *Alecto* (not certainly known?);[9] in India, *C. Nessus*, Drury, and *Lucasii*, Boisduval;[10] and an unnamed species from Port Natal.

In the species belonging to this group the subdorsal line may be more or less retained. Thus, *C. Celerio*, according to Hübner's figure, has a broad yellow line extending from the horn to the sixth segment, whilst it is completely absent on the three front segments. In the unnamed species from Port Natal[11] the subdorsal line extends to the front edge of the fifth segment, and on the fourth segment only is there a perfect eye-spot, whilst on the succeeding segments traces of such markings can be recognized as dark spots similar

[9] [Eng. ed. Dr. Staudinger has since obtained the caterpillar of *C. Alecto* from Beyrout; it possesses "a very distinct subdorsal line, and on the fourth segment a beautiful eye-spot, which is repeated with gradual diminution to segments 7—8.]

[10] Figured in "A Catalogue of Lepidopterous Insects in the Museum of the East India Company," by Thomas Horsfield and Frederick Moore. London, 1857. Vol. i., Pl. XI.

[11] Figured in Trans. Ent. Soc., New Series, vol. iv., Pl. XIII.

to those in *Elpenor* and *Porcellus*. The transition to the third group is through another unnamed species from Mozambique,[12] in which rather large eye-spots have become developed on the fourth and fifth segments and these are followed by a subdorsal line, which only appears distinctly at certain places. On this broken subdorsal line, and not completely separated from it, there are small, roundish eye-spots, situated near the front edge of each segment; these being, therefore, a somewhat more perfect repetition of the front eye-spots.[13]

[12] *Ibid.*

[13] [The following species figured by Butler (*loc. cit.* Pls. XC. and XCI.) appear to belong to the second group—*Chærocampa Japonica*, Boisd., which is figured in two forms, one brown, and the other green. The former has two distinct ocelli on the fourth and fifth segments, and a distinct rudiment on the sixth, whilst the subdorsal line extends from the second eye-spot to the caudal horn, and beneath this line the oblique lateral stripes stand out conspicuously in dark brown on a lighter ground. The ocelli are equally well developed on the fourth and fifth segments in the green variety, the subdorsal line commencing on the sixth segment, and extending to the caudal horn; there is no trace of a third eye-spot, nor are there any oblique lateral stripes; the insect is almost the exact counterpart of *C. Elpenor* in its fourth stage. (See Fig. 21, Pl. IV.) *Pergesa Mongoliana*, Butl., is brown, without a trace of the subdorsal line except on the three front segments, and with only one large eye-spot on the fourth segment. *Chærocampa Lewisii*, Butl., from Japan, is likewise figured in two forms. The brown variety has the subdorsal line on the three front segments only, distinct ocelli on the fourth and fifth segments, and gradually diminishing rudiments on the remaining seg-

GROUP 3.—In the species of this group the eye-spots are repeated on all the segments. I am acquainted with seven such *Chærocampa* larvæ, of which *C. Bisecta*, Horsfield,[14] shows some affinity to the foregoing group, since the eye-spots on segments 6—11 have not yet attained

ments. The green form appears to be transitional between the present and the third group, as it possesses a distinct, but rudimentary eye-spot on the third segment, besides the fully developed ones on the fourth and fifth, and very conspicuous, but gradually decreasing repetitions of rudimentary ocelli on segments 6—10. To this group may be added *Chærocampa Aristor*, Boisd., the caterpillar of which is figured by Burmeister (Lép. Rép. Arg., Pl. XV., Fig. 4) in the characteristic attitude of alarm, with the front segments retracted, and the ocelli on the fourth segment prominently exposed. The subdorsal line is present in this species. Burmeister also figures two of the early stages (Pl. XV., Fig. 7, A and B), and describes the complete development of *Philampelus Labruscæ*, another species belonging to the subfamily *Chærocampinæ*. The earliest stage (3—4 days old) is simple green, with no trace of any marking except a black spot on each side of the fourth segment, the position of the future ocelli. A curved horn is present both in this stage and the following one, during which the caterpillar is still green, but now has seven oblique red lateral stripes. The caudal horn is shed at the second moult, after which the colour becomes darker, the adult larva (figured by Madame Mérian, in her work on Surinam, pl. 34 and Sepp., pl. 32) being mottled brown. In addition to the ocellus on the fourth segment, there is another slightly larger on the eleventh segment, so that this species may perhaps be another transition to the third group; but our knowledge is still too imperfect to attempt to generalize with safety. R.M.]

[14] Cat. Lep. Ins. East Ind. Comp., Pl. XIII. [Figured also by Butler (=*Chærocampa Silhetensis*, Walker), *loc. cit.* Pl. XCII., Fig. 8. R.M.]

full perfection. In *C. Oldenlandiæ*, Fabr.,[15] and in *C. Alecto* from India,[16] the eye-spots appear to be perfectly alike on all the segments; whilst in *C. Acteus*, Cram.,[17] and in the North American *C. Tersa*[18] (Pl. IV., Fig. 35) they are smaller on the other segments than on the fourth; and in *C. Celerio*, Linn., from India,[19] the size of the spots diminishes from the head to the tail.

In this group also the subdorsal line is retained in a very variable degree. In some species it appears to have completely vanished (*C. Acteus*, *Celerio*); in others it is present as a light stripe extending along all the segments (*C. Alecto*); whilst in others it is retained as a broad white stripe, which extends only to the fourth segment (*C. Tersa*, Fig. 35). In species possessing eye-spots, the subdorsal line is thus a very variable character. It is, however, an interesting fact that even in the present group, which has made the greatest step forward, the subdorsal line is of general occurrence, because the eye-spots in all these species may have almost a similar development to those of *Elpenor* and *Porcellus*. The ontogeny of the tropical species would alone

[15] Cat. Lep. Ins. East Ind. Comp., Pl. XIII. [Figured also by Butler, *loc. cit.* Pl. XCI., Fig. 1. R.M.]

[16] Horsfield and Moore, *loc. cit.* Pl. X.

[17] *Ibid.* [= *Pergesa Acteus*, Walker. R.M.]

[18] [Figured also by Burmeister, *loc. cit.* Pl. XV., Fig. 3. R.M.]

[19] Horsfield and Moore, *loc. cit.*, Pl. XI.

give a definite reply on this point, but unfortunately we are not acquainted with any of the young forms, so that we can but presume that some of them at least would show only in the first stage the simple subdorsal line without eye-spots; that in the second stage the primary pairs of eye-spots would be formed on the fourth and fifth segments, whilst the transference of these spots to the remaining segments would take place in the last stage.

The foregoing assumption is based immediately on the ontogeny of *Elpenor* and *Porcellus;* it is supported by the considerable size attained by the eye-spots in many species of the third group, and would receive additional confirmation by observations on the Indian *C. Celerio*, supposing that Horsfield's statements do not arise from a confusion of species. This skilful observer, who was the first to breed systematically a large number of tropical larvæ, has given a figure of the Indian caterpillar of *C. Celerio,* according to which this species possesses eye-spots on all the segments from the fourth to the tenth. The European form of this same species has eye-spots only on segments four and five, a fact which does not appear to have been known to Horsfield, as no mention of it is made in his notice of the Indian species. If the caterpillar figured is really that of *Celerio*, which I consider to be by no means improbable, not only is it thus shown that in the species of the third group the ocelli on the hind segments have a secondary origin through a repetition of the primary ones of the front segments, but we can also establish that the same species in two different regions may arrive at two different phyletic stages.

If, finally, we sum up the facts taught by the ontogeny of the two German species, and the adult forms of the other species, we can form therefrom a tolerably complete picture of the course of development of the genus *Chærocampa*. Of the four phyletic stages indicated by the ontogeny of *Elpenor* and *Porcellus*, three still form the terminus of the development of existing species. The great differences among the caterpillars of this genus can be very simply explained on the view that they stand at different levels of phyletic development; some species having remained far behind (Group 1), others having advanced further (Group 2), and others having reached the highest point of development (Group 3). The fact that the species of the third group are only tropical accords well with this view, since many facts prove that phyletic development proceeds more rapidly in the tropics than in temperate climates.

The striking markings of the *Chærocampa* larvæ may, in brief, be stated to originate from a local transformation of two portions of the subdorsal line into eye-spots, and the subsequent transference of these two primary ocelli to the other segments. The eye-spots always originate on segments four and five, and from these the transference mostly occurs backwards, although in certain cases it takes place at the same time forwards. Herein, *i. e.* in the origin of the eye-spots, there lies a great distinction between the genus *Chærocampa* and the genus *Deilephila*, with which it was formerly associated, and in which the origin of a very similar kind of marking can be traced to quite another source.

The Genus Deilephila, Ochsenheimer.

I am acquainted with the caterpillars of nine European and one North American species, these differing in marking to such a wonderful extent that they appear to offer at first sight but little hope of being able to trace them to a common form. These ten species can be separated, according to their markings, into five groups, which I will briefly define before entering upon their ontogeny.

The first group consists of three species, and comprises the commonest and most widely-ranging of all the European species, *Deilephila Euphorbiæ*, as well as *D. Dahlii* from Sardinia and Corsica, and *D. Nicæa*, a species of very restricted range, which appears to occur only in one small district on the French coast of the Mediterranean. These three species agree in marking to the extent of their possessing in the adult form two rows of ring-spots on each side, whilst the subdorsal line is completely absent.

The second group, consisting also of three species, shows a great resemblance to *Euphorbiæ*, but has only one row of ring-spots. It contains *D. Vespertilio*, *D. Galii*, and the Algerian *D. Mauritanica*.

For the third group I only know one representative, *D. Livornica*, Esp., which possesses a single row of ring-spots connected by a subdorsal line.

Another group is composed of *D. Zygophylli*, which occurs on the shores of the Caspian Sea, and the North American *D. Lineata*; these species possessing a strongly-marked subdorsal line, associated with more or less distinct ring-spots, which I shall designate as "open rings," because their black border does not intersect the subdorsal line, but has the form of an arch above and below it.

In the last group, represented by *D. Hippophaës*, which occurs at the foot of the Alps (Wallis), and southward as far as Andalusia, there is only a broad subdorsal line, generally without any trace of a row of spots.

The important differences of marking displayed by these five groups are not in any way accidental, but they represent different stages of phyletic development; or, in other words, the five groups are of different ages, the first (*Euphorbiæ*, &c.) being the youngest, and the last (*Hippophaës*) the oldest of the genus.

According to their phyletic age, the groups follow each other in inverse order, the first being *Hippophaës*, the second that of *Zygophylli*, the third that of *Livornica*, the fourth that of *Galii*, and the fifth and youngest that of *Euphorbiæ*. Only in this last am I acquainted with the *complete* development of one species, for which reason I commence with this group, thus proceeding from the youngest to the oldest forms, instead of taking the more natural course from the simplest and oldest to the youngest and most complicated.

DEILIPHILA EUPHORBIÆ, LINN.

Some captured females were at once placed in an enclosure about the size of a small sitting-room. It was evident that they did not feel quite at home under these conditions, frequently beating their heads and wings against the tarlatan, but some of them nevertheless laid eggs at the base of the leaves of *Euphorbia Cyparissias*. The eggs much resemble those of *Chærocampa Elpenor*, being spheroidal in form, but rather smaller, and of a somewhat darker green. They were laid in small clusters composed sometimes of as many as seven, the single eggs being placed near together, but never touching, and seldom at the point of the leaf, but generally near the end of a twig, where young shoots are in close proximity. During the embryonic development the eggs become coloured, first yellow and partly blackish, and finally completely black.

First Stage.

The young caterpillars (Fig. 37, Pl. V.) immediately after hatching measure four millimeters in length; they are at first rather light, but in the course of half-an-hour they are seen by the naked eye to become of a deep velvety black; later, on increasing in size, they again become paler, appearing of a greenish-black, and subsequently blackish-green. On further increasing in size (Fig. 38), they are blackish-green, with the horn, head, legs, and a crescent-shaped chitinous plate on the back of the prothorax black. There are also on the last segment a double and two single black chitinous plates. Of the later marking of the caterpillar there is scarcely anything present. The spiracles appear as white spots, and on each segment there are a number (mostly ten) of small warts, each of which emits a single bristle.

When the young larvæ have attained a length of seven millimeters they are olive-green, and do not contrast so brilliantly with the green of the *Euphorbia* leaves as before; neither do they as yet possess any markings.

Second Stage.

The first ecdysis occurs after five days, and with this there appears quite suddenly a very complicated pattern. The ground colour is now a light yellowish-green (Fig. 39), and on each of the twelve segments, near the front border, there is a pure white round spot in the middle of a large black transverse spot. I shall designate these, in accordance with the nomenclature employed for *Chærocampa*, as the white "mirrors" on black "ground-areas," both together constituting "ring-spots," as distinguished from "eye-spots" proper, in which a "nucleus," the pupil of the eye, is also added. In many, but not in all specimens, very distinct traces of a subdorsal line can be seen as a light whitish stripe connecting the white spots. The horn, the thoracic and prolegs, and some spots on the head, are black.

The caterpillars remain unaltered till after four days, when, having a length of 17 millimeters, the second moult takes place, bringing with it changes quite as great as those which occurred with the first.

Third Stage.

The caterpillar now assumes the shagreened appearance which it possesses in the adult state. Small white warts are arranged in rows from the dorsal to the spiracular line, and again underneath this line on the abdominal legs. These dots are not only of value as a character for differentiating the genera *Deilephila* and *Chærocampa*, but they also play a part in the peculiar spot-marking which will be shown later on. The ground-colour of the caterpillar is now light green (Fig. 40), replaced by black on certain parts. From the black "ground-area" of the ring-spots, two black triangles extend towards the posterior borders of the segments, but usually without reaching them.

The ring-spots are not essentially changed, although it may be observed that in most specimens the shagreen-dots under each ring-spot are somewhat larger, and stand closer together than in other places. In the following stage they become fused into a second white "mirror," so that two ring-spots stand one above the other, their black ground-areas meeting. The formation of the second ring-spot sometimes takes place in the present stage (Fig. 42).

The subdorsal line has now completely vanished, whilst the spiracular-line [20] appears as a broad stripe above the legs. The horn is yellow with a black point, and the black spots on the head have increased in size.

[20] To be accurate this should be designated the infra-spiracular line; but this term cannot be well applied except in cases where there is also a supra-spiracular line, as, for instance, in *Anceryx* (*Hyloïcus*) *Pinastri*.

Fourth Stage.

The third moult, which again occurs after four days, is not accompanied by such important changes. The green ground-colour has now completely disappeared, and is replaced by a dull black. The caterpillars are now, as also in the previous stage, extremely variable. Thus, for example, a triangular patch of the green ground-colour may be retained on the posterior edge of the segments (Fig. 41), those specimens which possess this character generally having their markings retarded in development, as shown by the absence of the second "mirror" of the ring-spots.

In Fig. 41 the shagreen-dots from which this second "mirror" is subsequently formed, are distinctly larger than the others, and on the eleventh segment two of them have already coalesced.

Fifth Stage.

After another period of four days, the fourth moult takes place. The marking remains the same, but the colours become more vivid; the brick-red of the head, horn, dorsal line and legs, changing into a fiery red. The spiracular line, formerly green alternating with yellow, generally becomes resolved into a row of reddish-yellow spots. Ten days later the caterpillar (8·5 centimeters in length), ceases to feed, and prepares for pupation.

In this last stage also there is great variability of colour, but although each particular character is subject to fluctuation, the individuals of the same brood show but little variation among themselves.[21] Thus, the dorsal line is sometimes black, and sometimes red, or again, this colour interrupted with black, so that only small red spots mark its course. The head may be entirely red, or this colour mixed with black. On the under side of the caterpillar, red generally predominates, but in some specimens this is replaced by black. The ground-colour is also variable, being generally a shining brownish-black, but sometimes dull coaly black. The shagreen-dots are sometimes white and sometimes yellow, and the "mirrors" of the ring-spots are also often yellowish.

The most interesting variation, however, appears to me to be the following:—In many specimens from Kaiserstuhl (Breisgau), the red was unusually vivid, and was not limited to the ordinary places, but occupied also the triangles on the posterior edges of the segments (Fig. 44), which are green in the third and fourth stages (Fig. 42). This variety has also been figured by Hübner. In

[21] Upon this fact obviously depends the statement of that extremely accurate observer Rösel, that the caterpillar of *Euphorbiæ* is but very slightly variable ("Insektenbelustigungen," Bd. iii. p. 36). I formerly held the same opinion, till I convinced myself that this species is very constant in some localities, but very variable in others. It appears that local influences make the caterpillar variable.

one individual (Fig. 43), the under ring-spots were wanting, whilst the upper ones possessed a beautiful red nucleus fading away anteriorly, and showing the first step in the formation of a complete eye-spot.

I cannot positively assert that a fifth moult occurs in the last ten days, although I am very doubtful whether this is the case. It is certain, however, that some time before pupation, and whilst the larva is still feeding, the striking colours fade out, and become replaced chiefly by black.

The ontogeny of this species is obviously but a very incomplete representation of its phyletic development This is at once apparent from the large gap between the first and second stages. It is not possible that a row of ring-spots can have arisen suddenly; in all probability they have been developed from a subdorsal line, which in *Euphorbiæ* is now only indicated in the second stage by a faint line. This conjecture is raised to a certainty when we call in the aid of the remaining species of *Deilephila*.

Deilephila Nicæa, De Prunner.

I only know this species from blown larvæ in Staudinger's collection, and Duponchel's figure, of which Fig. 51, Pl. VI. is a copy. The adult insect possesses two perfectly separated rows of ring-spots. Duponchel figures also two younger stages, of which the youngest is probably the third stage.

The larva is 18 millimeters in length, of a leaf-green colour, and shows no trace of a subdorsal line, but possesses the two rows of ring-spots, which only differ from those of the succeeding stages in the green colour of the "mirror."

Deilephila Dahlii, Treitschke.

I am familiar with numerous specimens in various stages, collected in Sardinia by Dr. Staudinger, and preserved by inflation.

The first stage is blackish, and shows no kind of marking; thus agreeing with the corresponding stage of *Euphorbiæ*. The second stage is unfortunately not represented in Staudinger's collection.

The third stage shows a row of ring-spots, which are, however, connected by a very distinct and sharply defined subdorsal line. In the fourth stage a second row of (under) ring-spots is added, whilst the subdorsal line generally at the same time disappears.

The caterpillar remains unchanged during the fifth stage, when it shows a great resemblance in marking to *Euphorbiæ;* neither does it appear to differ essentially from this species in colour, so far as can be judged from preserved specimens and single figures (in Duponchel and Hübner). I have, moreover, seen several larvæ in the last stage, and the subdorsal could be distinctly recognized as a broad light stripe.

Of the four groups, the second (that of *Galii*),

appears to me to be of but very little importance, as I shall now proceed to show from the development of *D. Vespertilio*.

DEILEPHILA VESPERTILIO, FABRICIUS.

Hitherto I have unfortunately been unable to obtain fertile eggs of this species, so that I can say nothing about the first stage. The latter would have been of interest, not only because of the marking, but also because of the presence of a residual caudal horn.

I am likewise only acquainted with the end of the second stage, having found, at the end of June 1873, a single caterpillar on *Epilobium Rosmarinifolium*, just previous to its second ecdysis. In the case of such young caterpillars, however, the new characters which appear in the succeeding stage are generally perceptible through the transparent chitinous skin at the end of the preceding stage, so that the markings of the insect are thus caused to change. The caterpillar found was about 16 millimeters long, and of a beautiful smooth and shining grass-green (Fig. 13). A broad white subdorsal line extended from the first to the penultimate segment, from which the horn was completely absent. On close inspection the first traces of the ring-spots could be detected near the anterior edge of each segment as feeble, round, yellow, ill-defined spots, situated on the subdorsal line itself (Fig. 13). On the first segment only there is no spot, and here no ring-spot is afterwards formed. Besides these markings, there was only to be seen a yellowish-white spiracular line.

This solitary specimen unfortunately buried itself before the moult for which it had prepared itself had occurred; but this ecdysis is associated with a very important transformation. This statement is founded on a blown specimen in Staudinger's collection; it is only 18 millimeters in length, but already shows the later grey colouring in place of the beautiful green. In this, the third stage, the broad white subdorsal line bears on each segment a red spot enclosed between black crescents above and below (Fig. 49 A). In the fourth stage, during which I have seen many living caterpillars, the subdorsal line is still distinctly present in some individuals (Fig. 14), but the spots ("mirrors") are now completely surrounded by a narrow black ring ("ground-area"), which sharply separates them from the subdorsal line (Fig. 49 B). In the fifth stage this ring becomes a somewhat irregularly formed black "ground-area," whilst the subdorsal line completely vanishes (Figs. 51 and 49 C). The mirrors are white, but generally have a reddish nucleus, which obviously corresponds to the primary yellow spots from which the whole development of the ring-spots originates. This character is, however, sometimes absent; and many other variations also occur in the earlier stages, all of which can be easily explained as cases of arrested, or retarded development. Thus, the subdorsal line often disappears earlier, and is only present in the fourth stage as a feeble light stripe.

Deilephila Galii, Fabricius.

The markings of this species appear to be developed in a precisely similar manner to those of *D. Vespertilio*. The adult larva, as in the last species, shows no trace of a subdorsal line. A row of large black spots, each having an irregular round, yellowish-white nucleus, is situated on an olive-green, blackish-brown, brown, or dirty yellow ground. I have, unfortunately, also in this case been unable to procure fertile eggs. There is, however, one figure of a caterpillar, 2.5 centimeters long, by Hübner, which is of a light-green colour, and has five longitudinal lines; one dorsal, two subdorsal, and a spiracular line. The subdorsal is white, and bears in the place of the ring-spots small red dots, whilst the line itself is bordered with black where the red spots are situated. Hübner has probably figured the third stage, so that we may venture to conclude that in the second stage there is a subdorsal line either quite free from spots, or only showing such feeble rudiments as are to be seen in the second stage of *Vespertilio*.

I found two specimens in the fourth stage in the Upper Engadine. One of these (Fig. 45) was already of a dark, blackish-green ground-colour[22] with a broad, greenish-white subdorsal line sharply defined throughout its entire length, and containing ring-spots of a sulphur-yellow with an orange-red nucleus; the black "ground-area" did not encroach upon the subdorsal line, but was confined to two faint crescents situated above and below the "mirror." Only the two foremost "mirrors" (on the second and third segments) were without nuclei.

The remaining peculiarities of coloration are shown in the figure. I may here only point out the shagreening present on the sides and a portion of the under surface.

The specimen figured was 3.3 centimeters long; a second example measured 2.8 centimeters in length, and was essentially similar, but showed that a considerable amount of variability must prevail at this stage of development. It was pitchy black, with a very indistinct subdorsal line and a few ring-spots, the "mirrors" of which were also sulphur-yellow, with the orange-red nucleus. The shagreening was quite as strong as in the first specimen, the dots being yellow instead of white. It is specially to be observed, because of its important theoretical bearing, that in this larva the ring-spots were absent on the three front segments, and on the fourth only, a faint indication of one could be perceived. In the caterpillar figured

[22] The green is considerably too light in Fig. 45.

the ring-spots increase also in distinctness from the tail to the head.

Fifth Stage.

The two specimens just mentioned, after moulting, acquired the well-known markings of the adult caterpillar already briefly described above. The fifth is the last stage.

The larva is known to occur in several variations, Rösel having figured it in three forms; light green, olive-green, and dirty yellow. It has not been since considered worth the trouble to attend to the subject of caterpillar coloration. Thus, Wilde,[23] in his well-known work, takes no notice of Rösel's observation, but simply describes the caterpillar of *Galii* as "blackish olive-green."

Having had an opportunity of observing twenty-five adult specimens of this somewhat scarce species at one time, I am able to state that it is not in this instance di- or polymorphism, but a case presenting a great degree of variability, with which we have to deal. There are not several sharply-defined types of coloration; but the extremes are connected by numerous intermediate forms. The extreme forms, however, certainly preponderate.

I have never met with Rösel's light-green form; neither was there a dark-green specimen among

[23] "Die Pflanzen und Raupen Deutschlands." Berlin, 1860, p. 83.

the twenty-five mentioned, and I only know this variety from single individuals, found at a former period. Among the twenty-five caterpillars, all gradations of colour occurred, from pitchy black to light clay-yellow, and even to an almost whitish-yellow; some were brownish-black, others of a beautiful chestnut-brown, and others yellowish brown, dark clay-yellow, or brownish-red. Out of twenty-one specimens of which the ground-colours were noted, there were nine black, nine clay-yellow, and three brown; each of the three groups again showing various minor modifications of colour. The other colours also varied somewhat. Thus, the "mirrors" were sometimes white, sometimes strong yellow, and occasionally they also contained a reddish nucleus.

The variations in the shagreening were especially interesting, inasmuch as these appeared to have a striking connection with the general colouring of the caterpillar. Black specimens seldom show such sparse shagreening as that represented in Pl. V., Fig. 46, but are generally thickly scattered with large shagreen-dots right up to the dorsal line (Fig. 47, Pl. VI.), then strikingly resembling the adult larva of *D. Euphorbiæ*. The light ochreous-yellow individuals, on the other hand, were sometimes entirely without shagreening (Fig. 48, Pl. VI.), being smooth, and much resembling the light ochreous-yellow or yellowish-red caterpillar of *D. Nicæa* (Fig. 51, Pl. VI.).

I have never seen a caterpillar of *Galii* which showed traces of the subdorsal line in the last stage, nor have I ever met with one which possessed a second row of "mirror" spots; so that retrogression or a sudden advance in development does not appear to occur.

Of the North African *D. Mauritanica*, which likewise belongs to the *Galii* group, I have not been able to obtain specimens or figures of the younger stages. The adult caterpillar is very similar to that of *Euphorbiæ*, but differs in the absence of the second row of ring-spots. For this reason it must be regarded as a retarded form at an older stage of phyletic development.

I now proceed to the *Livornica* group.

Deilephila Livornica, Esper.

This, the only European species here to be considered, possesses almost the same markings as *Galii* in its fourth stage, *i.e.*, a subdorsal line with interpolated ring-spots. The species is known to be rare, and I have not been able to obtain living specimens, but I have examined several blown larvæ, all of which agree in having the ring-spots sharply distinct from the whitish subdorsal line, so that the latter is thereby interrupted. Figures of the adult larva are given in the works of Hübner, Boisduval, and Duponchel. In most specimens the ground-colour is brown, although Boisduval[24] also figures a light-green specimen; from which it may be inferred, from analogy with *Galii* and *Vespertilio*, that the first stages are green. In Dr. Staudinger's collection there is a young larva, probably in the fourth stage, the ground-colour of which is light ash-grey. The dorsal and subdorsal lines are white, the latter showing in the positions where the ring-spots subsequently appear, small white "mirrors" with red nuclei, exactly corresponding to the stage of *Vespertilio* represented in Fig. 49 A, Pl. VI. The "mirrors" are nothing more than dilatations of the subdorsal line, which is not therefore interrupted by them. The black "ground-area" does not surround the "mirrors" completely, but borders them only above and below, and is much more strongly developed above, extending in this direction to the dorsal line.

The fourth group comprises the two species *D. Lineata*, Fabr., and *D. Zygophylli*, Ochs., the former being the North American representative of our *D. Livornica*, but differing in remaining permanently at the fourth stage of this last species. I am acquainted with *D. Lineata* only through the figure of the adult larva given by Abbot and Smith, which figure, judging from the position and form of the spots, I am compelled to believe is not quite correct, notwithstanding the excellence

[24] Fig. 62, Pl. VII., is copied from Boisduval.

of the other illustrations. The ground-colour of the caterpillar is green; the subdorsal yellow, bordered with black, slightly curved, arched lines, which nowhere interrupt its continuity. This North American species appears therefore to be an older form than our *Livornica*.

DEILEPHILA ZYGOPHYLLI, OCHSENHEIMER.

This species, which is the next allied form to *D. Lineata*, is an inhabitant of Southern Russia. I have seen four specimens of the caterpillar in Dr. Staudinger's collection, three of which are certainly in the last ontogenetic stage. The ground-colour appears ash-grey, ash-brown, or blackish with whitish granulations. A broad white subdorsal line extends to the base of the black caudal horn, this line in one specimen appearing at first sight not to possess a trace of spot rudiments (Fig. 50). On closer investigation, however, there could be observed, in the same position where the ring-spots stand in the other species of *Deilephila*, small black crescents above and below the subdorsal line. In other specimens the white subdorsal line had also become expanded in these positions into distinct spots; indeed, in one individual light white mirror-spots, bordered above and below by black crescents, stood on the subdorsal line (Fig. 50 A).

It is thus in this distinguishing character that the caterpillar is extremely variable, and we may suppose either that this species is now in a state of transition to a higher stage of phyletic development, or else that the ring-spots were formerly more strongly developed, and are now degenerating. The developmental history of the larva could alone decide which of these two views is correct. There would be no difficulty in procuring materials for this purpose if one of the numerous and zealous Russian naturalists would take up the subject.

Deilephila Hippophaës, Esper.

This is the only representative of the fifth and oldest group known to me. The moth resembles *D. Euphorbiæ* to the extent of being sometimes confounded with it, a circumstance which is made the more remarkable by the fact that the caterpillars are so completely different.

The adult larva of this local moth has been made known by the figures, more or less exact, in the works of Hübner, Boisduval, and Duponchel. Wilde also gives a description of it, although from a foreign source. I will not here delay myself by criticizing the different descriptions and figures; they are partly correct, partly inexact, and sometimes altogether erroneous; they were of no avail for the question which here primarily concerns us, and new observation had to be undertaken.

I have been able to compare altogether about forty caterpillars, thirty-five of which were living. All these specimens possessed nearly the same greyish-green ground-colour, and most of them had exactly the simple marking as represented, for instance, in Hübner's figure, *i. e.*, a rather broad greenish-white subdorsal line, somewhat faded at the edges, and without a trace of spots on any of the segments with the exception of the eleventh, on which there was a yellowish, black-bordered mirror-spot, with a broad, diffused, vivid orange-red nucleus. Specimens also occur, and by no means uncommonly, in which no other markings are to be seen than those mentioned ; there were nine among twenty-eight examples compared from this point of view.

In many other individuals of this species small red spots appear on the subdorsal line, exactly in the positions where the ring-spots are situated in the other species of the genus (Fig. 60), so that these spots are thus repetitions of the single ring-spot—a fact which must appear of the greatest interest in connection with the development of the markings throughout the whole genus. But this is not all, for again in other specimens, these red spots stand on a large yellow " mirror," and in one individual (Fig. 59), they had become developed into well-formed ring-spots through the addition of a black border. We have thus presented to us in one and the same stage of a species, the complete development of ring-spots from a subdorsal line.

These facts acquire a still greater interest, as showing how new elements of marking are produced.

The spots on the subdorsal line decrease from the posterior to the anterior segments, so that they must undoubtedly be regarded as a repetition or transference of the ring-spot previously developed on the eleventh segment. I will now proceed to furnish proofs in support of this statement.

I have never met with any specimens having ring-spots on all the segments—in the most prominent instances these spots were present on segments 10—5. This was the case in three out of the twenty-eight caterpillars minutely examined. On all these segments, however, the ring-spots were not equally developed, but increased in perfection from the posterior towards the anterior segments. In the larva represented in Fig. 59 for example, there is a completely developed ring-spot on segment 10, which, although possessing but a feeble black "ground-area," is still distinctly bordered; on segment 9 this border is less sharp, and not so dark, and it is still less sharp and much lighter on segments 8 and 7, whilst it has completely disappeared from segment 6, the yellow "mirror" having at the same time lost in size. On segment 5, only two small contiguous reddish spots, the first rudiments of the nucleus,[25] can be recognized on close inspection.

Specimens in which the spots extend from the eleventh to the seventh segment are of more

[25] The fading of the red anteriorly has not been represented in the figure.

frequent occurrence, five having been found among the twenty-eight. In these the spots diminish anteriorly in size, perfection, and intensity of colour. Still more frequently (in eleven specimens) are the ring-spots or their rudiments restricted to the tenth and ninth segments, the spot on the latter being without exception less developed than that on the former segment.

An anteriorly progressing formation of ring-spots thus undoubtedly occurs, the spots generally diminishing in perfection very suddenly towards the front segments; and specimens, such as that represented in Fig. 60, Pl. VII., in which traces of ring-spots are to be seen on all the segments from the tenth to the fifth, are of rare occurrence.

From what elements of marking are these *secondary* ring-spots resulting from transference developed? They do not, as in the case of the *primary* eye-spots of the *Chærocampinæ*, originate in the separation of one portion of the subdorsal line, and the subsequent formation of this detached spot into a "mirror;" but they arise from the formation of a nucleus, first one and then two of the shagreen-dots on the subdorsal line acquiring a yellowish or reddish colour (Fig. 61, Pl. VII., segments 6 and 7). The ground on which these two spots are situated then becomes yellow (Fig. 61, Pl. VII., segment 8), and a more or less distinct black border, having the form of two small crescents, is afterwards formed. At a later period these two crescents and also the two primary nuclei coalesce, producing a ring-spot which, as in Fig. 61, Pl. VII., segment 9, can be distinctly resolved into two portions.

It certainly cannot be denied that these facts may also be theoretically interpreted in a reverse sense. We might interpret the phenomena in this case, as also in that of *D. Zygophylli*, as a gradual disappearance from the front towards the hind segments of ring-spots formerly present, a view which could only be refuted by the ontogeny of the species. I have not been fortunate enough to procure eggs of *D. Hippophaës*, so that the younger stages are unknown to me. Among my caterpillars, however, there were two in the fourth stage of development, but these did not show ring-spots on all the segments, as we should expect on the above view; on the contrary, no trace of such spots could be seen on any of the segments with the exception of the eleventh, on which there was a ring-spot less perfectly developed than in the last stage.

In this fourth stage the larva of *D. Hippophaës* is of a lighter green (Fig. 58), the subdorsal yellowish with sharp boundaries, and the infraspiracular line pure white, as in the next stage. The shagreening is present, but none of the shagreen-dots are red or reddish, and no trace of a ring-spot can be detected on the subdorsal line with the exception of that on the eleventh segment.

In this last position this line is somewhat widened, and a long, diffused, rose-red spot can there be recognized upon it (Fig. 58 A). The black "ground-area" present in the fifth stage is as yet absent, and the spot is not so sharply separated anteriorly from the subdorsal line as it becomes later.

From these observations we might venture to expect that in the third stage of *Hippophaës*, the subdorsal line would also be free from this spot on the eleventh segment, and it is possible that in the second stage this line is itself absent.

The Genus Deilephila: Summary of Facts and Conclusions.

Regarding only the *adult* larvæ of the species of *Deilephila*, these represent in their five groups, five stages in the phyletic development of the genus; but if we also take into consideration the developmental history, two more stages must be added, viz., that in which the caterpillar possesses no particular marking, as was found to be the case in the first stage of the development of *D. Euphorbiæ* and *D. Dahlii*; and a second stage with a subdorsal line, but without any ring-spot formations. Seven stages of phyletic development must therefore be distinguished.

Stage 1.—No species with entire absence of marking in the adult form now occurs.

Stage 2.—A subdorsal, accompanied by a spiracular line, extends from the caudal horn to the first segment. This also no longer forms the final stage of the ontogeny, but is, however, undoubtedly retained in the second stage of several species (*D. Vespertilio*, *Livornica*, *Lineata*, and perhaps also *Galii*).

Stage 3.—The subdorsal line bears a ring-spot on the penultimate segment; the other markings as in the last stage. *D. Hippophaës* only belongs to this stage, a small number of specimens, however, showing a transition to the following stage by the transference of ring-spots from the posterior to the anterior segments.

Stage 4.—Open ring-spots appear on the subdorsal line on all the segments from the eleventh to the first. *D. Zygophylli* and the North American *D. Lineata* belong here.

Stage 5.—Closed ring-spots are situated on the subdorsal line. Of the known species, only *D. Livornica* concludes its development at this phyletic stage.

Stage 6.—A single row of ring-spots replaces the subdorsal line. *D. Galii*, *Vespertilio*, and *Mauritanica* represent this stage at the conclusion of their ontogeny.[26]

Stage 7.—A double row of ring-spots. Only *D. Dahlii*, *Euphorbiæ*, and *Nicæa* attain to this highest stage of *Deilephila* marking, the two first

[26] [The caterpillar of *Deilephila Euphorbiarum*, figured by Burmeister (Lép. Rép. Arg., Pl. XVI, Fig. 1) belongs to this stage. R.M.]

species in the fourth stage, and *Nicæa* in the third stage of its ontogeny.

Although our knowledge of the history of the development of the individual species is still so fragmentary, we may conclude with certainty that the development of the markings has been uniform throughout—that it has proceeded in the same manner in all species. All the species appear to be making for the same goal, and the question thus arises whether there may not be an innate force urging their phyletic development. The rigorous examination of this conception must be reserved for a later section. Here, as we are only occupied essentially in establishing facts, it must be remarked that retrogression has never been observed. The young larval forms of a species never show the markings of a later phyletic stage than the older larval forms; the development takes the same course in all species, only making a greater advance in the same direction in some than in others.

Thus, *Nicæa* and *Euphorbiæ* have advanced to the seventh phyletic stage, *Zygophylli* and *Hippophaës* only to the third, and some specimens of *Zygophylli* to the fourth. But at whatever phyletic stage the ontogeny of a species may terminate, the young larval stages always display the older phyletic stages. Thus, *Galii* in its last ontogenetic stage reaches the *sixth* phyletic stage; in its penultimate stage it reaches the *fifth* phyletic stage; and in its third stage; the *fourth* phyletic stage is represented, so that little imagination is required to anticipate that in the second stage the *third* or *second* phyletic stage would be pictured.

If we tabulate the development of the various species, indicating the ontogenetic stages by Arabic numerals, and the stages of the phylogeny which are reached in each stage of the ontogeny by Roman numerals, we obtain a useful synopsis of the series of developments, and, at the same time, it shows how many gaps still remain to be filled up in order to complete our knowledge even of this small group of species.

TABLE OF DEVELOPMENT OF THE SPECIES OF DEILEPHILA.

Deilephila.	Ontogeny Stage 1.	Ontogeny Stage 2.	Ontogeny Stage 3.	Ontogeny Stage 4.	Ontogeny Stage 5.
1. Hippophaës	?	?	?	III.	III.—IV.
2. Zygophylli	?	?	?	?	III.—IV.
3. Lineata	?	?	?	?	IV.
4. Livornica	?	?	?	IV.	V.
5. Galii	?	?	IV.	V.	VI.
6. Vespertilio	?	II. (?)	IV.	V.	VI.
7. Mauritanica	?	?	?	?	VI.
8. Dahlii	I.	?	VI.	VII.	VII.
9. Euphorbiæ	I.	V.	VI.	VII.	VII.
10. Nicæa	?	?	VII.	VII.	VII.

From this very incomplete table we perceive that, in certain instances, the stages can be represented as a continuous series of phyletic steps, as in the case of *D. Galii*; that in others certain steps may be omitted, as with *D. Euphorbiæ*, in which grade I. of stage 1 is immediately followed by grade V. in stage 2. In reality the gap caused by this omission is still greater than would appear, as grade V. is only indicated, and not actually reached, the subdorsal not being present as a sharply-defined line, but only as a faint stripe. The suppression of phyletic steps increases with the advancement in phyletic development. The higher the step to which a species finally attains, the greater is the tendency of the initial stages to be compressed, or omitted altogether.

From what has thus far been seen with respect to the development of *D. Hippophaës*, there may be drawn what to me appears to be a very important conclusion, viz. that the ring-spots of *Deilephila* first originated on the segment bearing the caudal horn, and were then gradually transferred as secondary spots to the preceding segments. Complete certainty would be given to this conclusion by a knowledge of the young forms of other phyletically retarded species, especially those of the American *D. Lineata*, and perhaps also those of *Zygophylli* and *Livornica*. The few observations on the development of *D. Galii* already recorded give support to this view, since the absence of ring-spots on the three front segments in the young caterpillar (one instance), or their less perfect formation on these segments (second instance), indicates a forward transference of the spots.

If the foregoing view be accepted, there follows from it a fundamental difference between the development of the genera *Chærocampa* and *Deilephila.* In the former the formation of the eye-spots proceeds from a subdorsal line, but they first appear on two of the front segments, and are then transferred to the *posterior* segments. In *Deilephila*, on the other hand, a single ring-spot is formed on the penultimate segment bearing the caudal horn, and this is repeated on the *anterior* segments by secondary transference. With respect to the origination of the ring-spot also, there is a distinction between this genus and *Chærocampa*, inasmuch as the first step towards the eye-formation in the latter consists in the separation of a curved portion of the subdorsal line, whilst in *Deilephila* the nuclear spot first seems to originate and the separation of the mirror-spot from the subdorsal line appears to occur secondarily. It is difficult here to draw further conclusions, since the first appearance of the primary ring-spot has not yet been observed, and no more certain inference respecting the history of the formation of the *primary* ring-spots can be drawn from the manner in which the *secondary* ring-spots are formed.

Because in *Hippophaës* the formation of the secondary ring-spots begins with the red coloration of one or two shagreen-dots, it does not follow that the primary spot on the eleventh segment also originated in this manner; and this is not without importance when we are concerned with the causes which underlie the formation of ring-spots. In *Chœrocampa* also, the formation of the primary eye-spots appears to differ from that of the secondary—in the latter the black "ground-area" first appearing, and in the former the "mirror-spot." The secondary eye-spots certainly remain rudimentary in this last genus, so that the evidence in support of this conclusion is thus much weakened; but it must be admitted that we are here on ground still too uncertain to admit of wider conclusions being based thereon.

As a final result of the investigation, we may advance the opinion that the existing species of the genus *Deilephila* have reached five different phyletic stages, and that their very different external appearance is explained by their different phyletic ages; the appearance from these caterpillars of moths so extremely similar, can otherwise be scarcely understood.

It may appear almost unnecessary to bring forward additional proofs in support of this interpretation of the facts, but in a field where the data are so scanty, no argument which can be drawn from them should be considered as superfluous. The variations which occasionally occur in the larvæ, however, to a certain extent furnish a proof of the correctness of the theoretical interpretation offered.

When, in the ontogeny of these species, we actually see before us a series of stages of phyletic development, we must admit that ordinary reversion may occur, causing an adult caterpillar to show the characters of the young. Forms reverting to an earlier phyletic stage must, on the whole, occur but seldom, as this stage is removed further back in the ontogeny. Thus, indications of the subdorsal line must occur but rarely in the *adult* larvæ of *Euphorbiæ*, and still less frequently in *Nicæa*, whilst they must be expected to be of more common occurrence in *Vespertilio*, and also, as has already been seen, in *Dahlii*. In this last species, as also in *Vespertilio*, the completely-developed subdorsal line is still present in the third stage, whilst it is possessed by *Euphorbiæ* only in the second stage, and then in a rudimentary condition.

The state of affairs may in fact be thus described: Among several hundred adult larvæ of *Dahlii* found in Sardinia by Dr. Staudinger, there were some which did not actually possess a distinct subdorsal line, but in place thereof, and as its last indication, a feeble light stripe. One of Dr. Staudinger's caterpillars showed also a distinct line between the closed eye-spots. In the last stage of *Vespertilio* this line appears still more frequently, whilst in *Euphorbiæ* it is extremely rare, and when present it only appears as a faint indication. This is the case with one of the specimens figured in Hübner's work as an "aberration," and also with one in Dr. Staudinger's collection. Of *Nicæa* I have at most seen only eight specimens, none of which showed any trace of the long-vanished subdorsal line.

It must be expected that any ontogenetic stage would most readily revert to the preceding phyletic stage, so that characters present in the preceding stage are consequently those which would most commonly arise by reversion. This postulate of the theory also finds confirmation in the facts. Caterpillars which, when full grown, belong to the *seventh* phyletic stage, *e.g.* *D. Euphorbiæ*, not unfrequently show variations corresponding to the *sixth* stage, *i.e.* only one instead of two rows of ring-spots—the upper and first-appearing series. On the other hand, forms reverting to the *fifth* phyletic stage (ring-spots with connecting subdorsal line) occur but very rarely. I have never met with such cases in adult living caterpillars of *D. Euphorbiæ*, although in one instance such a larva was found in the fourth ontogenetic stage; but the strikingly dark, brownish subdorsal line which connected the otherwise perfectly developed ring-spots, completely disappeared in the fifth stage of the ontogeny. Those larvæ which, in the adult state, belong to the *sixth* phyletic stage, not unfrequently show the characters of the *fifth* stage more or less developed, as, for example, *D. Vespertilio.*[27]

THE GENUS SMERINTHUS, LATREILLE.

The caterpillars of this genus are very similar in appearance, and all possess extremely simple

[27] [In concluding this account of the *Chærocampinæ* I may call attention to the following species, which have since been figured by Burmeister :—*Pachylia Ficus*, Linn. (*loc. cit.* Pl. XIV., Figs. 1 and 2); during the three first stages the larva is uniformly green, with a yellow subdorsal line, and below this ten oblique yellow stripes slanting away from the head; after the third moult the colour completely changes, the whole area of the body being divided into two distinct portions by the subdorsal line, above which the colour is red, and underneath of a pale green; the oblique stripes have almost disappeared; no occelli nor annuli are present. *Pachylia Syces*, Hübn. (*loc. cit.* Fig. 3); very similar to the last species in its young stages (figured also by Mérian, Surin. pl. 33). *Philampelus Vitis*, Linn. (*loc. cit.* Figs. 4 and 5); two stages represented; between first and second moults green, with oblique paler stripes slanting in same direction as in *Pachylia*, and each one containing a red streak surrounding the spiracle. When adult, the ground-colour is yellow above and green beneath, the whole surface being mottled with deep black and red transverse markings; the oblique stripes whitish, bordered with black at their lower extremities (figured also by Mérian, pls. 9 and 39). *Philampelus Anchemolus*, Cram. (*loc. cit.* Pl. XV., Fig. 1; Mérian, pl. 47); green when young, with seven oblique red stripes; when adult, uniformly brown, with seven pale yellow lateral markings, the first four of which are spots, and the remainder broad oblique stripes slanting forwards. *Philampelus Labruscæ*, (see note 13, p. 195). R.M.]

markings. The occurrence of numerous stages of development of these markings is thus excluded, and the study of the ontogeny therefore promised to furnish less information concerning the phyletic development of the genus than in the case of the preceding genera. This investigation has nevertheless also yielded interesting results, and the facts here recorded will be found of value in likewise throwing light on the causes which have produced the markings of caterpillars.

I shall commence, as in former cases, with the developmental history. I have easily been able to obtain fertile eggs of all the species of *Smerinthus* known to me. Impregnated females laid large numbers of eggs in confinement, and also bred females of the commoner species can readily be made to copulate, when pinned, and exposed in a suitable place in the open air. A male soon appears under these circumstances, and copulation is effected as readily as though the insect were not fastened in the way indicated.

SMERINTHUS TILIÆ, LINN.[28]

The light green eggs are nearly spherical, and

[28] [*Mimas Tiliæ* of Butler's revision. The author states that this genus is "easily distinguished from *Laothoë* by the form of the wings, the outer margin of secondaries deeply excavated below the apex, and the secondaries narrow and not denticulated." Here again we have a clashing of the results arrived at by a study of the ontogeny of the larvæ, on the one hand, and the founding of genera on the characters of the imagines

after fourteen days (beginning of July) the young larvæ emerge. These are also of a light green colour, and are conspicuous for the great length of the caudal horn, which is nearly half as long as the body. This horn is likewise of a light green at first, but becomes dark violet in the course of an hour. No trace of any markings can be detected at this stage.

As soon as the caterpillars are hatched they commence to nibble the empty egg shells; then they run about with great activity, and after several hours take up their position on the largest vein on the under side of the lime leaves, where they remain for a long period. In this situation they have the same form and colour as the leaf-vein, and are very difficult to discover, which would not be the case if they reposed obliquely or transversely to the vein. In about 4—5 days the caterpillars undergo their first moult, and enter upon the second stage. On each side of the segments 11—4, there now appear seven oblique whitish stripes on a somewhat darker

only, on the other. Of the three species discussed by Dr. Weismann, Mr. Butler, following other authors, refers *Tiliæ* to the genus *Mimas*, *Populi* to *Laothoë*, and *Ocellatus* to *Smerinthus*. It is to be hoped that when our knowledge of the developmental history of larvæ is more complete in all groups, a reconciliation between the results of the biological investigator and the pure systematist will be brought about, so that a genus may not, as at present, have such very different values when regarded from these two points of view. R.M.]

green ground; these slope in the direction of the caudal horn. Owing to the transparency of the skin, a dark green dorsal line appears in the position of the underlying dorsal vessel, the green contents of the alimentary canal being distinctly visible through the absence of adipose matter in the tissues. The larvæ possess also a fine whitish subdorsal line, which extends from the horn to the head. The horn at this stage becomes black with a yellowish red base.

In the third stage, which occurs after six or seven days, the oblique stripes appear darker, and the subdorsal line disappears.

Fourth Stage.

After another period of 4—5 days the third moult takes place, and there now commences a dimorphism which will perhaps be better designated as variability, since the two extremes are connected by transitional forms. The majority of the larvæ have, as in the preceding stage, pure white oblique stripes, but many of them possess a blood-red spot on the anterior side of the stripes, this spot showing all gradations in size and depth of colour between maximum development and a mere trace. Special interest attaches to these spots, as they are the first rudiments of the coloured border of the oblique stripes which occurs in so many *Sphinx* caterpillars.

In the fifth stage—the last of the larval development—the red spots become more strongly pronounced. Among eighty caterpillars from one brood there were about twenty without any red whilst the remainder were ornamented with more or less vivid blood-red spots, often large and irregular in form. In some specimens the spots had become drawn out into lines,[29] forming a coloured edge to the oblique white stripes, similar to that possessed by the larva of *Sphinx Ligustri*. The caterpillar is thus represented in many figures, but generally the coloured stripe is made too regular, as in reality it is always irregularly defined above, and never so sharp and even as in *Sphinx Ligustri*. The character is here obviously not yet perfected, but is still in a state of development.

Smerinthus Populi, Linn.

From green spherical eggs there emerged larvæ 6.5 millimeters in length without any markings. They were of a light greenish-white, the large head and long caudal horn being of the same colour. The posterior boundary of the segments appears as a light shining ring (Pl. VI. Fig. 55).

The characteristic markings of the genus appear on the following day without the occurrence of any moult: seven oblique white stripes arise from near the dorsal line, and extend along the sides in a direction parallel to that of the horn. On the three front segments they are represented only by

[29] The caterpillar is thus figured by Rösel.

three small white spots (Fig. 56). The caterpillar likewise possesses a marking of which the adult species of the genus retain only a trace, viz., a well-developed, pure white subdorsal line, which is crossed by the six anterior oblique stripes, and uniting with the upper part of the seventh extends to the caudal horn.

I long believed that the markings described were first acquired in the second stage, as I was possessed with the generally accepted idea that the changes of form and colour in insects could only occur at the period of ecdysis. I at first thought that the moult had escaped my notice, and I was only undeceived by close observation of individual specimens.

Second Stage.

The first moult took place after five days, the larvæ being 1.4 centimeters in length. Only unimportant changes of marking are connected therewith. The subdorsal line loses much in thickness and definition, and the first and last of the oblique stripes become considerably broader than the intermediate ones (Fig. 57). The green ground colour and also the stripes acquire a yellowish hue.

On the other hand, there occur changes in form. The head, which was at first rounded, becomes of the characteristic triangular shape, with the apex upwards, common to all the species of the genus, and at the same time acquires two white lines, which unite above at the apex of the angle. The shagreening of the skin now also takes place, and the red spot at the base of the horn is formed.

There appears to be at this stage a general tendency for the suffusion of red, the thoracic legs also becoming of this colour.

Third Stage.

The second moult occurs after six or eight days, the marking only changing to the extent of the subdorsal line becoming still more indistinct. This line can now only be distinctly recognized on the three front segments in a few individuals, whilst in the majority it is completely absent. Sometimes the ferruginous red spots on the oblique stripes now appear, but this character is not completely developed till the fifth stage. Out of about ninety bred specimens in which I followed the entire development, only one possessed such spots, and these were situated on both sides of the sixth segment.

Fourth Stage.

The third moult, which takes place after another period of six days, is not associated with any change of marking.

In this stage also I observed in one specimen (not the one just mentioned) the ferruginous spots, and again only on the sixth segment. On account of the theoretical conclusions which may be drawn from this localization of the spots—supposing it to be of general occurrence—it becomes of importance to institute observations with different broods, so as to investigate their first appearance, frequency, and local limitation. It appears to me very probable that, with respect to frequency and time of appearance, there would be great differences, since, in the last stage, it is just this character which shows a great variability. It would be more remarkable if it should be established that the first appearance of the spots was always limited to a certain segment; and there would then be a great analogy with the first appearance of the eye-spots in *Chærocampa* and the ring-spots in *Deilephila*.

Fifth Stage.

The adult caterpillar does not differ in marking to any considerable extent from the preceding stages. The first and last stripes do not appear larger than the intermediate ones, as the latter now increase in size. Many specimens were entirely without red spots; in others they were present, but were small and inconspicuous, whilst in others again there were two spots, one above the other, of a vivid ferruginous red, these coalescing in some cases, and thus forming one spot of a considerable size. I have never seen these spots formed into a regular, linear, coloured border to the white oblique stripes—as occasionally happens in *Tiliæ*—either in living specimens, blown larvæ, or in figures.

Smerinthus Ocellatus, Linn.

The green eggs much resemble those of *Populi*, as also do the newly hatched caterpillars, which, as in the case of this last species, are entirely without markings. As with *Populi*, the markings are formed in the course of the first stage, and are distinctly visible before the first moult. The long caudal horn is of a red colour.

After two to three days the caterpillars moult, their length then being one centimeter ; the seven beautiful oblique white stripes, and the fine white subdorsal line, are more strongly pronounced, the latter becoming broader in front. They differ from *Populi* in having the oblique stripes united in the dorsal line.

The second moult occurs after another three days, and brings no important change ; only the fine subdorsal line becoming somewhat fainter. Neither is the third moult, which takes place four days later, associated with the appearance of any essentially new character. The oblique stripes remain as before, but their upper portions now stand on a somewhat darker green ground colour, whilst the subdorsal line vanishes, leaving distinct traces only on the three or four front segments.

The fourth moult follows after a period of seven days, and my bred larvæ underwent scarcely any alteration in marking. Only small differences in coloration became perceptible in the head and horn, these changing to bluish. Specimens occur, although but rarely, which show in this last stage red spots in the vicinity of the oblique stripes, just in the same manner as with *Populi*, in which species, however, they occur more commonly. I only once found an adult larva of *Ocellatus* possessing reddish-brown spots above and below the oblique stripes,[30] exactly as in one of the specimens figured by Rösel.[31]

[30] [In 1879 Mr. E. Boscher found about thirty full-grown caterpillars of this species in the neighbourhood of Twickenham, ten to twelve of which were feeding on *Salix viminalis*, and the remainder, from a locality not far distant, on *Salix triandra.* The whole of the specimens taken on the plant first named, had the red-brown spots above and below the oblique stripes more or less completely developed, as I myself had an opportunity of observing. In these spotted specimens the ground-colour was bright yellowish-green, and in the others this colour was dull whitish-green above, passing into bluish-green below. Should these observations receive wider confirmation, it would be fair to conclude that this species is now in two states of phyletic development, the more advanced stage being represented by the brighter spotted variety. (See also Proc. Ent. Soc. 1879, p. xliv.). Mr. Peter Cameron has recently suggested (Trans. Ent. Soc. 1880, p. 69) that the reddish-brown spots on the *Smerinthus* caterpillars may serve for purposes of disguise, as they closely resemble, both in colour and form, certain galls (*Phytoptus*) of the food-plants of these species. If this view be admitted, these spots must be considered as a new character, now being developed by natural selection. The variation in the ground-colour of the two forms of *S. Ocellatus* may possibly be phytophagic, but this can only be decisively settled by a series of carefully conducted experiments. R. M.]

[31] "Insekten-Belustigungen," Suppl. Pl. 38, Fig. 40.

In this stage also there remains almost always on the three to six front segments, a more or less distinct residue of the subdorsal, which extends backwards from the head as a whitish line intersecting the foremost oblique stripes. (Fig. 70, Pl. VII.)

Results of the Developmental History of Smerinthus Tiliæ, Populi and Ocellatus.

From the meagre materials furnished by these three obviously nearly related species, we may at least conclude that, with respect to marking, three stages of development can be distinguished:—(1) Simple (green) coloration without marking; (2) subdorsal lines crossed by seven pairs of oblique stripes; (3) more or less complete absence of the subdorsal lines, the oblique stripes remaining, and showing a tendency to become edged with a red border.

Which of the three species is the oldest I will not attempt to decide. If we might venture to form any conclusion from the frequency of the red spots, *Tiliæ* would be the youngest, *i. e.*, the species which has made the farthest advance. But this does not agree with the fact that the oblique stripes appear somewhat later in this species. Both these distinctions are, however, too unimportant to enable us to build certain conclusions on them. Neither does a comparison of the adult larvæ with other species of *Smerinthus* furnish any further information of importance.

Of the genus *Smerinthus*, Latr., thirty species were catalogued by Gray,[32] of which I am only acquainted with the larvæ of eight (five European, and three North American). None of these in the last stage possess a complete subdorsal line together with oblique stripes. Neither, on the other hand, do any of them show a more advanced stage of development in having the red spots constantly formed into coloured border-stripes. We must therefore admit that they have all reached nearly the same stage of phyletic development. On turning to the doubtfully placed genus *Calymnia*, Boisduval, which is represented in Gray by only one species, figured by Westwood[33] as a *Smerinthus*, we first meet with an older stage of development of the genus.

The adult caterpillar of *C. Panopus*, from the East Indies, possesses, in addition to the oblique stripes, a completely developed subdorsal line,[34]

[32] "Catalogue of Lepidop." British Museum. [Butler divides the subfamily *Smerinthinæ* into 17 genera, containing 79 species, viz. *Metamimas*, 2; *Mimas*, 4; *Polyptychus*, 7; *Lophostethus*, 1; *Sphingonæpiopsis*, 1; *Langia*, 2; *Triptogon*, 23; *Laothoë*, 2; *Cressonia*, 3; *Paonias*, 2; *Calasymbolus*, 5; *Smerinthus*, 5; *Pseudosmerinthus*, 2; *Daphnusa*, 4; *Leucophlebia*, 5; *Basiana*, 10; *Cæquosa*, 1. R.M.]

[33] "Cabinet Orient. Entom.," p. 13, Pl. VI., Fig. 2. [Butler places this species doubtfully among the *Sphinginæ*. R.M.]

[34] "Catalogue of the Lepidop. Insects of the E.I. Co.," by Horsfield and Moore. Pl. VIII., Fig. 6.

and thus corresponds to the first stage of *S. Populi*. This species may possibly retain in its ontogeny a stage in which the oblique stripes are also absent, whilst the subdorsal line is present. From the early disappearance of the subdorsal line in the species of *Smerinthus*, we may venture to conclude that this character appeared at an early stage of the phylogeny, whilst the oblique stripes represent a secondary form of marking, as shall be further established subsequently.[35]

[35] [The larvæ of four other species of this subfamily have since been made known through Mr. Butler's figures. *Smerinthus Tatarinovii*, Ménetriés (*loc. cit.* Pl. XC., Fig. 16), from Japan, is "pale sea-green, tuberculated with white, with seven lateral, oblique, crimson-edged white stripes." There is no trace of the subdorsal line shown in the figure, so that this species thus appears to be in the third phyletic stage of development. *Smerinthus Planus*, Walker, from China (*loc. cit.* Pl. XCII., Fig. 11), is "pale green, with white or yellow lateral stripes." A trace of the subdorsal line remains on the front segments, thus showing that the species is in the second phyletic stage of development. *Triptogon Roseipennis*, Butler, from Hakodadi (*loc. cit.* Pl. XCI., Fig. 6), is represented as yellow, with seven oblique white stripes, with large irregular triangular red spots extending from the anterior edge of the stripes, nearly across each segment. It is probably in the third phyletic stage. The Indian *Polyptychus Dentatus*, Cramer (*loc. cit.* Pl. XCI., Fig. 10), is "bluish-green at the sides, with oblique purple stripes, with a broad, dorsal, longitudinal, golden-green band, bordered by subtriangular purple spots, one above each stripe." The dorsal band is bordered by coloured stripes, which may be the subdorsal lines; but the position in which it is figured, and its very different mode of coloration, make it very difficult to compare satisfactorily with the foregoing species. The genus *Ambulyx* is closely allied to the *Smerinthinæ*, and the two

THE GENUS MACROGLOSSA, OCHSENHEIMER.

The adult larvæ of five species are known, and to these I can now add a sixth. In Gray the genus contains twenty-six species.[36] I cannot find any figures or descriptions of the young stages of these caterpillars, and I have myself only observed the complete ontogeny of one species.

By placing a captured female *M. Stellatarum* in a capacious breeding-cage, in the open air, I was enabled to procure eggs. The moth hovered about over the flowers, and laid its small, grass-green, spherical eggs (partly when on the wing), singly, on the leaves, buds, and stalks of *Galium Mollugo*. Altogether 130 were obtained in three days.[37]

following species may be here mentioned: *A. Gannascus*, Stål, figured by Burmeister (*loc. cit.* Pl. XIII., Fig. 5), is green, with a yellow subdorsal line, and seven oblique white lateral stripes, edged with red. *A. Liturata*, Butl. (*loc. cit.* Pl. XCI., Fig. 2), is yellowish-green above, passing into bluish-green below. The subdorsal is present on the three front segments, and is followed by a row of white, elongated patches, one on each segment, these being the upper portions of a row of lateral oblique stripes. The thickened upper extremities of the latter are edged with red, and their arrangement is very suggestive of their having arisen from the breaking up of a subdorsal line. R.M.]

[36] [Butler catalogues 43 species of this genus. R.M.]

[37] The deposition of eggs was accomplished by the insect laying hold of the point of a twig with its legs during flight, and curving its abdomen upwards against a leaf, the wings being kept vibrating. The egg is instantaneously fastened to the leaf. This operation is repeated from twice to four times

First Stage.

After about eight days the caterpillars emerge. They are only two millimeters in length, and are at first yellowish, but soon become green, set with small single bristles, and they possess a short greenish caudal horn, which afterwards becomes black. The head is greenish-yellow. The young larvæ are entirely destitute of marking. (Pl. III., Fig. 1).

Second Stage.

The first moult takes place after four days, the caterpillar now acquiring the marking which it essentially retains to pupation.

Fine white subdorsal and spiracular lines appear, and at the same time a dark green dorsal line, which, however, does not arise from the deposition of pigment, as is generally the case, but from a division in the folds of the fatty tissue along this position. (Fig. 2, Pl. III.)

The colour is now dirty green in all specimens, the skin being finely shagreened.

Third Stage.

The second moult, occurring after another period of four days, does not bring any change of

successively, the moth then hovering over and sucking at the flowers for some time. The eggs exactly resemble in colour the young green buds of *Galium*.

marking, the colour only becoming somewhat darker. Length, twelve millimeters.

Fourth Stage.

The third moult (after another four days) likewise brings only a change of colouring, which is of such a nature that the caterpillar becomes dimorphic. At the same time that peculiar roughening of the skin takes place which, in the case of *Chærocampa*, was designated as "shagreening." The colour is now light grass-green in some specimens, and dark green in others; in these last the subdorsal line is edged above with dark brown, and the spiracles are also of this colour. Length, seventeen millimeters.

Fifth Stage.

Four days later, after the fourth ecdysis, the dimorphism becomes a polymorphism. Five chief types can be distinguished :—

Variety I.—Light green (Fig. 7, Pl. III.); dorsal line, blackish-green, strongly marked; subdorsal line broad, pure white, edged above with dark green; spiracular line, chrome-yellow; horn, black, with yellow tip and blue sides. Spiracles, blackish-brown, with narrow yellow border; legs, and extremities of prolegs, vermilion-red.

Variety II.—Blackish-brown (Fig. 6, Pl. III.); head and prothorax, yellowish-brown; markings the same as above.

Variety III.—Blackish-green or greenish-black (Figs. 10 and 11, Pl. III.); subdorsal line with blackish-green border above, gradually passing into a light green ground-colour; spiracular line, chrome yellow; head and prothorax, greenish-yellow.

Variety IV.—Light green (Figs. 4 and 12, Pl. III.); dorsal line quite feeble; subdorsal broad, only faintly edged with dark green; subspiracular line, faint yellowish; head and prothorax, green.

Variety V.—Brownish-violet (Fig. 8, Pl. III.); the black dorsal line on a reddish ground either narrow or broad.

From these five varieties we see that the different types do not stand immediately next to one another; they are, in fact, connected by numerous transitional forms, the ground-colour varying greatly, being dark or light, yellowish or bluish. (Compare Figs. 4, 5, 7, and 12.) The markings remain the same in all, but may be of very different intensities. The dorsal line is often only very feebly indicated, and the subdorsal line is frequently but faintly edged; the latter is also sometimes deep black above and bordered rather darkly beneath, the sides then being of a dark green, often with blackish dots on the yellow spiracular line (Fig. 5, Pl. III.), this likewise being frequently edged with black. Only the horn and legs are alike in all forms. The green ground-colour passes into blackish-green, greenish or brownish-black, and again, from reddish-brown to lilac (Fig. 3), this last being the rarest colour.

The designation "polymorphism" may here appear very inapplicable, since we have no sharply distinct forms, but five very variable ground-colours connected by numerous intermediate modes of coloration. Should, however, the term "variability" be suggested, I am in possession of an observation which tends to show that the different colours have to a certain extent become fixed. I found a brown caterpillar, the five front segments of which were light green on the left side, and the fifth segment brown and green mixed (Fig. 9, Pl. III.). Such parti-coloration can evidently only appear where we have contending characters which cannot become combined; just as in the case of hermaphrodite bees, where one half of a segment is male and the other half female, the two characters never becoming fused so as to produce a truly intermediate form.[38] From this observation, I conclude that some of the chief

[38] [Figures of a remarkable case of gynandromorphism in a butterfly (*Cirrochroa Aoris*, Doubl.) have recently been published by Prof. Westwood (Trans. Ent. Soc. 1880, p. 113). On the right fore and hind-wings of a male specimen there are patches of female colouring, thus bearing out in a very striking manner the above views concerning the non-fusibility of characters (in this case sexual) which have been long fixed. Complete (*i. e.* half-and-half) gynandromorphism is not uncommon in butterflies. R.M.]

varieties of *Stellatarum* have already become so far removed from one another that they must be regarded as intermediate fixed forms, the colours of which no longer become fused together when they occur in one individual, but are developed in adjacent regions. Other facts agree with this conclusion. Thus, among the 140 adult larvæ which I bred from the batch of eggs above mentioned, the transition forms were much in the minority. There were forty-nine green and sixty-three brown caterpillars, whilst only twenty-eight were more or less transitional.

On these grounds I designate the phenomenon as "polymorphism," although it may not yet have reached, as such, its sharpest limits. This would be brought about by the elimination of the intermediate forms.[39]

[39] [I have long held the opinion that the di- and trimorphism displayed by certain butterflies has originated through polymorphism from ordinary variability. I will not here enter into details, but will only cite a few instances indicating the general direction of the arguments. The phenomenon to which I refer is that so ably treated of by Mr. A. R. Wallace (see Part I., p. 32, note 18) and others. One male has often two or more distinctly coloured females, and in such cases one form of the female generally resembles the male in colour. Cases of polymorphic *mimetic* females may for the present be excluded, in order to reduce the argument to its greatest simplicity. Thus, in the case of native species, *Colias Edusa* has two females, one having the orange ground-colour of the male, and the other the well-known light form, var. *Helice*. So, also, *Argynnis Paphia* has a normal female and the dark melanic form var. *Valezina*. Numerous other cases might be mentioned

Immediately before pupation, all the caterpillars, both green and brown, acquire a lilac coloration. The fifth stage lasts seven days, and the whole larval development twenty-three days, the period from the deposition of the eggs to the appearance of the moth being only thirty-one days.

I have treated of the polymorphism of *Stellatarum* in detail, not only because it has hitherto remained unknown, and an analysis of such cases has been completely ignored,[40] but more particularly because, it appears to me, that important conclusions can be drawn therefrom. Moreover,

among exotic species; and, looking at the phenomenon as a whole, it is seen to be one of *gradation.* For instance, our common "Blues" (*Plebeius Icarus, P. Thetis,* &c.) have females showing a complete gradation between the ordinary blue male and the brown female coloration. In a large number of specimens of *Callosune Eupompe* in my cabinet, collected in Arabia by the late J. K. Lord, there is a completely graduated series of females, varying from individuals having the scarlet tips of the fore-wings as strongly developed as in the males, to specimens without a trace of such colouring; and the same is the case with other species of this and allied genera. In such instances it is only necessary for the intermediate female forms to become extinct, in order to have true cases of dimorphism. It is significant that in 1877, when *Colias Edusa* appeared in this country in such extraordinary profusion, large numbers of *intermediate forms* were captured, these forming an uninterrupted series connecting the normal female and the var. *Helice.* R.M.]

[40] [Many of our best describers of caterpillars, such as the late Edward Newman, Messrs. Hellins and Buckler, &c., have described the various forms of numerous polymorphic species, but not from the point of view of the comparative morphology and ontogeny of the markings. R.M.]

such an extreme multiplicity of forms is interesting, since, so far as I know, polymorphism to this extent has not been observed in any insect.

The theoretical bearing of this polymorphism will be treated of subsequently. It is not in any way connected with a more advanced development of the markings, since *M. Stellatarum* shows in this respect a very low state of development. This species displays only two stages :—(1), complete absence of all markings; and (2), a simple subdorsal, with dorsal and spiracular lines. We must therefore admit that the phyletic development of the markings has for a long time remained at a standstill, or, what expresses the same thing, that the marking which the adult larva now possesses is extremely old.

In order to complete my observations on *M. Stellatarum*, I now add some remarks on the pupa, the colour variations of which it appeared of importance to investigate, owing to the extraordinary variability of the caterpillar, The pupa varies but very slightly; the ochreous yellow ground-colour sometimes passes into reddish, and sometimes into greenish; the rather complicated blackish-brown marking of streaky lines is very constant, especially on the wing portions, being at most only more or less strongly pronounced. The minute colour variations of the pupa therefore have no connection with the colour of the caterpillar, both green and brown larvæ furnishing sometimes reddish-yellow and sometimes greenish-yellow pupæ.

The comparison of *M. Stellatarum* with the other known species of the genus, brings scarcely any addition to our knowledge of the phyletic development. Thus, the two European species of which the caterpillars are known, viz. *M. Fuciformis* and *Bombyliformis*,[41] show essentially the same markings as *Stellatarum*, the chief element being a well-developed subdorsal line. The Indian *M. Gilia*, Herrich-Schäf., possesses also this line,[42] and, together with the East Indian *M. Corythus*, Walk.,[43] has oblique stripes in addition; the stripes do not, however, cross this line, but commence underneath it, and probably originated at a later period than the subdorsal line. Should this be the case, we must regard *M. Corythus* as representing a later phyletic stage. According to Duponchel's figures, in both *M. Fuciformis* and *Bombyliformis* small oblique stripes (red) occur near the spiracles, but these have nothing to do with the oblique stripes of *M. Gilia* just mentioned, as they run in a contrary direction. Of the two European species, I have

[41] [In Butler's revision both these species are placed in the genus *Hemaris*. R.M.]

[42] [This species is figured also by Butler (*loc. cit.* Pl. XC., Fig. 9), who represents it with seven oblique green lines between the spiracles and below the subdorsal line. R.M.]

[43] "Cat. E. Ind. Co. Mus.," Pl. VIII., Fig. 2. [Walker, Lepidop. Heter. VIII., p. 92, No. 14, 1856; this species is strictly confined to Java. R.M.]

only seen the living caterpillar of *Fuciformis*, and this possessed no oblique stripes.

To these five species I am now enabled to add a sixth, viz. *Macroglossa Croatica*,[44] a species inhabiting Asia Minor and Eastern Europe, of which a specimen and notice were kindly forwarded to me by Dr. Staudinger. The adult caterpillar much resembles that of *M. Stellatarum* in form and marking, but the subdorsal line appears much less distinctly defined, and the dorsal and spiracular lines seem to be entirely absent. The colour is generally green, but varies to red, and the subdorsal is more distinct and sharper in the young than in the adult larva. The markings of this species do not therefore in any way surpass those of *Stellatarum*, but are, on the contrary, much simpler.[45]

[44] [Eng. ed. The caterpillar is described and figured by Millière, "Iconographie des Chenilles et Lépidoptères inédits," tome iii., Paris, 1869; also in the Annales, Soc. Linn. de Lyon, 1871 and 1873.] [This sp. = *Hemaris Croatica*, Esper., of Butler's revision. R.M.]

[45] [The following additional species of the subfamily *Macroglossinæ* have been figured by Butler:—*Lophura Hyas*, Walk. (*loc. cit.* Pl. XC., Figs. 1 and 2), Hong-Kong, Silhet, and Java. The larva is apparently figured in two stages, the younger being red-brown with oblique white stripes, and the head and three front segments green. The larger specimen is green, mottled with red-brown, and no oblique stripes. In both figures the subdorsal line is indicated. The whole colouring is very suggestive of protective resemblance. *Hemaris Hylas*, Linn., from China, Japan, Ceylon, India, Australia, and Africa (*loc. cit.* Pl. XC., Fig. 4). The upper part of the body is light blue, and the lower part

THE GENUS PTEROGON, BOISD.[46]

Although I am acquainted with only a small portion of the developmental history of a single species of this genus, I will here proceed to record this fragment, since, taken in connection with two other species, it appears to me sufficient to determine, at least broadly, the direction of development which this genus has taken.

green, the two areas being separated by a white subdorsal line bordered above with brown. The dorsal line is feebly represented. *Macroglossa Belis*, Cram., N. India (*loc. cit.* Pl. XC., Fig. 6), is figured with the ground-colour deep indigo; a conspicuous white subdorsal, and a yellow spiracular line is present; on the side of each segment, between the two lines mentioned, there is a large red spot with a yellow nucleus (? eye-spots), the spots decreasing in size towards the head and tail; these probably confer upon this species some special protective advantage. *Macroglossa Pyrrhosticta*, Butler, China and Japan (*loc. cit.* Pl. XC., Fig. 8), is greenish-white with dorsal and subdorsal lines, and seven dark oblique stripes along the sides, below the subdorsal line. Of the foregoing species *Hemaris Hyas* appears to be in the same phyletic stage as *M. Stellatarum* and *M. Croatica*, &c., whilst *M. Pyrrhosticta* is probably, together with *M. Corythus* and *M. Gilia*, in another and more advanced stage, which is also passed through by *Lophura Hyas* in the course of its ontogenetic development. This last species (adult) and *M. Belis* may represent phyletic stages still further advanced. *Caliomma Pluto*, Walk., of which the caterpillar is figured by Burmeister (*loc. cit.* Pl. XIII., Fig. 1), appears to be a case of special protective resemblance to a twig or branch of its food-plant. Figured also by Chavannes; Bull. Soc. Vadoise des Sci. Nat., Dec. 6th 1854. R.M.].

[46] [Genus *Pterogon*, Boisd., = *Proserpinus* and *Lophura* (part). Butler, *loc. cit.* p. 632. The species above treated of = *Proserpinus Œnotheræ*, Fabr. R.M.]

PTEROGON ŒNOTHERÆ, FABR.

The adult larva, as made known by many, and for the most part good figures, has very complicated markings, which do not seem derivable from any of the elements of marking in the *Sphingidæ* hitherto considered. I was therefore much surprised at finding a young caterpillar of this species, only twelve millimeters in length, of a light green colour, without any trace of the subsequent latticed marking, and with a broad white subdorsal line extending along all the twelve segments. (Pl. VII., Fig. 63). Judging from the size and subsequent development, this caterpillar was probably in the third stage.

The same colouring and marking remained during the following (fourth) stage; but in the position occupied by the caudal horn in other *Sphingidæ*, there could now be observed the rudiment of a future ocellus in the form of a round yellowish spot (Pl. VII., Fig. 64). The subdorsal line disappears suddenly in the fifth stage, when the larva becomes dark green (rarely) or blackish brown; the latticed marking and the small oblique stripes are also acquired, together with the beautifully developed eye-spots, consisting of a yellow mirror with black nucleus and ground-area (Pl. VII., Fig. 65).

The North American *Pterogon Gauræ* and *P. Abboti*[47] also show markings precisely similar to

[47] [These species = *Thyreus Abboti* and *Proserpinus Gauræ*

those of this European species in the adult state; but in the two former the markings are of special interest as indicating the manner in which the primary Sphinx-marking has become transformed into that of the apparently totally different adult *P. Œnotheræ*. *P. Gauræ* is green, with a complicated latticed marking, which closer observation shows to arise from the dorsal line being resolved into small black dots, whilst the subdorsal line is broken up into black, white-bordered triangles. This caterpillar therefore gives fresh support to the remarkable phenomenon that the animals as well as the plants of North America are phyletically older than the European fauna and flora, a view which also appeared similarly confirmed by *Deilephila Lineata*, the representative form of *D. Livornica*. In entire accordance with this is the fact that the larva of *P. Gauræ* is without the eye-spot on the eleventh segment, and instead thereof still shows the original although small caudal horn. The perfect insect also resembles our *P. Œnotheræ* in colour and marking, but not in the form of the wings.

That the caterpillars of the genus *Pterogon* originally possessed the caudal horn we learn

of Butler's revision. Of the former he states:—"Transformations described, and larva and imago figured, Am. Ent. ii. p. 123, 1870; the larva is also figured by Scudder in Harris's 'Correspondence,' Pl. III., Fig. 1 (1869), and by Packard in his 'Guide,' p. 276, Fig. 203." R.M.]

from *P. Gorgoniades*, Hübn.,[48] a species now inhabiting south-east Russia, and for a knowledge of which I am indebted to Dr. Staudinger's collection. There are in this about eight blown specimens, from 3.7 to 3.9 centimeters in length, which show a marking, sometimes on a red and sometimes on a green ground, which unites this species with the young form of *P. Œnotheræ*, viz., a broad white subdorsal line, extending from the small caudal horn to the head. In addition to this, however, the caterpillar possesses an extraordinarily broad white red-bordered infra-spiracular line, a fine white dorsal stripe, and a similar line between the subdorsal and spiracular, *i.e.* a supra-spiracular line.

The caterpillars in Staudinger's collection, notwithstanding their small size, all belong to the last stage, as the moth itself does not measure more than 2.6 centimeters in expanse, and is therefore among the smallest of the known *Sphingidæ*. This species has therefore in the adult condition a marking very similar to that of *Œnotheræ* when young—it bears to *Œnotheræ* the same relationship that *Deilephila Hippophaës* does to *D. Euphorbiæ*, only in the present case the interval between the two species is greater. *Gorgoniades* is obviously a phyletically older species, as we perceive from the marking and from the possession of a horn.

[48] [*Proserpinus* (*Sphinx*) *Gorgon*, Esp. R.M.]

We certainly do not yet know whether *Œnotheræ* possesses a horn in its earliest stages, although in all probability it does so; in any case the ancestor of *Œnotheræ* had a horn, since the closely allied *P. Gauræ* now possesses one.

We thus see that also in the genus *Pterogon* the marking of the caterpillars commences with a longitudinal line formed from the subdorsal; an infra-spiracular or also a supra-spiracular line (*Gorgoniades*) being added. A latticed marking is developed from the linear marking by the breaking up of the latter into spots or small patches, which finally (in *Œnotheræ*) become completely independent, their connection with the linear marking being no longer directly perceptible.

THE GENUS SPHINX, LINN.

Of this genus (in the narrow sense employed by Gray) I have only been able, in spite of all trouble, to obtain fertile eggs of one species. The females cannot be induced to lay in confinement, and eggs can only be obtained by chance.

I long searched in vain the literature of this subject for some account of the young stages of these caterpillars, and at length found, in a note to Rösel's work, an observation of Kleemann's on the young forms of *Sphinx Ligustri*, which, although far from complete, throws light on certain points.

From a female of *S. Ligustri* Kleemann obtained 400 fertile eggs. The caterpillars on emerging are "at first entirely light yellowish-green, but become greener after feeding on the fresh leaves;" the horn is also at first light green, and then becomes "darker." The young larvæ spin webs, by which they fasten themselves to the leaves of their food-plant (this, so far as I know, has not been observed in any species of *Sphingidæ*). They moult four times, the border round the head and the purple stripes appearing after the third moult, these stripes "having previously been entirely white." The ecdyses follow at intervals of about six days, increasing to about ten days after the fourth moult.[49]

From this short account we gather that in the third stage the marking consists of seven oblique white stripes, which acquire coloured edges in the fourth stage, a fact which I have myself frequently observed. On the most important point Kleemann's observations unfortunately give no information—the presence or absence of a subdorsal line in the youngest stages. That he does not mention this character, can in no way be considered as a proof of its actual absence. I am rather inclined to believe that it is present in the first, and perhaps also in the second stage. There occur, however, species of the genus *Sphinx* (*sensû strictiori*) which possess a subdorsal line when young, as I think may be certainly inferred from the fact

[49] Rösel, *loc. cit.* vol. iii., p. 26, note.

that the remains of such a line are present in the adult larva of *S. Convolvuli.*

This conclusion becomes still more certain on comparing the markings with those of a nearly allied genus; without such comparison the separation of the genus *Macrosila*, Boisd., from *Sphinx* is scarcely justifiable. If to these two genera we add *Dolba*, Walk., and *Acherontia*, Ochs., we must be principally struck with the great similarity in the markings, which often reaches to such an extent that the differences between two species consist entirely in small shades of colour, while the divergence of the moths is far greater.

Of the genera mentioned, I am acquainted altogether with fourteen species of caterpillars:—*Macrosila Hasdrubal*, *Rustica*,[50] and *Cingulata*;[50] *Sphinx Convolvuli*, *Ligustri*, *Carolina*,[50] *Quinquemaculata*,[50] *Drupiferarum*,[50] *Kalmiæ*,[50] and *Gordius*;[50] *Dolba Hylæus*;[50] *Acherontia Atropos*, *Styx*,[51] and *Satanas*.[51] With one exception all these caterpillars possess oblique stripes of the nature of those of the *Smerinthus* larvæ, and most of them are without any trace of a subdorsal line; one species—the North American *M. Cingulata*—has a completely developed subdorsal; and the typical European species, *S. Convolvuli*, has a rudimentary

[50] Figured and described by Abbot and Smith. [*Macrosila* (*Sphinx*) *Cingulata* is figured also by Burmeister, *loc. cit.* Pl. XII., Fig. 1. R.M.]

[51] Figured in "Cat. Lep. E. Ind. Co."

subdorsal line. The ground-colour in most of these species is of the same green as that of the leaves of their food-plants; some are brown, *i.e.* earth-coloured, and in these the markings do not appear so prominently; others again possess very striking colours (*A. Atropos*), the oblique stripes in these cases being very vivid. Only *M. Hasdrubal* [52] separates itself completely from this

[52] See the figure in Sepp's Surinam Lepidoptera, P. 3, Pl. CI., 1848. A specimen in alcohol of the adult caterpillar is in the Berlin Museum. [The following is the synonymy of the above mentioned species:—*Macrosila Hasdrubal*, Walk.= *Pseudosphinx* (*Sphinx*) *Tetrio*, Linn.; *M. Cingulata* = *Protoparce* (*Sphinx*) *Cingulata*, Fabr.; *M. Rustica* = *Protoparce* (*Sphinx*) *Rustica*, Fabr.; *Sphinx Convolvuli*, Linn. = *Protoparce Convolvuli; S. Carolina*, Linn. = *P. Carolina;* the other species remain in the genera, as given above. The following additional species of *Sphinginæ* and *Acherontiinæ* have been figured by Butler:—*Pseudosphinx Cyrtolophia*, Butl., from Madras (*loc. cit.* Pl. XCI., Figs. 11 and 13); *Protoparce Orientalis*, Butl., from India, China, Java, &c. (Pl. XCI., Fig. 16); *Diludia Vates*, Butl. from India, &c. (Pl. XCI., Fig. 18); *Nephele Hespera*, Fabr., from India, Australia, &c. (Pl. XCI., Fig. 20); *Acherontia Morta*, Hübn., from Java, China, India, &c. (Pl. XCII., Fig. 9); and *A. Medusa*, Butl., from nearly the same localities as the last (Pl. XCII., Fig. 10). Most of these species fall under Dr. Weismann's general remarks, so that it is unnecessary to give detailed descriptions. The most divergent marking is that of *P. Cyrtolophia*, which has a broad white dorsal line bordered with pink, and two large pink ovals on the back of the four anterior segments, the hindmost and larger of these being bisected by the dorsal line. In *N. Hespera* the subdorsal line is present on segments 6 to 11 only, and it is highly significant that the oblique stripes are absent from these segments, but are present on the anterior segments, where the subdorsal line

system of classification, since this species is deep black with narrow yellow rings, the horn and last segment being red.

The large and most striking caterpillar of *M. Hasdrubal* is the same which Wallace has made use of for his theory of the brilliant colours of caterpillars. The explanation of the origin of this widely divergent mode of marking could only be furnished by the ontogeny, in which one or another of the older phyletic stages will certainly have been preserved.

fails. With reference to the larva of *A. Atropos*, Mr. Mansel Weale states (Proc. Ent. Soc. 1878, p. v.) that in S. Africa the ordinary form feeds generally on *Solanaceæ*, whilst the darker and rarer variety is found only on species of *Lantana*. The following species of these subfamilies are figured by Burmeister: *Amphonyx Jatrophæ* (*loc. cit.* Pl. XI., Fig. 1); *Protoparce* (*Diludia*) *Florestan*, Cram. (Fig. 2); *Sphinx Justiciæ*, Walk. (Fig. 3); *Protoparce* (*Diludia*) *Lichenea*, Walk. (Fig. 4); *Sphinx* (*Protoparce*) *Cingulata*, Fabr. (Pl. XII., Fig. 1); and *Sphinx Cestri* (Fig. 5). All these species have the characteristic Sphinx-like markings. *Dilophonota Ello*, Linn. (Pl. XII., Fig. 2), is greenish-brown with a yellow subdorsal line, and *D. Hippothöon* (Fig. 4), yellow with a whitish subdorsal. Neither of these has oblique stripes. *D. Œnotrus*, Cram. (Fig. 3), has neither stripes nor subdorsal, but is uniform brown above, passing into green beneath. *Protoparce Albiplaga*, Walk. (Pl. XIII., Fig. 2, also Mérian, Pl. III., and Abbot and Smith, I., Pl. XXIV.), pale green with large yellow, black-bordered patches surrounding the spiracles. *Pseudosphinx Tetrio*, Linn. (Pl. XIII., Fig. 3), and *P. Scyron* (Fig. 4) are black with broad transverse belts, yellow and white respectively, encircling the middle of each segment. These light bands serve very effectively to break up the uniform surface of the large bodies of these insects. R.M.]

Strictly speaking the same should be said of the other species—nevertheless their comparison with the so similarly marked *Smerinthinæ*, together with the circumstance that in certain species a subdorsal line can be traced, makes it appear correct to suppose that here also the subdorsal was the primary marking, this line being subsequently entirely replaced by the oblique stripes. The *Sphinginæ* would therefore be a younger group than the *Smerinthinæ*, a conclusion which is borne out by the fact that in the former the oblique stripes have reached a higher development, being always of two, and sometimes even of three colours (*S. Drupiferarum*, white, red, black), whilst in the species of *Smerinthus* they only occasionally possess uniformly coloured borders.

THE GENUS ANCERYX, BOISD.

Although this genus is not admitted into most of the European catalogues—the solitary European species representing it being referred to the genus *Sphinx*, Linn.[53] — its separation from *Sphinx* appears to me to be justified, not because of the striking differences presented by the moths, but because the caterpillars, judging from the little we know of them, likewise show a similar degree of difference.

[53] [The species referred to is placed by Butler in Hübner's genus *Hyloicus*. R.M.]

I have frequently succeeded in obtaining fertile eggs of *Anceryx Pinastri* and I will now give the developmental history of this caterpillar, which has already been figured with great accuracy in Ratzeburg's excellent work on forest insects. Rösel was acquainted with the fact that the "pine moth" laid its eggs singly on the needles of the pine in June and July, and he described them as "yellowish, shining, oval, and of the size of a millet seed."

On emerging, the caterpillars are six millimeters in length, of a light yellow colour, the head shining black with a yellow clypeus. The caudal horn, which is forked at the tip, is also at first yellowish, but soon becomes black. No particular marking is as yet present, but a reddish stripe extends along the region of the dorsal vessel, and the course of the spiracles is also marked by an orange-red line. (Fig. 53, A & B, Pl. VI.)

As soon as the young larvæ are filled with food they acquire a greenish streak. The first moult occurs after four days, and immediately after this there is still an absence of distinct markings, with the exception of a greenish-white spiracular line. In the course of some hours, however, the original light green ground-colour becomes darker, and at the same time a sharp, greenish-white subdorsal line appears, together with a parallel line extending above the spiracles, which, in *Pterogon Gorgoniades*, has already been designated as the "supra-spiracular." The dorsal line is absent;

the head is light green, with two narrow blackish brown lines surrounding the clypeus; the horn and thoracic legs are black; claspers, reddish green; length, twelve to thirteen millimeters. (Fig. 54.)

Third Stage.

After another period of four days the second moult occurs, neither colour nor marking being thereby affected. Only the horn, now no longer forked, becomes brownish with a black tip. The young caterpillars are now, as before, admirably adapted to the pine needles, on which they feed by day, and from which they can only be distinguished with difficulty.

Fourth Stage.

The third moult also brings no essential change. The ground-colour and marking remain the same, only the spiracles, which were formerly dull yellowish, are now of a vivid brick-red. The horn becomes yellowish-red at the base.

Fifth Stage.

The marking is only completely changed in the fifth and last stage. A broad reddish-brown dorsal line replaces the subdorsal, more or less completely. The supra-spiracular line also becomes broken up into numerous short lengths, whilst the green ground-colour in some specimens becomes more or less replaced by a brownish shade extending from the back to the sides. Horn, black; the upper part of the first segment with a corneous plate, similar to that of the *Deilephila* larvæ.

This stage is very variable, as shown by the figures in various works. The variations arise on the one hand from the struggle between the green ground-colour and the reddish-brown extending from above, and, on the other hand, from a more or less complete disappearance of the associated longitudinal lines. The latter are sometimes completely retained, this being the case in a caterpillar figured by Hübner (*Sphinges*, III., *Legitimæ* C, b), where both the subdorsal and supra-spiracular lines are continuous from segment 11 to segment 1, an instance which may perhaps be regarded as a reversion to the primary form.

The entire change of the marking from the fourth to the fifth stage depends upon the fact that the young larvæ resemble the *needles* of the pine, whilst the adults are adapted to the *branches*. I shall return to this later.

The ontogeny of *A. Pinastri* makes us acquainted with three different forms of marking: (1) simple coloration without marking; (2) a marking composed of three pairs of parallel longitudinal lines; (3) a complicated marking, arising from the breaking up of the last and the addition of a darker dorsal line.

Of the fourteen species placed by Gray in the

genus *Anceryx*, I find, in addition to the one described, notices of only two caterpillars :—

A. Coniferarum,[54] a North American species, lives on *Pinus Palustris*, and was figured by Abbot and Smith. Colour and marking very similar to *A. Pinastri.*

A. Ello, Linn.,[55] according to the authority of Mérian, is described by Clemens[56] as dark brown, "with a white dorsal line, and irregular white spots on the sides." It lives on a "species of *Psidium* or *Guava*."

Most of the species of *Anceryx* appear to live on *Coniferæ*, to which they show a general and decided adaptation. In the absence of decisive information, I partly infer this from the names, as *Anceryx Juniperi* (Africa). It has long been known that in our *A. Pinastri* the mixture of brown and fir-green, interspersed with conspicuous irregular light yellowish and white spots, causes the adult larva to present a very perfect adaptation to its environment. Of this caterpillar Rösel states :—"After eating it remains motionless, and is then difficult to see, because it is of the same colour as its food, since its brown dorsal line has almost the colour of the pine twigs ; and who

[54] [= *Ellema Coniferarum*, of Butler's revision. R.M.]

[55] [= *Dilophonota Ello* of Butler's revision. R.M.]

[56] "Synopsis of the North American Sphingides." Philadelphia, 1859.

is not familiar with the fact that beneath the green needles there is also much yellow to be found?"

This adaptation to the needles and twigs obviously explains why this caterpillar in the adult condition is so far removed from those of the genus *Sphinx*, while the moths are so nearly related that they were only separated as a distinct genus when we became acquainted with a large number of species.

II.

Conclusions from Phylogeny.

The considerations previously set forth are entirely based on Fritz Müller's and Haeckel's view, that the development of the individual presents the ancestral history *in nuce*, the ontogeny being a condensed recapitulation of the phylogeny.

Although this law is generally true—all recent investigations on development having given it fresh confirmation—it must not be forgotten that this "recapitulation" is not only considerably abbreviated, but may also be "falsified," so that a searching examination into each particular case is very desirable.

The question thus arises, in the first place, as to whether the markings of caterpillars, so distinct at the different stages of growth, are actually to be regarded as residual markings inherited from the parent form; or whether their differences do not depend upon the fact that the caterpillar, in the course of growth, is exposed to different external conditions of life, to which it has adapted itself by assuming a different guise.

The former is undoubtedly the case. It can

by no means be denied that the conditions of life in young caterpillars are *sometimes* different to those of the adults. It will, in fact, be shown later on, that in certain cases the assumption of a new guise at an advanced age actually depends upon adaptation to new conditions of life; but as a rule, the external conditions remain very similar during the development of the larva, as follows from the fact that a change of food-plant never takes place.[1] We should therefore rather expect a complete similarity of marking throughout the entire larval period, instead of the great differences which we actually observe.

Different circumstances appear to me to show that the markings of young larvæ are only exceptionally due to a new adaptation, but that as a rule they depend upon heredity. In the first place, there is the fact that closely allied species, exposed to precisely similar external conditions, as, for instance, *Chærocampa Elpenor* and *Porcellus*, possess exactly the same markings when young, these markings nevertheless appearing at different stages of growth. Thus, the subdorsal line first

[1] [The larvæ of many moths which feed on deciduous trees during the autumn and hibernate, are stated to feed on low-growing plants in the spring, before the buds of their food-trees open. On the other hand, low-plant feeders, such as *Triphæna Fimbria*, &c., are stated to sometimes feed at night in early spring on the buds of trees. The habits and ontogeny of these species are of special interest in connection with the present researches, and are well worthy of investigation. R.M.]

appears in *Elpenor* in the second stage, whilst in *Porcellus* it is present during the first stage. If this line were acquired by the young larva for adapting it at this age to special conditions of life, it should appear in both species at the same stage. Since this is not the case, we may conclude that it is only an inherited character derived from the adult ancestor of the two species, and now relegated to the young stages, being (so to speak), pushed further back in one species than in the other.

But the strongest, and, as it appears to me, the most convincing proof of the purely phyletic significance of the young larval markings, is to be found in the striking regularity with which these are developed in a similar manner in all allied species, howsoever different may be their external conditions of life. In all the species of the *Chærocampa* group (the genera *Chærocampa* and *Deilephila*) the marking—no matter how different this may be in later stages—arises from the simple subdorsal line. This occurs even in species which live on the most diverse plants, and in which the markings can be of no biological importance as long as the larvæ are so small as to be only visible through a lens, and where there can be no possible imitation of leaf-stalks or veins, the leaves and caterpillars being so very distinct.

Moreover, when in the *Macroglossinæ* (the genera *Macroglossa*, *Pterogon*, and *Thyreus*) we see precisely the same simple marking (the subdorsal) line retained throughout all the stages in two genera, whilst in the *Smerinthinæ* this line vanishes at a very early stage, and in the *Sphinginæ* is only present in traces, we can give but one explanation of these facts. We have here a fragmentary series representing the phyletic development of the Sphinx-markings, which latter have arisen from one original plan—the simple subdorsal line—and have then undergone further development in various directions. As this subsequent development advanced, the older phyletic stages would always be relegated to younger ontogenetic stages, until finally they would be but feebly represented even in the youngest stage (*D. Euphorbiæ*), or else entirely eliminated (most of the species of the genus *Sphinx*). I believe that no other sufficient explanation of these facts can be adduced. Granting that the correctness of the above views can no longer be doubted, we may now take up the certain position that the ontogeny of larval markings reveals their phylogeny, more or less completely, according to the number of phyletic stages omitted, or, in some exceptional cases, falsified. In other words, the ontogeny of larval markings is a more or less condensed and occasionally falsified recapitulation of the phylogeny.

Considering this to be established, we have next to deal with the uniformity of the developmental phenomena, from which we may then attempt to trace out the inciting causes underlying this development.

The law, or, perhaps better, the line of direction followed by the development, is essentially the following :—

1. The development commences with a state of simplicity, and advances gradually to one of complexity.

2. New characters first make their appearance in the last stage of the ontogeny.

3. Such characters then become gradually carried back to the earlier ontogenetic stages, thus displacing the older characters, until the latter disappear completely.

The first of these laws appears almost self-evident. Whenever we speak of development, we conceive a progression from the simple to the complex. This result therefore does nothing but confirm the observation, that we have actually here before us a development in the true sense of the word, and not simply a succession of different independent conditions.

The two following laws, on the other hand, lay claim to a greater importance. They are not now enunciated for the first time, but were deduced some years ago by Würtemberger[2] from a study

[2] "Neuer Beitrag zum geologischen Beweise der Darwin'schen Theorie." 1873, Nos. 1 and 2. [This principle, in common with many others which have only been completely worked out

of the ammonites. In this case also the new characters predominate in the later periods of life, and are then transferred back to the younger ontogenetic stages in the course of phyletic development. "The change in the character of the shell in ammonites, first makes itself conspicuous in the last chamber; but in the succeeding generations this change continually recedes towards the beginning of the spiral chambers, until it prevails throughout the greater part of the convolutions."

In the same sense must also be conceived the case which Neumayr and Paul have recently made known respecting certain forms of *Melanopsis* from the West Sclavonian *Paludina* bed. In *M. Recurrens* the last convolutions of the shell are smooth, this being a new character; the small upper convolutions, however, are delicately ribbed, as is also the case with the last convolution of the immediate progenitor. The embryonic con-

of late years, is foreshadowed by Darwin. Thus, he states when speaking of inheritance at corresponding periods of life: "I could give a good many cases of variations (taking the word in the largest sense) which have supervened at an earlier age in the child than in the parent" ("Origin of Species," 1st ed., 1860, p. 444). In the case of inherited diseases also: "It is impossible to . . . doubt that there is a strong tendency to inheritance in disease at corresponding periods of life. When the rule fails, the disease is apt to come on earlier in the child than in the parent; the exceptions in the other direction being very much rarer." ("Variation of Animals and Plants under Domestication," 1st ed., 1868, vol. ii., p. 83.) R.M.]

volutions again are smooth, and the author believes (on other grounds) that the more remote progenitor possessed a smooth shell.

In this case therefore, and in that of the ammonites, every shell to a certain extent proclaims the ancestral history of the species ; in one and the same shell we find different phyletic stages brought into proximity. The markings of caterpillars do not offer similar facilities ; nevertheless I believe that by their means we are led somewhat further, and are able to enter more deeply into the causes underlying the processes of transformation, because we can here observe the living creature, and are thus enabled to study its life-history with more precision than is possible with a fossil species.

When, in 1873, I received Würtemberger's memoir, I was not only struck with the agreement of his chief results with those which I had arrived at by the study of larval markings, but I was almost as much astonished at the great difference in the interpretation of the facts. The latter indicate the gradual backward transference of a new character from the latest to the earlier ontogenetic stages. Without further confirmation Würtemberger assumes that it is to a certain extent self-evident that the force producing this backward transference is the same as that which, according to his view, first called forth the character in question in the last stage, viz., natural selection. "Variations acquired at an advanced age of the organism may, when advantageous, be inherited by the succeeding generations, in such a manner that they always appear a little earlier than in the preceding generations."

It is certainly theoretically conceivable that a newly acquired character, when also advantageous to the earlier stages, might be gradually transferred to these stages, since in this case those individuals in which this character appeared earliest would have the greatest chance of surviving. In the case of the development of larval markings, however, there are facts which appear to me to show that such backward transference of a new character is, in a certain measure, independent of the principle of utility, and that it must therefore be referred to another cause—to the innate law of growth which rules every organism.

When, in the larva of *C. Elpenor*, we perceive that the two eye-spots which are first formed on the fourth and fifth segments appear subsequently on the other segments as faint traces of no biological value whatever, we cannot explain this phenomenon by natural selection. We should rather say that in segmented animals there is a tendency for similar characters to be repeated on all the segments; and this simply amounts to the statement, that an innate law of growth is necessary for the repetition of such newly acquired characters.

The existence of such a law of growth, acting

independently of natural selection, may therefore be considered as established, and indeed cannot be disputed (Darwin's "correlation of growth"). In the present case it appears to me that an innate law of this kind, determining the backward transference of new characters, is deducible from the instances already quoted in another sense, viz., from the fact that in many cases characters which are decidedly advantageous to the adult are transferred to the younger stages, where they are at most of but indifferent value, and can certainly be of no direct advantage. This is the case with the oblique stripes of *Smerinthus*, which, in the adult larvæ, resemble the leaf ribs, as will be shown more fully later on, and, in conjunction with the green coloration, cause these caterpillars to be very difficult of detection on their food-plants. The insects are easily overlooked, and can only be distinctly recognized on close inspection.

Now these oblique stripes appear, in all the *Smerinthus* caterpillars known to me, in the second, and sometimes even in the first stage, *i. e.* in larvæ of from 0·7 to 1 centimeter in length. The stripes are here much closer together than the ribs of any of the leaves of either willow, poplar, or lime, and can therefore have no resemblance to these leaves. The young caterpillars are certainly not rendered more conspicuous by the oblique stripes, since they can only be recognized on close inspection. It is for this reason that the stripes have not been eliminated by natural selection.

The remarkable phenomenon of the backward transference of newly acquired characters may therefore be formulated as follows :—Changes which have arisen in the later ontogenetic stages have a tendency to be transferred back to the younger stages in the course of phyletic development.

The facts of development already recorded furnish numerous proofs that this transference occurs gradually, and step by step, taking the same course as that which led to the first establishment of the new character in the final ontogenetic stage.

Did this law not obtain, the ontogeny would lose much of the interest which it now possesses for us. It would then be no longer possible, from the ontogenetic course of development of an organ or of a character, to draw a conclusion as to its phylogeny. If, for instance, the eye-spots of the *Chærocampa* larvæ, which must have been acquired at a late age, were transferred back to the younger ontogenetic stages in the course of phyletic development, as eye-spots already perfected, and not showing their rudimentary commencement as indentations of the subdorsal line, the phenomenon would then give us no information as to the manner of their formation.

It is well known to all who have studied the developmental history of any group of animals, that no organ, or no character, however complex, appears suddenly in the ontogeny; whereas, on the other hand, it appears certain that new, or more advanced, but simpler characters, predominate in the last stage of development. We are thus led to the following modification of the foregoing conclusion:—Newly acquired characters undergo, as a whole, backward transference, by which means they are to a certain extent displaced from the final ontogenetic stage by characters which appear later.

This must be a purely mechanical process, depending on that innate law of growth, the action of which we may observe without being able to explain fully. Under certain conditions the operation of this law may be prevented by natural selection. Thus, for instance, if the young caterpillars of *Anceryx Pinastri* have not acquired the characteristic marking of the adults, it is probably because they are better protected by their resemblance to the green pine-needles than they would be if they possessed the pattern of the larger caterpillars in their last stage.

The backward transference of newly acquired characters may also possibly be accelerated when these characters are advantageous to the younger stages; but this transference takes place quite independently of any advantage if the characters are of indifferent value, being then entirely brought about by innate laws of growth.

That new characters actually predominate in

the last stage of the ontogeny, may also be demonstrated from the markings of caterpillars. It is, of course, not hereby implied, that throughout the whole animal kingdom new characters can only appear in the last ontogenetic stage. Haeckel is quite correct in maintaining that the power of adaptation of an organism is not restricted to any particular period. Under certain circumstances transformations may occur at any period of development ; and it is precisely insects undergoing metamorphosis that prove this point, since their larvæ differ so widely from their imagines that the earlier stages may be completely disguised. It is here only signified that, with respect to the development of caterpillars, new characters first appear in the adult. The complexity of the markings, which so frequently increases with the age of the caterpillar, can scarcely bear any other interpretation than that the new characters were always acquired in the last stage of the ontogeny. In certain cases we are able, although with some uncertainty, to catch Nature in the act of adding a new character.

I am disposed to regard the blood-red or rust-red spots which occur in the last stage of the three species of *Smerinthus* larvæ in the neighbourhood of the oblique stripes as a case in point. It has already been shown that these red spots must be regarded as the first rudiments of the linear coloured edges which reach complete development in the genus *Sphinx.* In some specimens of *Smerinthus Tiliæ* the spots coalesce so as to form an irregular coloured edge to the oblique stripes. In *S. Populi* they occur in many individuals, but remain always in the spot stage; whilst *S. Ocellatus* is but seldom, and *S. Quercus* appears never to be spotted.

The spots both of *S. Tiliæ* and *Populi* certainly do not show themselves exclusively in the fifth (last) stage, but also in the fourth, and sometimes in *Populi* even as early as the third stage, from which we might be disposed to conclude that the new character did not first appear in the last stage. But the majority of the spotted individuals first acquire their spots in the fifth stage, and only a minority in the fourth; so that their occasional earlier appearance must be ascribed to the backward transference of a character acquired in the fifth stage. Moreover, the fourth and fifth stages of the caterpillars are closely analogous both in size, mode of life, and marking, and are therefore analogous with reference to the environment, so that it is to be expected that new characters, when depending on adaptation, would be rapidly transferred from the fifth stage to the fourth.[3] We should

[3] [If the reddish-brown spots on the larva of *S. Populi* have the protective function assigned to them by Mr. Peter Cameron (Trans. Ent. Soc. 1880, p. 69), it can be readily understood that they would be of service to the insect in the fourth stage, and the backward transference of this character might thus be accelerated by natural selection, in accordance with the above principles. (See, also, note [30], p. 241.) R.M.]

thus have a case of the acceleration by natural selection, of processes determined by innate causes. Why changes should predominate in the last stage, is a question closely connected with that of the causes of larval markings in general, and may therefore be investigated later.

But if we here assume in anticipation that all new markings depend on adaptation to the conditions of life, and arise through natural selection, it will not be difficult to draw the conclusion that such new characters must prevail in the last stage. There are two conditions favouring this view; the size of the insect, and the longer duration of the last stage. As long as the caterpillar is so small as to be entirely covered by a leaf, it only requires a good adaptation in colour in order to be completely hidden; independently of which, it is also possible that many of its foes do not consider it worth attacking at this stage. The last stage, moreover, is of considerably longer duration than any of the four preceding ones; in *Deilephila Euphorbiæ* this stage lasts for ten days, whilst the remaining stages have a duration of four days; in *Sphinx Ligustri* the last stage also extends over ten days, and the others over six days.

In its last stage, therefore, a caterpillar is for a longer period exposed to the danger of being discovered by its foes; and since, at the same time, its enemies become more numerous, and its increased size makes it more easy of detection, it is readily conceivable that a change in the conditions of life, such, for instance, as removal to a new food-plant, would bring about the adaptation of the *adult larva* as its chief result.

I shall next proceed to show how far the assumption here made—that all markings depend on natural selection—is correct.

III.

Biological Value of Marking in General.

Having now described the development of larval markings, so far as possible from their external phenomena, and having traced therefrom the underlying law of development, I may next proceed to the main problem—the attempt to discover those deeper inciting causes which have produced marking in general.

The same two contingencies here present themselves as those which relate to organic life as a whole; either the remarkably complex and apparently incomprehensible characters to which we give the name of markings owe their origin to the direct and indirect gradual action of the changing conditions of life, or else they arise from causes entirely innate in the organism itself, *i. e.* from a phyletic vital force. I have already stated in the Introduction why the markings of caterpillars appear to me such particularly favourable characters for deciding this question, or, more precisely, why these characters, above any others, appear to me to render such decision more easily possible; repetition is here therefore unnecessary.

The whole of the present investigation had not been planned when I joined with those who, from the first, admitted the omnipotence of natural selection as an article of faith or scientific axiom. A question which can only be solved by the inductive method cannot possibly be regarded as settled, nor can further evidence be considered unnecessary, because the first proofs favour the principle. The admission of a mysteriously working phyletic power appears very unsatisfactory to those who are striving after knowledge; the existence of this power, however, is not to be considered as disproved, because hundreds of characters can be referred to the action of natural selection, and many others to that of the direct action of the conditions of life. If the development of the organic world is to be considered as absolutely dependent on the influence of the environment, not only should we be able here and there to select at pleasure characters which appeared the most accessible for elucidating this point, but it becomes in the first place necessary to attempt to completely refer *all* characters belonging to any particular group of phenomena, however small this group might be, to known transforming factors. We should then see whether this were possible, or whether there would remain residual phenomena not explicable by known principles and compelling us to admit the existence of a force of development innate in the organism. In any case the "phyletic vital force" can only be got rid of by a process of elimination —by proving that all the characters generally occurring throughout the group of phenomena in question, must be attributed to other causes, and that consequently nothing remains for the action of the supposed phyletic vital force, which would in this manner be negatived, since we cannot infer the presence of a force if the latter exerts no action whatever.

I shall here attempt such an investigation of the group of phenomena displayed by larval markings, with special reference to those of the *Sphingidæ*. The alternatives upon which we have to decide are the following:—Are the markings of caterpillars purely morphological characters, produced entirely by internal causes? or, are they simply the response of the organism to external influences?

The solution of these questions will be arrived at by seeking to refer all the markings present to one of the known transforming factors, and the success or failure of this attempt will give the required decision. The first question to be attacked is obviously this,—whether the Sphinx-markings are actually, as they appear at first sight, purely morphological characters. If it can be shown that all these markings were originally of biological value, they must be attributed to the action of natural selection.

Did I here at once proceed to establish the biological value of larval markings—and especially of those of the *Sphingidæ*—so as to arrive in this manner at a conclusion as to their dependence upon natural selection, it would be impossible to avoid the consideration of the total coloration of the caterpillars, since the marking frequently consists only of a local strengthening of the colour, and cannot be comprehended without coming to this understanding. The action of the markings also often appears to be opposed to that of the colouring, making the caterpillar again conspicuous; so that the two factors must necessarily be considered together. I shall therefore commence the investigation with colour in general, and then proceed to treat of marking.

IV.

BIOLOGICAL VALUE OF COLOUR.

THE general prevalence of protective colouring among caterpillars has already been so frequently treated of that it is not here my intention to recall particular instances. In order to judge of the effect of marking, however, it will be well to bear in mind that these insects, being generally defenceless and thus requiring protection, have acquired the most diverse means of rendering themselves in some measure secure from their foes.

The sharp spines which occur on the caterpillars of many butterflies (*Vanessa, Melitæa, Argynnis*), and the hairs on those of many moths, serve for protective purposes. Among other means of protection—although in a different sense—we have in all the species of the great family of the *Papilionidæ* the strikingly coloured (yellowish red) odour-emitting tentacles concealed near the head, and suddenly protruded for terrifying foes; and likewise the forked horn at the tail of the caterpillars of the genus of moths *Harpyia*, the tentacles of which can be suddenly protruded in a similar manner. Adaptive colours and forms combined with certain habits [1] are, however, much

[1] [For cases of correlation of habit with protective resemblance in larvæ, see a paper in "Ann. and Mag. of Nat. Hist.," Feb., 1878, pp. 159, 160. Also Fritz Müller on a Brazilian Cochliopod larva, Trans. Ent. Soc. 1878, p. 223. Mr. Mansel Weale states, with reference to S. African *Sphingidæ* (Proc. Ent. Soc. 1878, p. vi.), that many species when seized "have a habit of doubling up the body, and then jumping a considerable distance with a spring-like action. This is especially the case with species having eye-like markings; and it is probable that if attacked by birds in a hesitating manner, such species might effect their escape amid the grass or foliage." Many of the defensive weapons and habits of larvæ are doubtless means of protection from ichneumons and other parasitic foes. In the case of saw-flies, Mr. Peter Cameron has shown (Trans. Ent. Soc. 1878, p. 196) that the lashing about of the posterior part of the body may actually frighten away such enemies. The grotesque attitude and spider-like appearance and movements of the caterpillar of *Stauropus Fagi* are considered by Hermann Müller ("Kosmos," Nov., 1879, p. 123) to be means of protection from ichneumons. Among the most remarkable means of defence possessed by larvæ is that of secreting a liquid, which Mr. W. H. Edwards has shown, in the case of certain North American *Lycænidæ* ("Canadian Entomologist," vol. x., 1878, pp. 3—9 and 131—136), to be attractive to ants, who regularly attend these caterpillars, in the same manner and for the same purpose as they do our aphides. The mutual advantage derived by the ants and larvæ was discovered in the case of *Lycæna Pseudargiolus*. Mr. Edwards states that the *mature* larva of this species is singularly free from Hymenopterous and Dipterous parasites:—"Why this species, and doubtless many other *Lycænæ*, are thus favoured will, perhaps, in some degree appear from a little incident to be related. On 20th June, in the woods, I saw a mature larva on its food-plant; and on its back, facing towards the tail of the larva, stood motionless one

more common than defensive weapons. Thus, the caterpillars of the *Noctuæ* belonging to the genus *Catocala* and its allies, feed only at night on the green leaves of various forest-trees; by day they rest in crevices of the bark on the tree trunk, which they resemble so perfectly in the colour of their peculiar glossy dull grey or brownish skin beset with small humps, that only sharp eyes can

of the larger ants. . . . At less than two inches behind the larva, on the stem, was a large ichneumon-fly, watching its chance to thrust its ovipositor into the larva. I bent down the stem, and held it horizontally before me, without alarming either of the parties. The fly crawled a little nearer and rested, and again nearer, the ant making no sign. At length, after several advances, the fly turned its abdomen under and forward, thrust out its ovipositor, and strained itself to the utmost to reach its prey. The sting was just about to touch the extreme end of the larva, when the ant made a dash at the fly, which flew away, and so long as I watched—at least five minutes—did not return. The larva had been quiet all this time, its tubes out of sight, and head buried in a flower-bud, but the moment the ant rushed and the fly fled, it seemed to become aware of the danger, and thrashed about the end of its body repeatedly in great alarm. But the tubes were not protruded, as I was clearly able to see with my lens. The ant saved the larva, and it is probable that ichneumons would in no case get an opportunity to sting so long as such vigilant guards were about. It strikes me that the larvæ know their protectors, and are able and willing to reward them. The advantage is mutual, and the association is friendly always." Those who are familiar with Mr. Belt's description of the standing armies of ants kept by the "bull's-horn thorn" ("Naturalist in Nicaragua," pp. 218—222) and by certain *Cecropiæ* and *Melastomæ*, will be struck with the analogy between these and the foregoing case. R.M.]

detect them, even when we are familiar with their habits.[2]

The striking resemblance of many moths to splinters of wood is well known, and to this is added a habit which helps their disguise, viz., that of remaining stiff and motionless on the approach of danger, just like a splinter projecting from the branch.[3] Among the moths coming under this category may be mentioned *Cucullia Verbasci*, and particularly those of the genus *Xylina*, which, when at rest, closely resemble a broken splinter of wood in the colour and marking of their fore wings, and when touched, have a habit of drawing in their legs and falling without opening their wings as though dead.

That simple adaptive colouring prevails widely

[2] [The adaptive resemblance is considerably enhanced in *Catocala* and in *Lasiocampa Quercifolia* by the row of fleshy protuberances along the sides of these caterpillars, which enables them to rest on the tree trunks by day without casting a sharp shadow. The hairs along the sides of the caterpillar of *Pœcilocampa Populi* doubtless serve the same purpose. (See a paper by Sir John Lubbock, Trans. Ent. Soc. 1878, p. 242; also Peter Cameron, *ibid.*, 1880, p. 75.) It is well known to collectors that one of the best methods of finding the caterpillars of the *Catocalæ* is to *feel* for them by day on the barks of their respective food-trees, or to beat for them at night. R.M.]

[3] [See Wallace's "Contributions to the Theory of Natural Selection," 1st ed., p. 62. Also a paper in "Ann. Mag. Nat. Hist." Feb. 1878, p. 159, for cases in point. Rösel in 1746 mentioned this habit in *Calocampa Exoleta*. Hermann Müller has recorded many other similar instances on the authority of Dr. Speyer; see "Kosmos," Nov., 1879, p. 114. R.M.]

among caterpillars is shown by the large number of green species.[4] It may be fairly said that all caterpillars which possess no other means of protection or defence are adaptively coloured. These facts are now well known; so also is the explanation of the varied and striking colours of many caterpillars given by Wallace.[5] There is, how-

[4] [Andrew Murray called attention to this fact in 1859 ("Edinburgh New Philos. Journ.," Jan., 1860, p. 9). This view is also corroborated by the fact that no internal feeders are green; see note 2, p. 310 and Proc. Zoo. Soc. 1873, p. 159. R.M.]

[5] [Proc. Ent. Soc. March 4th, 1867; and "Contributions to the Theory of Natural Selection," 1st ed., pp. 117—122; also Darwin's "Descent of Man," 2nd ed., p. 325. Among the most important recent additions to the subject of the colours, spines, and odours of caterpillars, I may call attention to a paper by Fritz Müller ("Kosmos," Dec., 1877), the following abstract of which I communicated to the Entomological Society (Proc. 1878, pp. vi, vii) :—"The larvæ of *Dione Juno* and *Acræa Thalia* live gregariously, and are brown in colour; they are covered with spines, but, being of dull colours, their spiny protection (which in the case of *D. Juno* is very imperfect) would not preserve them unless they were distinguished as inedible at the right time, and not after being seized, in accordance with the principles laid down by Mr. Wallace. It is suggested that the social habits of the larvæ, which lead them to congregate in large numbers, make up for their want of colour, since their offensive odour then gives timely warning to an approaching enemy. The caterpillars of *Colænis Julia* and *Dione Vanillæ* are equally wanting in bright colours, but are solitary in their habits, and these species rest on the under side of the leaf when feeding. On the other hand, the caterpillars of *Heliconius Eucrate*, *Colænis Dido*, and *C. Isabella*, which are of solitary habits, and which freely expose themselves, are very gaudily coloured, and there-

ever, novelty in the proof contained in the fore-going descriptions of larval development, as to

fore most conspicuous. As examples of nearly allied larvæ, of which some species are gregarious and others solitary, Fritz Müller mentions *Morpho* and *Brassolis*, which are gregarious; while *Opsiphanes* and *Caligo* are solitary. The larva of *Papilio Pompeius* also is gregarious, and those of *P. Nephalion*, *P. Polydamas*, and *P. Thoas* are solitary. . . . . Fritz Müller sums up his observations by remarking that those caterpillars which live alone, and lack the bright colouring as a sign of offensiveness, must hide themselves; as those of *C. Julia* and *D. Vanillæ.* The spiny covering is much less a protection against birds than against smaller enemies; and they may, by the protective habit of living together, diffuse around themselves an offensive atmosphere, even to man, and thus gradually becoming shorter (as with *D. Juno*), the spines of these caterpillars become useless, and finally are altogether dropped." See also Sir John Lubbock's "Note on the Colours of British Caterpillars," Trans. Ent. Soc. 1878, p. 239. Mr. Peter Cameron finds (Trans. Ent. Soc. 1880, pp. 71 and 75) that these remarks are also applicable to the larvæ of certain saw-flies. In 1877 Mr. J. W. Slater published a paper "On the Food of gaily-coloured Caterpillars" (Trans. Ent. Soc. 1877, p. 205), in which he suggested that such caterpillars might derive their distasteful qualities from feeding on plants containing poisonous or otherwise noxious principles. A much larger number of observations will be required, however, before this view can be accepted as of general application. A beautiful illustration of the theory of warning colours is given by Belt in his "Naturalist in Nicaragua," p. 321. All the frogs found in the woods round St. Domingo are, with one exception, protectively coloured; they are of nocturnal habits, and are devoured by snakes and birds. The exception was a species of bright red and blue colours, which hopped about by day and made no attempt at concealment. From these facts Mr. Belt concluded that this species was inedible, and on trying the experiment with ducks and fowls this was found to be the case. R.M.]

the manner in which the di- and polymorphism of caterpillars can be explained from the external phenomena which they present, these phenomena being well adapted for showing the great importance of protective colouring to the larvæ. We have here presented a double adaptation, although not quite of the nature of that which I formerly admitted on hypothetical grounds.[6] In the first place, from the developmental history there results the conclusion that all Sphinx-larvæ which, in the adult state, are di- or polymorphic, are unicolorous when young. Thus, the caterpillars of *Chærocampa Elpenor* all remain green till the fourth stage, when they mostly become light or dark brown, and only very seldom retain their green colour. *Chærocampa Porcellus* behaves in a precisely similar manner; as also does *Pterogon Œnotheræ*, which inhabits the same localities, and is found on the same food-plant, but is not very closely related to the *Chærocampa*. In this species also (*P. Œnotheræ*) the brown is more common than the green form in the adult state, both varieties showing a complicated marking. The young larvæ possess only a light green colour, and a pure white subdorsal line as the only marking; they are so well adapted to the leaves of their food-plants, *Epilobium Hirsutum*, and *E. Rosmarinifolium*, that they can only be detected

[6] See the essay "Über den Einfluss der Isolirung auf die Artbilding." Leipzig, 1872, p. 22.

with great difficulty. After the third moult they become brown, and can be easily seen when at rest on their food-plant.

Now in all known caterpillars brown colours are adaptive, sometimes causing a resemblance to the soil, and at others to dead leaves or branches. As soon, therefore, as the caterpillars have attained a considerable size, they remain concealed by day.[7] The truth of this observation not only appears from various entomological notes, but I have frequently convinced myself of its accuracy. I well remember from the earliest times that *C. Elpenor*, especially when the larva is adult, always rests by day among the dead branches and leaves of its shrub-like food-plant, *Epilobium Hirsutum;* and even when this species lives on the low-growing *Epilobium Parviflorum*, it conceals itself by day on the ground, among the tangled leaves and branches. I have observed that *Sphinx Convolvuli* has a precisely similar habit, for which reason it is difficult to obtain, even in localities where it occurs very commonly.

In the neighbourhood of Basle I once found at mid-day a brown caterpillar of *Pterogon Œnotheræ* on an isolated dead branch of *Epilobium Rosmarinifolium*, and I was informed by Herr Riggenbach-Stähelina—collector of great experience who accompanied me—that these caterpillars always rest (by day) on withered plants as soon as they

[7] [See also preceding note 5, p. 294. R.M.]

become brown, but before this change they are only to be found on green plants.

Thus, it cannot well be doubted that the change of colour is associated with a change in the habits of life, and the question arises as to which has been the primary change.

If the view here entertained, that the later brown coloration is adaptive, be correct, the species must have first acquired the habit of concealing itself by day on the ground and among dead herbage, before the original green colour could have been changed into brown by natural selection. This must represent the actual facts of the case.

Nearly allied species which at an advanced age are not dimorphic, but are darkly coloured in all individuals, are especially calculated to throw some light on this point. For instance, the caterpillar of *Deilephila Vespertilio*, which comes under this denomination, is light green when young, and rests both by day and night on the leaves of the plant on which it feeds. As soon as it acquires its dark colour—after the third moult—it changes its habits, concealing itself by day on the ground and feeding only by night. For this reason collectors prefer seeking for it in the evening, or with a lantern by night.

The most instructive case, however, is that of *Deilephila Hippophaës*, in which no change of colour is associated with age, the caterpillar, throughout its whole life, remaining of a greyish green, which exactly matches the colour of the leaves of its food plant, *Hippophae Rhamnoides.* Nevertheless this species also possesses the habit of feeding only at night as soon as it has attained to a considerable size, hiding itself by day at the root of its food-plant. Collectors expressly state that this larva can scarcely be found by day, and recommend that it should be sought for at night with a lantern.

From the foregoing facts and considerations it may fairly be concluded, that the habit of hiding by day, possessed by these and other allied caterpillars, was acquired when they resembled the leaves in colour, and that the adaptation to the colour of the soil, or dead foliage and withered branches, ensued as a secondary consequence.

But why have these caterpillars acquired such a habit, since they appear to be perfectly protected by their resemblance in colour to the green leaves? The answer to this question is easily given when we consider in which species this habit generally occurs.

Does the habit prevail only among the species of the one genus *Deilephila*, and in all the species of this genus? This is by no means the case, since, on the one hand, many species of *Deilephila*, such as *D. Euphorbiæ*, *Galii*, *Nicæa*, and *Dahlii*, do not possess the habit, and, on the other hand, it occurs in species of other genera, such as *Macro-*

glossa Stellatarum, *Sphinx Convolvuli*, and *Acherontia Atropos*.

The habit in question must therefore be the result of certain external conditions of life common to all those species which rest by day. The mode of life common to them all is that they do not live on trees with large leaves or with thick foliage, but on low plants or small-leaved shrubs, such as the Sea Buckthorn.[8] I believe I do not err when I attribute the habit possessed by the adult larvæ, of concealing themselves by day, to the fact that the green colour is protective only so long as they are small—or, more precisely speaking, as long as their size does not considerably exceed that of a leaf or twig of their food-plant. When they become considerably larger, they must become conspicuous in spite of their adaptive colour, so that it would then be advantageous for them to conceal themselves by day, and to feed only by night. This habit they have acquired, and still observe, even when the secondary adaptation to the colour of the soil, &c., has not been brought about. We learn this from *D. Hippophaës*, which

[8] [Eng. ed. The habit of hiding by day occurs also in those caterpillars which resemble the bark of their food-trees. Thus *Catocala Sponsa* and *Promissa* conceal themselves by day in crevices of the bark, and are, under these circumstances, only found with difficulty. Dr. Fritz Müller also writes to me that in Brazil the caterpillars of *Papilio Evander* rest in this manner in large numbers, crowded together into dense masses, on the trunks of the orange-trees, which they resemble in colour.]

remains green throughout its whole larval existence; and no less from the green forms of the adult larvæ of *Sphinx Convolvuli*, *Chærocampa Elpenor*, and *Porcellus*, all of which conceal themselves by day in the same manner as their brown allies.

It may be objected that there are Sphinx-larvæ—instances of which I have myself adduced—which live on low small-leaved plants, and which nevertheless do not hide themselves by day. This is the case with the spurge-feeding *D. Euphorbiæ*, so common in many parts of Germany. This caterpillar must, however, be classed with those which, on account of their distastefulness, or for other reasons to be subsequently considered, are rejected by birds and other larger foes, and which, as Wallace has shown, derive advantage from being coloured as vividly as possible. I shall return to this subject later, when treating of the biological value of special markings.

On the other hand, it is readily conceivable that, from the conditions of life of caterpillars living on trees or shrubs with dense foliage, the habit of resting by day and descending from the tree for concealment would not have been acquired. Such larvæ are sufficiently protected by their green colour among the large and numerous leaves; and I shall have occasion to show subsequently that their markings increase this protective resemblance.

The di- or polymorphism of the larvæ of the

Sphingidæ does not therefore depend upon a *contemporaneous* double adaptation, but upon the replacement of an old protective colour by a new and better one, and therefore upon a *successive* double adaptation. The adult caterpillars of *C. Elpenor* are not sometimes brown and sometimes green because some individuals have become adapted to leaves and others to the soil, but because the anciently inherited green has not yet been completely replaced by the newly acquired brown coloration, some individuals still retaining the old green colour.

When, in another place,[9] I formerly stated "that a species can become adapted in this or that manner to given conditions of life, and that by no means can only one best adapted form be allowed for each species," this statement is theoretically correct speaking generally, but not in its application to the present class of cases. A comparison with one another of those caterpillars which repose by day, distinctly shows that they all possess a tendency to abandon the green and assume a dull colour, but that this process of replacement has advanced further in some species than in others. It will not be without interest to follow this operation in some detailed cases, since we may thus obtain an insight into the processes by which polymorphism has arisen, as well as into the con-

[9] "Über den Einfluss der Isolirung auf die Artbildung." Leipzig, 1872, p. 21.

nection between this phenomenon and simple variability.

In *D. Hippophaës* the process has either not yet commenced, or is as yet in its first rudiments. If we may trust the statements of authors, together with the ordinary green form there occurs, rarely, a silver-grey variety, which may be regarded as the beginning of a process of colour substitution. Among thirty-five living specimens of this scarce species which I was able to procure, the grey form did not occur, neither have I found it in collections.

In *Macroglossa Stellatarum* we see the transforming process in full operation. A large number of individuals (about thirty-five per cent.) are still green; the number of dark-coloured individuals reaches forty-six per cent., these, therefore, preponderating; whilst between the two extremes there are about nineteen per cent. of transition forms, showing all possible shades between light green and dark blackish-brown or brownish-violet, and even, in solitary individuals, pure violet (See Figs. 3—12, Pl. III.). The relatively small number of the intermediate forms, taken in connection with the fact that all the 140 specimens employed in my investigation were obtained from one female, leads to the conclusion that these forms owe their existence to cross-breeding. It would be superfluous to attempt to prove this last conclusion with reference to the before-mentioned case, in which a caterpillar was streaked with brown and green (Fig. 9, Pl. III.).

The process of transformation, as already mentioned, advances in such a manner that the intermediate forms diminish relatively to the dark individuals. This is found to be the case with *Sphinx Convolvuli*, and almost to the same extent with *Chærocampa Elpenor*, in both of which species the green caterpillars are the rarest.[10] Forms truly intermediate in colour between green and brown no longer occur, but apparently only different shades of light and dark brown, passing into brownish-black.

The process has again made a further advance in *Chærocampa Porcellus* and *Celerio* as well as in *Pterogon Œnotheræ*. In all these species the green form occurs,[11] but so rarely that very few collectors have seen it. The brown form has therefore in these cases nearly become the predominant type, and the solitary green specimens which occasionally occur, may be regarded as reversions to an older phyletic stage.

Deilephila Livornica appears to have reached a similar stage, but the caterpillar of this species has been so imperfectly observed, that it is

[10] I am unfortunately not able to give exact numbers showing the relative proportions of the different forms, since I have never bred *S. Convolvuli* from eggs, nor *C. Elpenor* in sufficient numbers.

[11] [With reference to *C. Porcellus*, see note 2, p. 188. R.M.]

difficult to determine, even approximately, the relative proportion of the brown to the green individuals. I have only seen one of the latter in Dr. Staudinger's collection (Compare Fig. 62, Pl. VII.).

In *Deilephila Vespertilio*, *Euphorbiæ*, *Dahlii*, *Mauritanica*, *Nicæa*, and *Galii*, the green form has completely disappeared. The blackish olive-green colour shown by many caterpillars of the two last species, can be considered as a faint retention of the light green colour which they formerly possessed, and which they both show at the present time in their young stages.

Beginning with the appearance of single darker individuals, we pass on in the first place to a greater variability of colouring, and from this, by the greater diminution of the intermediate forms, to polymorphism; the complete extermination of these forms ending in dimorphism. The whole process of transformation has been thus effected:—As the new colouring always prevailed over the old, the latter was at length completely displaced, and the caterpillars, which were at first simply variable, became polymorphic and then dimorphic, finally returning to monomorphism.

We thus see the process of transformation still going on, and no doubt can arise as to its inciting causes. When a character can with certainty be ascribed to adaptation, we can explain its origin in no other way than by the action of natural selection. If, as I believe, it can and has been shown, not only that caterpillars in general possess adaptive colours, but that these colours can change during the lifetime of one and the same species, in correspondence with external conditions, we must certainly gain a very high conception of the power which natural selection exerts on this group of living forms.[12]

[12] [In the class of cases treated of in the foregoing portions of this essay, the external conditions remain unaltered during the lifetime of the caterpillar, but change of habit, and in some cases of colour, occurs when the insect has attained a size sufficient to make it conspicuous. Cases are, however conceivable *à priori*, and are realized by observation, in which the environment itself may undergo change during the lifetime of the individual caterpillar. Thus, in the case of hibernating species, the colour which is adaptive to the autumnal colours of the foliage of their food-trees would not assimilate to that of the newly-opened leaves in the spring. I have already quoted (Proc. Zoo. Soc. 1873, p. 155) as instances of what may be called "seasonal adaptation," the larvæ of *Geometra Papilionaria*, *Acidalia Degeneraria*, and *Gnophos Obscurata*, and many more could be named. These species undergo a change of colour before or after hibernation, the change being always adaptive to the environment.

It has long been known that caterpillars which feed on flowers or on plants of variously-coloured foliage, in some cases partake of the colour of their food. See, for instance, Dr. L. Möller's memoir, "Die Abhängigkeit der Inseckten von ihrer Umgebung," 1867, and B. D. Walsh "On Phytophagic Varieties and Phytophagic Species," Proc. Ent. Soc. Philadelph., vol. iii., p. 403. In 1869 Mr. R. McLachlan published a paper entitled "Observations on some remarkable varieties of *Sterrha Sacraria*, Linn., with general notes on variation in Lepidoptera" (Trans. Ent. Soc. 1865, p. 453), in which he gave many illustrations of this phenomenon. The larva of *Heliothis*

Peltiger, according to Mr. Reading's description (Newman's "British Moths," p. 438), is another case in point. In 1874 a number of instances were published by Mr. Thomas G. Gentry in a paper entitled "Remarkable Variations in Coloration, Ornamentation, &c., of certain Crepuscular and Nocturnal Lepidopterous Larvæ" ("Canadian Entomologist," vol. vi., p. 85. See also W. H. Edwards' description of the summer and autumnal larvæ of *Lycæna Pseudargiolus; Ibid.*, vol. x., pp. 12, 13).

The caterpillars of the *Sphingidæ* appear also in some cases to vary in a manner very suggestive of phytophagic influences. The observations upon *S. Ocellatus* recorded in the previous note (p. 241) may perhaps be interpreted in this sense. In order to get experimental evidence upon this subject, I may add that Mr. E. Boscher was good enough at my request to repeat his observations, and conduct some breeding experiments during the present year (1880). In the same locality as that previously mentioned, seven larvæ were found feeding on *Salix viminalis*, all of which were the bright green spotted variety; and in the same osier-bed six more were found on another species of *Salix*, two of these being the bluish-green variety, and the other four the bright green form. Unless we have here a local race, these observations, in connection with those of last year, tend to show that the light green form is associated with *Salix viminalis*. When found in the natural state feeding on apple, the caterpillar of this species is generally, perhaps invariably, the bluish-green form. In order to try the effect of breeding the larvæ *ab ovo* on distinct food-plants, a large number of eggs laid by a female *Ocellatus* in July were divided into three batches, one being supplied with *Salix triandra*, another with *S. viminalis*, and the third lot with apple. The experiment unfortunately failed in great part, owing to most of the larvæ dying off, three from the third lot only surviving; but these were all of the bluish-green form, which colour was shown by all the caterpillars of this batch from their earliest stage. The observation is thus so far successful, as it goes to support the view that the variety mentioned is associated with apple (and *S. triandra ?*) My friend Mr. W. J. Argent informs me that he had a number of specimens of *Sphinx*

Ligustri in his possession this autumn, some of which had been found on lilac and others on laurestinus, and he states that all those on the latter plant had the ground-colour distinctly darker than in those feeding on lilac. I learn also from Mr. W. Davis, of Dartford, that he found a number of these larvæ this year feeding on ash, and that they were all differently coloured to those found on lilac or privet, being of a more greyish-green. Another case of colour-variation in larvæ is that *Emmelesia Unifasciata*, specimens of which I have recently had an opportunity of examining, through the courtesy of Mr. W. Davis. This species feeds on the seeds of a species of *Bartsia* when the capsules are in various stages of growth, and (omitting details of marking) those caterpillars found on the green capsules were green, whilst those on the brown capsules were of a corresponding colour.

On the whole I am inclined to believe that sufficient importance has not hitherto been given to phytophagic variability as a factor in determining larval coloration, and a large field for experimental investigation here lies open for future work. The obscure chemico-physiological processes which may perhaps be shown by such researches to lead to phytophagic variation, cannot, I am persuaded, produce any great divergence of character if unaided; but when such causes of variability play into the hands of natural selection variations of direct *protective advantage* to the species, we can easily see that this all-important agency would seize upon and perpetuate such a power of adaptability to a variable environment. (See Proc. Zoo. Soc. 1873, p, 158, and "Nature," vol. xiv., pp. 329 and 330. R.M.]

V.

Biological Value of Special Markings.

The following questions now present themselves: Have the markings of caterpillars any biological value, or are they in a measure only sports of nature? Can they be considered as partially or entirely the result of natural selection, or has this agency had no share in their production?

The problem here offers itself more distinctly than in any other group of living forms, because it presents an alternative without a third possibility. In other words, if it is not possible to show that larval markings have a distinct biological significance, there remains only for their explanation the assumption of a phyletic force, since the direct action of the environment is insufficient to account for such regularity of development throughout a series of forms. The explanation by sexual selection is excluded *ab initio*, since we are here concerned with larvæ, and not with reproductive forms.[1]

[1] [In 1879 Mr. George Francis, of Adelaide, forwarded from the latter place a number of moths (a species of *Anapæa*)

The biological significance of marking—if such significance it possess—will be most easily investigated by examining whether species with similar markings have any conditions of life in common which would permit of any possible inference as to the significance of the markings.

Among the *Sphingidæ* we find four chief forms of marking; (1) complete absence of all marking; (2) longitudinal stripes; either a simple subdorsal or this together with a spiracular and dorsal line; (3) oblique stripes; (4) eye-spots and ring-spots, single, paired, or in complete rows.

Now if we consider in which species these four kinds of marking are of general occurrence, not

together with their larvæ (in alcohol) and cocoons (Proc. Ent. Soc. 1879, p. xvi), and in an accompanying note he stated that the male larva when living is of "a bright emerald green, with red and pink markings on the back, and yellow, black, and white streaks on the sides." The male larva is described as being smaller than the female, and as possessing all the brilliant colours, the latter "having no red markings, but only white, yellow, and green, with a little black." I was at first disposed to think that we might be dealing here with two distinct species having differently marked larvæ; but Mr. Francis this present year (1880) forwarded a large number of the living cocoons of this species, which I separated according to size, and, on the emergence of the moths (August), I found that all those from the small cocoons were males, and those from the larger cocoons females. There can be no doubt, therefore, that we have but one species in this case, the larva of which presents the remarkable phenomenon of sexual difference of coloration. As an analogous fact I may here mention the well-known case of *Orgyia Antiqua*, the larva of which differs in the colour of the tufts of hair according to sex. R.M.]

only in the small group of the *Sphingidæ* but in the whole order Lepidoptera, we shall arrive at the following results :—

1. *Complete absence of marking*, so common in the larvæ of other insects, such as the Coleoptera, is but seldom found among Lepidopterous caterpillars.

To this category belong all the species of *Sesiidæ* (the genera *Sesia*, *Trochilia*, *Sciapteron*, *Bembecia*, &c.), the larvæ of which, without exception, are of a whitish or yellowish colour, and live partly in the wood of trees and shrubs and partly in the shoots of herbaceous plants. Subterranean larvæ also, living at the roots of plants, such as *Hepialus Humuli* at the roots of hop, and *H. Lupulinus* at those of *Triticum Repens*, possess neither colour nor marking. These, like the foregoing, are yellowish-white, evidently because they are deprived of the influence of light.[2] The larvæ of

[2] [I have already given reasons for suspecting that the colour of green caterpillars may be due to the presence of chlorophyll (or some derivative thereof) in their tissues (see Proc. Zoo. Soc. 1873, p. 159). This substance appears to be one of great chemical stability, and, according to Chautard, who has detected it in an unaltered state in the tissues of certain leaf-feeding insects by means of its absorption spectrum (" Comp. Rend." Jan. 13th, 1873), it resists the animal digestive processes (Ann. Ch. Phys. [5], iii., 1—56). If this view should be established by future observations, we must regard the green colour of caterpillars as having been produced, when protective, from phytophagic variability by the action of natural selection; and the absence of colour in internal feeders, above referred to,

certain small moths, such as *Tortrix Arbutana* and *Pomonana*, which live in fruit, and many case-bearing *Tineina*, are likewise without marking and devoid of bright colour, being generally whitish.

Many of the small caterpillars which feed exteriorly are also—so far as my experience extends—without definite markings, these being among the most minute, such as the greenish leaf-mining species of *Nepticula*. It is among the larger species that we first meet with longitudinal and oblique stripes. Eye-spots do not occur in any of these larvæ, a circumstance of the greatest importance for the biological significance of this character, as will be shown subsequently. The small size of the caterpillars cannot be the sole cause of the absence of such eye-spots, since in young *Smerinthus* caterpillars one centimeter long, the oblique stripes are beautifully developed, and the larvæ of many of the smaller moths considerably exceed this size. The surface of these caterpillars therefore, *i.e.*, the field on which markings are displayed, is not absolutely too small for the development of such a character.

is only secondarily due to the exclusion of light, and depends primarily on the absence of chlorophyll in their food. In connection with this I may adduce the fact, that some few species of *Nepticula* (*N. Oxyacanthella*, *N. Viscerella*, &c.) are green, although they live in leaf-galleries where this colour can hardly be of use as a protection; but their food (hawthorn and elm) contains chlorophyll. See also note 2, p. 293. Further investigations in this direction are much needed. R.M.]

Besides the larvæ of the Micro-lepidoptera and of those species living in the dark, there is also a complete absence of marking in the young stages of many caterpillars. Thus, all the *Sphingidæ* of which I have been able to observe the development, show no markings immediately after emergence from the egg; in many they appear very soon, even before the first moult, and, in other species, after this period.

2. *The second category of markings, longitudinal stripes*, is very widely distributed among the most diverse families. This character is found among the larvæ of butterflies, *Sphingidæ*, *Noctuæ*, Micro-lepidoptera, &c., but in all these groups it is absent in many species. This last fact is opposed to the view that this character is purely morphological, and leads to the supposition that it may have a biological value, being of service for the preservation of the individual, and therefore of the species.

I find that such marking is of service, stripes extending longitudinally along the upper surface of the caterpillar generally making the latter less conspicuous. This, of course, does not hold good under all circumstances, since there are many species with very striking colours which possesss longitudinal stripes. Let us consider, however, a case of adaptive colouring, such as a green caterpillar, which, on this account only, is difficult to see, since it accords with the colour of the plant on which it lives. If it is a small caterpillar, *i.e.*, if its length and thickness do not considerably exceed that of the parts of its food-plant, it can scarcely be better concealed—stripes would hardly confer any special advantage unless the parts of the plant were also striped. But the case is quite different if the caterpillar is considerably larger than the parts of the plant (leaves, stalks, &c.). The most perfect adaptive colouring would not now prevent it from standing out conspicuously as a larger body, among the surrounding parts of the plants. It must be distinctly advantageous therefore to such a caterpillar to be striped, since these markings to a certain extent divide the large body into several longitudinal portions—they no longer permit it to be seen as a whole, and thus act more effectively than mere assimilative colouring in causing it to escape detection. This protection would be the more efficacious if the stripes resembled the parts of the plant in colour and size, such, for instance, as the lines of light and shadow produced by stalks or by long and sharp-edged leaves,

If this view be correct, we should expect longitudinal stripes to be absent in the smallest caterpillars, and to be present more especially in those species which live on plants with their parts similarly disposed, *i. e.*, on plants with numerous thin, closely-growing stalks and grass-like leaves, or on plants with needle-shaped leaves.

It has already been mentioned that the smallest species are devoid of longitudinal striping. The larvæ of the Micro-lepidoptera show no such marking, even when they do not live in the dark, but feed either on the surface or in superficial galleries of the leaves (*Nepticula*, &c.), in which they must be exposed to almost as much light as when living on the surface. The fact that the subdorsal line sometimes appears in very young Sphinx-larvæ is explained, as has already been shown, by the gradual backward transference of adaptational characters acquired in the last stage of development.

It can easily be demonstrated that longitudinally striped caterpillars mostly live on plants, of which the general appearance gives the impression of a striped arrangement. We have only to consider in connection with their mode of life, any large group of adaptively coloured species marked in this manner. Thus, among the butterflies, nearly all the *Satyrinæ* possess larvæ conspicuously striped—a fact which is readily explicable, because all these caterpillars live on grasses. This is the case with the genera *Melanargia*, *Erebia*, *Satyrus*, *Pararge*, *Epinephele*, and *Cænonympha*, no species of which, so far as the larvæ are known, is without longitudinal stripes, and all of which feed on grasses. It is interesting that here also, as in certain *Sphingidæ*, some species are brown, *i. e.*, adapted to the soil, whilst the majority are green, and are therefore adapted to living grass. Just as in the case of the *Sphingidæ* also, the brown species conceal themselves by day on the earth, whilst some of the green species have likewise acquired this habit. I have already shown how this habit originates from the increasing size of the growing larva, which would otherwise become too conspicuous, in spite of adaptive colour and marking, A beautiful confirmation of this view is found in the circumstance that only the largest species of *Satyrus*, such as *S. Proserpinus*, *Hermione*, *Phædrus*, &c., possess brown caterpillars. I should not be surprised if a more exact investigation of these species, which have hitherto been but seldom observed, revealed in some cases a dimorphism similar to that of the *Sphingidæ*; and I believe that I may venture to predict that the young stages of all these brown larvæ—at present quite unknown—are, as in the last-named group, green.

Besides the *Satyrinæ*, most of the larvæ of the *Pierinæ* and *Hesperidæ* possess longitudinal stripes, which are generally less strongly pronounced than in the former subfamily. Some of the *Pierinæ* live on *Cruciferæ*, of which the narrow leaves and thin leaf- and flower-stalks present nothing but a linear arrangement; other species of this group, however, feed on *Leguminosæ* (*Lathyrus*, *Lotus*, *Coronilla*, *Vicia*), and some few on broad-leaved bushes (*Rhamnus*). This last fact may appear to be opposed to the theory; but light lateral stripes, such for example, as those possessed by *Gonepteryx Rhamni*, can never be disadvantageous, and may be of use, even on large leaves, so that if we consider them as an inherited character, there is no reason for natural selection to eliminate them. In the case of caterpillars living on vetch, clover, and other *Leguminosæ*, it must not be forgotten that, although their food-plants do not present any longitudinal arrangement of parts, they always grow among grasses, the species feeding on such plants always resting between grass stems, and very frequently on the grass itself, so that they can have no better protective marking than longitudinal stripes. The striping of the *Hesperidæ* larvæ, which partly feed on grasses but mostly on species of *Leguminosæ*, can be explained in a similar manner.

It is not here my intention to go through all the groups of Lepidoptera in this manner. The instances adduced are quite sufficient to prove that longitudinal stripes occur wherever we should expect to find them, and that they really possess the biological significance which I have ascribed to them. That these markings are occasionally converted into an adaptive imitation of certain special parts of a plant, is shown by the larvæ of many moths, such for example as *Chesias Spartiata*, which lives on broom (*Spartium Scoparium*), its longitudinal stripes deceptively resembling the sharp edges of the stems of this plant.[3]

3. *Oblique striping.* Can the lilac and white oblique stripes on the sides of a large green caterpillar, such as those of *Sphinx Ligustri;* or the red and white, or white, black, and red stripes of *Smerinthus Tiliæ* and *Sphinx Drupiferarum* respectively, be of any possible use? Have we not here just one of those cases which clearly prove that such a character is purely morphological, and worthless for the preservation of the individual? Does not Nature occasionally sport with purposeless forms and colours; or, as it has often been poetically expressed, does she not here give play to the wealth of her phantasy?

At first sight this indeed appears to be the case. We might almost doubt the adaptive importance of the green ground-colour on finding coloured stripes added thereto, and thus—as one might suppose—abolishing the beneficial action of this ground-colour, by making the insect strikingly conspicuous. But this view would be decidedly incorrect, since oblique stripes are of

[3] [The same applies to *Pseudoterpna Cytisaria*, also feeding on broom at the same time of the year. The most striking cases of adaptive resemblance brought about by longitudinal stripes are to be found among fir and pine feeders, species belonging to the most diverse families (*Hyloicus Pinastri*, *Trachea Piniperda*, *Fidonia Piniaria*, &c., &c.) all being most admirably concealed among the needle-shaped leaves. R.M.]

just the same importance as longitudinal stripes. The former serve to render the caterpillar difficult of detection, by making it resemble, as far as possible, a leaf; they are imitations of the leaf-veins.

Nobody who is in the habit of searching for caterpillars will doubt that, in cases where the oblique stripes are simply white or greenish-white, it is extremely difficult to see the insect on its food-plant, *e.g. S. Ocellatus* on *Salix;* not only because it possesses the colour of the leaves, but no less because its large body does not present an unbroken green surface, which would bring it into strong contrast with the leaves, and thus arrest the attention. In the case of the species named, the coloured area of the body is divided by oblique parallel stripes, just in the same manner as a willow leaf. In such instances of course we have not presented to us any special imitation of a leaf with all its details—there is not a perfect resemblance of the insect to a leaf, but only an arrangement of lines and interspaces which does not greatly differ from the division of a leaf by its ribs.

That this view is correct is shown by the occurrence of this form of marking. It is on the whole rare, being found, besides in many *Sphingidæ,* in isolated cases in various families, but is always confined to those larvæ which live on ribbed leaves, and never occurring in species which feed on grasses or on trees with needle-shaped leaves. This has already been shown with respect to the *Sphingidæ*, in which the oblique stripes are only completely developed in the subfamilies *Smerinthinæ* and *Sphinginæ*. The species of *Smerinthus* all live on trees such as willows, poplars, lime, oak, &c., and all possess oblique stripes. The genus *Anceryx* also belongs to the *Sphinginæ*, and these caterpillars, as far as known, live on trees with needle-shaped leaves. The moths of this last genus are very closely allied to the species of *Sphinx*, not only in form and colour, but also in many details of marking. The larvæ are however different, this distinction arising entirely from their adaptation to needle-shaped leaves, the *Sphinx* caterpillars being adapted to ordinary foliage. The species of *Anceryx*, as has been already shown, are brown mixed with green, and never possess even a trace of the oblique stripes, but have a latticed marking, consisting of many interrupted lines, which very effectively serves to conceal them among the needles and brown bark of the *Coniferæ*.

Of the *Sphinginæ* living on plants with ordinary foliage, not a single species is without oblique stripes. I am acquainted with ten species of caterpillars and their respective food-plants, viz. *Sphinx Carolina*, *Convolvuli*, *Quinquemaculata*, *Prini*, *Drupiferarum*, *Ligustri*; *Macrosila Rustica* and *Cingulata*; *Dolba Hylæus* and *Acherontia Atropos*.

Besides among the *Sphinginæ*, oblique stripes occur in the larvæ of certain butterflies, viz. *Apatura Iris*, *Ilia*, and *Clytie*, all of which live on forest trees (aspen and willows), and are excellently adapted to the leaves by their green colour. In addition to these, I am acquainted with the larvæ of some few moths, viz. of *Aglia Tau* and *Endromis Versicolora*, both of which also live on forest trees.

Oblique stripes also occasionally occur in the smaller caterpillars of *Noctuæ*, *Geometræ*, and even in those of certain *Pyrales*, in all of which they are shorter and differently arranged. In these cases also, my theory of adaptation holds good, but it would take us too far if I attempted to go more closely into them. I will here only mention the extraordinary adaptation shown by the caterpillar of *Eriopus Pteridis*. This little Noctuid lives on *Pteris Aquilina;* it possesses the same green colour as this fern, and has double oblique white stripes crossing at a sharp angle on each segment, these resembling the lines of *sori* of the fern-frond so closely, that the insect is very difficult to perceive.

After all these illustrations it can no longer remain doubtful that the oblique stripes of the *Sphingidæ* are adaptive. But how are the coloured edges bordering these stripes in so many species to be explained?

I must confess that I long doubted the possibility of being able to ascribe any biological value to this character, which appeared to me only conspicuous, and not protective. Cases may actually occur in which the brightly coloured edges of the oblique stripes make the caterpillar conspicuous—just in the same manner as any marking may bring about a conspicuous appearance by presenting a striking contrast of colour. I am acquainted with no such instance, however. As a rule, in all well-adapted caterpillars, considering their colour in its totality, this is certainly not the case. The coloured edges, on the contrary, enhance the deceptive appearance by representing the oblique shadows cast by the ribs on the under-side of the leaf; all these caterpillars rest underneath the leaves, and never on the upper surface.

This explanation may, perhaps, at first sight appear far-fetched, but if the experiment be made of observing a caterpillar of *Sphinx Ligustri* on its food-plant, not immediately before one's eyes in a room, but at a distance as under natural conditions, it will be found that the violet edges do not stand out brightly, but show a colour very similar to that of the shadows playing about the leaves. The coloured edges, in fact, produce a more effective breaking up of the large green surface of the caterpillar's body, than whitish stripes alone. Of course if the insect was placed on a bare twig in the sun, it would be easily visible at a distance; the larva never rests in such a position, however, but always in the deep shadow of the leaves, in which situation the coloured edges produce their peculiar effect. It may be objected that the oblique white stripes, standing simply on a dark-green ground-colour, would produce the same effect, and that my explanation therefore leaves the bright colouring of these edges still unaccounted for. I certainly cannot say why in *Sphinx Ligustri* these edges are lilac, and in *S. Drupiferarum*, *S. Prini*, and *Dolba Hylæus* red, nor why they are black and green in *Macrosila Rustica*, and blue in *Acherontia Atropos*. If we knew exactly on what plants these caterpillars fed originally, we might perhaps indulge in comparing with an artistic eye the shadows playing about their leaves, seeing in one case more red, and in another more blue or violet. The coloured stripes of the *Sphingidæ* must be regarded as the single strokes of a great master on the countenance of a human portrait. Looked into closely, we see red, blue, or even green spots and strokes; but all these colours, conspicuous when close, disappear on retreating, a general effect of colour being then produced, which cannot be precisely described by words.

Quite in accordance with this explanation, we see caterpillars with the brightest coloured stripes concealing themselves in the earth by day, and betaking themselves to their food-plants only in the dusk of the evening or dawn of morning and even during the night; *i.e.* in a light so faint that feeble colours would produce scarcely any effect. The bright blue of *Acherontia Atropos*, for example, would give the impression of oblique shadows without any distinctive colour.

It is precisely the case of this last caterpillar, which formerly appeared to me to present insurmountable difficulties to the explanation of the coloured stripes by adaptation, and I believed that this insect would have to be classed with those species which are brightly coloured because they are distasteful, and are avoided by birds. But although we have no experiments on this point, I must reject this view. Unfortunately, we know scarcely anything of the ontogeny of this caterpillar; but we know at least that the young larvæ (stage four) are greener than the more purely yellow ones of the fifth stage (which, however, are also frequently green), and we know further that some adults are of a dark brownish grey, without any striking colours. From analogy with the dimorphism of the species of *Chærocampa* and *Sphinx*, fully considered previously, it must therefore be concluded that in this case also, a new process of adaptation has commenced—that the caterpillar is becoming adapted to the soil in and on which it conceals itself by day.[4] An insect which

[4] The geographical distribution of the dark form indicates that in the case of this species also, the form referred to is replacing the yellow (green) variety. Whilst in the middle of

acquires undoubted protective colours cannot, however, be classed with those which possess an immunity from hostile attacks.

That the coloured edges are correctly explained as imitations of the oblique shadows of the leaf-ribs, may also be proved from another point of view. Let us assume, for the sake of argument, that these coloured stripes are not adaptive, and that they have not been produced by natural

Europe (Germany, France, Hungary) the dark form is extremely rare, in the south of Spain this variety, as I learn from Dr. Noll, is almost as common as the yellow one. I hear also from Dr. Staudinger that in South Africa (Port Natal) the dark form is somewhat the commoner, although the golden-yellow and, more rarely, the green varieties, occur there. I have seen a caterpillar and several moths from Port Natal, and these all agree exactly with ours. The displacement of the green (yellow) form by the dark soil-adapted variety, appears therefore to proceed more rapidly in a warm than in a temperate climate. [Eng. ed. Dr. Noll writes to me from Frankfort that the caterpillar of *Acherontia Atropos* in the south of Spain does not, as with us, conceal itself by day in the earth, but on the stems underneath the leaves. "At Cadiz, on the hot, sandy shore, *Solanum violaceum* grows to a height of three feet, and on a single plant I often found more than a dozen *Atropos* larvæ resting with the head retracted, . It can easily be understood why the lateral stripes are blue when one has seen the south European *Solaneæ*, on which this larva is at home. *Solanum violaceum* is scarcely green : violet tints alternate with brown, green, and yellow over the whole plant, and between these appear the yellow-anthered flowers, and golden-yellow berries of the size of a greengage. Thus it happens that the numerous thorns, an inch long, between which the caterpillar rests on the stem, pass from violet into shades of blue, red, green, and yellow."]

selection, but by a hypothetical phyletic force. We should then expect to see them appear at some period in the course of the phyletic development—perhaps at first only in solitary individuals, then in several, and finally in all; but we certainly could not expect that at first single, irregular, coloured spots should arise in the neighbourhood of the oblique white stripes—that these spots should then multiply, and fusing together, should adhere to the white stripes, so as to form an irregular spot-like edge, which finally becomes formed into a straight, uniformly broad stripe. The phyletic development of the coloured edges takes place, however, in such a manner, the species of *Smerinthus*, as has already been established, showing this with particular distinctness. In *S. Tiliæ* the course of development can be followed till the somewhat irregular red border is formed. In the species of *Sphinx* this border has become completely linear. It is very possible that the ontogeny of *S. Ligustri* or *Drupiferarum* would reveal the whole process, although it may also be possible that owing to the contraction of the development, much of the phylogeny is already lost.

I have now arrived at the consideration of the last kind of marking which occurs in the *Sphingidæ*, viz.:—

4. *Eye-spots and Ring-spots.*—These markings, besides among the *Sphingidæ*, are found only in a very few caterpillars, such as certain tropical

Papilionidæ and *Noctuæ*. I know nothing of the conditions of life and habits of these species, however, and without such knowledge it is impossible to arrive at a complete explanation.

With Darwin, I take an eye-spot to be "a spot within a ring of another colour, like the pupil within the iris," but to this central spot "concentric zones" may be added. In the *Chærocampa* larvæ and in *Pterogon Œnotheræ*, in which complete ocelli occur, there are alway three zones—a central spot, the pupil, or, as I have called it, the "nucleus;" then a light zone, the "mirror;" and, surrounding this again, a dark zone (generally black), the "ground-area."

As ring-spots I will consider those ocelli which are without the nucleus (pupil), and which are not therefore, strictly speaking, deceptive imitations of an eye, but present a conspicuous light spot surrounded by a dark zone.

Between these two kinds of markings there is, however, no sharp boundary, and morphologically they can scarcely be separated. Species with ring-spots sometimes have nuclei, and ocellated larvæ in some cases possess only a pale spot instead of a dark pupil. I deal here with the two kinds separately, because it happens that they appear in two distinct genera, in each of which they have their special developmental history. Ring-spots originate in a different position, and in another manner than eye-spots; but it must not, on this account, be assumed without further inquiry, that they are called into existence by the same causes; they must rather be investigated separately, from their origin.

Eye-spots are possessed by the genera *Chærocampa* and *Pterogon*; ring-spots by the genus *Deilephila*. In accordance with the data furnished by the above-given developmental histories, the origination of these markings in the two genera may be thus represented :—

In the genera named, eye-spots and ring-spots are formed by the transformation of single portions of the subdorsal line.

In *Chærocampa* the primary ocelli originate on the fourth and fifth segments by the detachment of a curved portion of the subdorsal, this fragment becoming the "mirror," and acquiring a dark encircling zone ("ground-area"). The nucleus (pupil) is added subsequently.

In *Deilephila* we learn from the development of *D. Hippophaës*, that the primary annulus arises on the segment bearing the caudal horn (the eleventh) by the deposition of a red spot on the white subdorsal line, which is somewhat enlarged in this region. The formation of a dark "ground-area" subsequently occurs, and with this, at first the partial, and then the complete, detachment of the mirror-spot from the subdorsal line takes place.

In both genera the spots arise at first locally on one or two segments, from which they are transferred to the others as a secondary character. In *Chærocampa* this transference is chiefly backwards, in *Deilephila* invariably forwards.

We have now to inquire whether complete eye-spots—such as those of the *Chærocampa* larvæ—have any significance at all, and whether they are of biological importance. It is clear at starting, that these spots do not belong to that class of markings which make their possessors more difficult of detection; they have rather the opposite effect.

We might thus be disposed to class ocellated caterpillars with those "brightly coloured" species which, like the *Heliconinæ* and *Danainæ* among butterflies, possess a disgusting taste, and which to a certain extent bear the signal of their distastefulness in their brilliant colours. But even if I had not found by experiment that our native *Chærocampa* larvæ were devoured by birds and lizards, and that they are not therefore distasteful to these insect persecutors, from the circumstance that these caterpillars are all protectively coloured, it could have been inferred that they do not belong to this category. It has been found that all adaptively coloured caterpillars are eaten, and one and the same species cannot possibly be at the same time inconspicuously (adaptively) and conspicuously coloured; the one condition excludes the other.

What other significance can eye-spots possess

than that of making the insects conspicuous? Had we to deal with sexually mature forms, we should, in the first place, think of the action of sexual selection, and should regard these spots as objects of taste, like the ocelli on the feathers of the peacock and argus-pheasant. But we are here concerned with larvæ, and sexual selection is thus excluded.

The eye-spots must therefore possess some other significance, or else they are of no importance at all to the life of the insect, and are purely "morphological characters;" in which case, supposing this could be proved, they would owe their existence exclusively to forces innate in the organism itself—a view which very closely approaches the admission of a phyletic vital force.

I am of opinion, however, that eye-spots certainly possess a biological value as a means of terrifying—they belong to that numerous class of characters which occur in the most diverse groups of animals, and which serve the purpose of making their possessors appear as alarming as possible.

The caterpillars of the *Sphingidæ* are known to behave themselves in different manners when attacked. Some species, such, for instance, as *Sphinx Ligustri* and *Smerinthus Ocellatus*, on the approach of danger assume the so-called Sphinx attitude; if they are then actually seized, they dash themselves madly to right and left, by this means not only attempting to get free, but also to terrify their persecutor. This habit frequently succeeds with men, and more especially with women and children; perhaps more easily in these cases than with their experienced foes, birds.

The ocellated *Chærocampa* larvæ behave differently. They remain quiet on being attacked, and do not put on a Sphinx-like attitude, but only withdraw the head and three small front segments into the large fourth segment, which thus becomes much swollen, and is on this account taken for the head of the insect by the inexperienced.[5] Now the large eye spots are situated on the fourth segment, and it does not require much imagination to see in such a caterpillar an alarming monster with fiery eyes, especially if we consider the size which it must appear to an enemy such as a lizard or small bird. Fig. 28 represents the larva of *C. Porcellus* in an attitude of defence, although but imperfectly, since the front segments can be still more withdrawn.

These facts and considerations do not, however, amount to scientific demonstration, and I therefore made a series of experiments, in order to determine whether these caterpillars did actually frighten small birds. The first experiment proved but little satisfactory. A jay, which had been domesticated for years, to which I threw a caterpillar of *Chæro-*

[5] [For Mr. J. P. Mansel Weale's remarks on the habits of certain ocellated S. African Sphinx-larvæ see note 1, p. 290. R.M.]

campa Elpenor, did not give the insect any time for manœuvring, but killed it immediately by a strong blow with its bill. This bird had been tame for years, and was in the habit of pecking at everything thrown to him. Perhaps a wild jay (*Garrulus Glandarius*) would have treated the insect differently, but it is hardly possible that such a large and courageous bird would have much respect for our native caterpillars. I now turned to wild birds. A large brown *Elpenor* larva was placed in the food-trough of an open fowl-house from which the fowls had been removed. A flock of sparrows and chaffinches (*Fringilla Domestica* and *Cœlebs*) soon flew down from the neighbouring trees, and alighted near the trough to pick up stray food in their usual manner. One bird soon flew on to the edge of the trough, and was just about to hop into it when it caught sight of the caterpillar, and stood jerking its head from side to side, but did not venture to enter. Another bird soon came, and behaved in a precisely similar manner; then a third, and a fourth; others settled on the perch over the trough, and a flock of ten or twelve were finally perched around. They all stretched their heads and looked into the trough, but none flew into it.

I now made the reverse experiment, by removing the caterpillar and allowing the birds again to assemble, when they hopped briskly into the trough.

I often repeated this experiment, and always with the same result. Once it could be plainly seen that it was really fear and not mere curiosity that the birds showed towards the caterpillar. The latter was outside the trough amongst scattered grains of food, so that from one side it was concealed by the trough. A sparrow flew down obliquely from above, so that at first it could not see the caterpillar, close to which it alighted. The instant it caught sight of the insect, however, it turned in evident fright and flew away.

Of course these experiments do not prove that the larger insectivorous birds are also afraid of these caterpillars. Although I have not been able to experiment with such birds, I can certainly prove that even fowls have a strong dislike to these insects. I frequently placed a large *Elpenor* larva in the poultry yard, where it was soon discovered, and a fowl would run hastily towards it, but would draw back its head just when about to give a blow with the bill, as soon as it saw the caterpillar closely. The bird would now run round the larva irresolutely in a circle—the insect in the meantime assuming its terrifying attitude—and stretching out its head would make ten or twenty attempts to deal a blow with its bill, drawing back again each time. All the cocks and hens acted in a similar manner, and it was often five or ten minutes before one particularly courageous bird would give the first peck, which would soon be followed by a second and third, till the caterpillar, appearing palatable, would finally be swallowed.

These experiments were always made in the presence of several persons, in order to guard myself against too subjective an interpretation of the phenomena; but they all invariably considered the conduct of the birds to be as I have here represented it.[6]

If it be admitted that the ocelli of caterpillars are thus means of exciting terror, the difficulty of their occurring in protectively coloured species at once vanishes. They do not diminish the advantage of the adaptive colouring, because they do not make the caterpillars conspicuous, or at least any more easily visible at a distance, excepting when the insects have assumed their attitude of alarm. But these markings are of use when, in spite of protective colouring, the larva is attacked by an enemy. The eye-spots accordingly serve the caterpillar as a second means of defence, which is resorted to when the protective colouring has failed.

By this it must not be understood that the ocelli of the *Chærocampa* larvæ invariably possess only this, and no other significance for the life of the insect. Every pattern can be conceived to render

[6] [Some experiments with the caterpillar of *C. Elpenor*, confirming these results, have been made by Lady Verney. See "Good Words," Dec. 1877, p. 838. R.M.]

its possessor in the highest degree conspicuous by strongly contrasted and brilliant colouring, so that it might be anticipated that perfect eye-spots in certain unpalatable species would lose their original meaning, and instead of serving for terrifying become mere signals of distastefulness. This is perhaps the case with *Chærocampa Tersa* (Fig. 35), the numerous eye-spots of which make the insect easily visible. Without experimenting on this point, however, no certain conclusion can be ventured upon, and it may be equally possible that in this case the variegated ocelli with bright red nuclei resemble the blossoms of the food-plant (*Spermacoce Hyssopifolia*).[7] I here mention this possibility only in order to show how an inherited form of marking, even when as well-defined and complicated as in the present case, may, under certain circumstances, be turned in quite another direction by natural selection, for the benefit of its possessor. Just in the same manner one and the same organ, such, for instance, as the limb of a crustacean, may, in the course of phyletic development, perform very different functions—first serving for locomotion, then for respiration, then for reproduction or oviposition, and finally for the acquisition of food.

[7] [The eye-spots on *Ch. Nerii* have thus been supposed by some observers to be imitations of the flowers of the periwinkle, one of its food-plants. See, for instance, Sir John Lubbock's "Scientific Lectures," p. 51. R.M.]

I now proceed to the consideration of the biological value of incomplete eye-spots, or, as I have termed them, ring-spots. Are these also means of terrifying, or are they only signals of distastefulness ?

I must at the outset acknowledge that on this point I am able to offer but a very undecided explanation. The decision is only to be arrived at by experiments conducted with each separate species upon which one desires to pronounce judgment. It is not here legitimate to draw analogical inferences, and to apply one case to all, since it is not only possible, but very probable, that the biological significance of ring-spots changes in different species. Nothing but a large series of experiments could completely establish this. Unfortunately I have hitherto failed in obtaining materials for this purpose. I would have deferred the publication of this essay for a year, could I have foreseen with certainty that such materials would have been forthcoming in sufficient quantity during the following summer ; but this unfortunately depends very much upon chance, and I believed that a preliminary conclusion would be preferable to uncertainty. Perhaps some entomologist to whom materials are more easily accessible, may, by continuing these experiments, accomplish this object.

The experiments hitherto made by other observers, are not sufficient for deciding the question under consideration. Weir,[8] as is well known, showed that certain brightly coloured and conspicuous larvæ were refused by insectivorous birds; and Butler[9] proved the same for lizards and frogs. These experiments are unfortunately so briefly described, that in no case is the species experimented with mentioned by name, so that we do not know whether there were any Sphinx caterpillars among them.[10] I have likewise experimented in this direction with lizards, in order to convince myself of the truth of the statement that (1) there are caterpillars which are not eaten on account of their taste, and (2) that such larvæ possess bright colours. I obtained positive, and on the whole, very decided results. Thus, the common orange and blue striped caterpillars of *Bombyx Neustria* enjoyed complete immunity from the attacks of lizards, whilst those of the nearly allied *Eriogaster Lanestris* and *L. Pini* were devoured, although not exactly relished. That the hairiness is not the cause of their being unpalatable, is shown by the fact that *L. Pini* is much more hairy than *B. Neustria.* The very conspicuous yellow and black ringed caterpillar of *Euchelia Jacobææ* gave also most decided results. I frequently placed this

[8] "On Insects and Insectivorous Birds," Trans. Ent. Soc. 1869, p. 21.

[9] *Ibid.*, p. 27.

[10] [Messrs. Weir and Butler inform me that they have not experimented with Sphinx-larvæ. R.M.]

insect in a cage with *Lacerta Viridis*, but they would never even notice them, and I often saw the caterpillars crawl over the body, or even the head of the lizards, without being snapped at. On every occasion the larvæ remained for several days with the lizards without one being ever missed. The reptiles behaved in a precisely similar manner with respect to the moth of *E. Jacobææ*, not one of which was ever touched by them. The yellow and black longitudinally striped caterpillars of *Pygæra Bucephala* were also avoided, and so were the brightly coloured larvæ of the large cabbage white (*Pieris Brassicæ*), which when crushed give a disagreeable odour. This last property clearly shows why lizards reject this species as distasteful. Both caterpillar and butterfly possess a blood of a strong yellow colour and oily consistency, in which, however, I could not detect such a decided smell as is emitted by that of the *Heliconinæ* and *Danainæ*.[11]

I next made the experiment of placing before a lizard a caterpillar as much as possible like that of *E. Jacobææ*. Half grown larvæ of *Bombyx Rubi* likewise possess golden yellow (but narrower)

[11] [It appears that the nauseous character of these last butterflies is to a certain extent retained after death, as I found that in an old collection which had been destroyed by mites, the least mutilated specimens were species of *Danais* and *Euplœa*, genera which are known to be distasteful when living, and to serve as models for mimicry. See Proc. Ent. Soc. 1877, p. xii. R.M.]

transverse rings on a dark ground, and they are much more hairy than those of *E. Jacobææ*. The lizard first applied its tongue to this caterpillar and then withdrew it, so that I believed it would also be avoided; nevertheless it was subsequently eaten. The caterpillars of *Saturnia Carpini* were similarly devoured in spite of their bristly hairs, and likewise cuspidate larvæ (*Dicranura Vinula*), notwithstanding their extraordinary appearance and their forked caudal horn.[12] These lizards were by no means epicures, but consumed large numbers of earth-worms, slugs, and great caterpillars, and once a specimen of the large and powerfully biting Orthopteron, *Decticus Verucivorus*. Creatures which possessed a strongly repugnant odour were, however, always rejected, this being the case with the strongly smelling beetle, *Chrysomela Populi*, as also with the stinking centipede, *Iulus Terrestris*, whilst the inodorous *Lithobius Forficatus* was greedily eaten. I will call particular attention to these last facts, because they favour the supposition that with rejected caterpillars a disgusting odour—although perhaps not always perceptible by us—is the cause of their being unpalatable.

Striking colours are of course only signals of distastefulness, and the experiment with *Bombyx*

[12] [This bears out the view expressed in a previous note 1, p. 290, that the grotesque attitude and caudal tentacles are more for protection against ichneumons than against larger foes. R.M.]

Rubi shows that the lizards were from the first prejudiced against such larvæ, the prejudice only being overcome on actually trying the specimen offered. A subsequent observation which I made after arriving at this conclusion, is most noteworthy. After the lizard had learnt by experience that there might be not only distasteful caterpillars (*E. Jacobææ*), but also palatable ones banded with black and yellow (*B. Rubi*), it sometimes tasted the *Jacobææ* larvæ, as if to convince itself that the insect was actually as it appeared to be, viz., unpalatable!

A striking appearance combined with a very perceptible and penetrating odour is occasionally to be met with, as in the caterpillar of the common Swallow-tail, *Papilio Machaon*. I have never seen a lizard make the slighest attempt to attack this species. I once placed two large specimens of this caterpillar in the lizard vivarium, where they remained for five days, and finally pupated unharmed on the side of the case.

I have recorded these experiments, although they do not thus far relate to Sphinx-caterpillars, with the markings of which we are here primarily concerned, because it appeared to me in the first place necessary to establish by my own experiments that signals of distastefulness did occur in caterpillars.

I now come to my unfortunately very meagre experience with *Deilephila* larvæ, with only two species of which have I been able to experiment, viz., *D. Galii* and *Euphorbiæ*.

The first of these was constantly rejected. Two large caterpillars, one of the black and the other of the yellow variety, were left for twelve hours in the lizard vivarium, without being either examined or touched. It thus appears that *D. Galii* is a distasteful morsel to lizards; and the habits of the caterpillar are quite in accordance with this, since it does not conceal itself, but rests fully exposed by day on a stem, so that it can scarcely escape being detected. It is almost as conspicuous as *D. Euphorbiæ*.

I was much surprised to find, however, that this last species was not rejected by lizards. On placing a large caterpillar, six to seven centimeters long, in the vivarium, the lizard immediately commenced to watch it, and as soon as it began to crawl about, seized it by the head, and, after shaking it violently, commenced to swallow it. In spite of its vigorous twisting and turning, the insect gradually began to disappear, amidst repeated shakings; and in less than five minutes was completely swallowed.[13] With regard to lizards, therefore, the prominent ring-spots of this larva are not effective as a means of alarm, nor are they considered as a sign of distastefulness.

[13] These experiments, as already mentioned above, were not made with the common German lizard (*Lacerta Stirpium*), but with the large South European *Lacerta Viridis*.

Unfortunately I have not hitherto been able to make any experiments with birds. It would be rash to conclude from the experience with lizards that ring-spots were of no biological value. There is scarcely any one means of protection which can render its possessor secure against *all* its foes. The venom of the most poisonous snakes does not protect them from the attack of the secretary bird (*Serpentarius Secretarius*) and serpent eagle (*Spilornis Cheela*); and the adder, as is well known, is devoured by hedgehogs without hesitation. It must therefore be admitted that many species which are protected by distastefulness, may possess certain foes against which this quality is of no avail. Thus, it cannot be said that brightly coloured caterpillars, which are not eaten by birds and lizards, are also spared by ichneumons. It is readily conceivable therefore, that the larva of *D. Euphorbiæ* may not be unpalatable to lizards, because they swallow it whole; whilst it is perhaps distasteful to birds, because they must hack and tear in order to swallow it.

From these considerations it still appears most probable to me that *D. Euphorbiæ*, and the nearly allied *D. Dahlii* and *Mauritanica*, bear conspicuous ring-spots as signs of their being unpalatable to the majority of their foes. The fact that these species feed on poisonous *Euphorbiaceæ*, combined with their habit of exposing themselves openly by day, so as to be easily seen at a distance, may perhaps give support to this view. As these insects are not protectively coloured, this habit would long ago have led to their extermination; instead of this, however, we find that in all situations favourable to their conditions of life they are among the commonest of the *Sphingidæ*.

Thus, *D. Euphorbiæ* occurs in large numbers both in South and North Germany (Berlin); and Dr. Staudinger informs me that in Sardinia the larvæ of *D. Dahlii* were brought to him by baskets full.

But if the conspicuous ring-spots (combined of course with the other bright colours) may be regarded as signals of distastefulness in many species of *Deilephila*, this by no means excludes the possibility that in some species these markings play another part, and are effective as a means of alarm. It even appears conceivable to me that in one and the same caterpillar they may play both parts against different foes, and it would certainly be of interest to confirm or refute this supposition by experiment.

In the light yellow variety of the caterpillar of *D. Galii* the ring-spots may serve as means of alarm, and still more so in that of *D. Nicæa*, the resemblance of which to a snake has struck earlier observers.[14]

[14] Thus, Boisduval states of this caterpillar, which in Provence lives on *Euphorbia esula* and allied species:—"Its resemblance to a serpent, and its brilliant colour, permit of its being

In those species of *Deilephila* which conceal themselves by day, the ring-spots cannot be considered as signals of distastefulness, and they must therefore have some other meaning. As examples of this class may be mentioned *D. Vespertilio*, which is protectively coloured both in the young and in the adult stages; and likewise *D. Hippophaës*, in which this habit of concealment is associated with adaptive colouring. In the case of the first-named species, it appears possible that the numerous large ring-spots may serve to alarm small foes, but the truth of this supposition could only be decided by experiment. In *D. Hippophaës*, on the other hand, such an interpretation must be at once rejected, since most individuals possess but a single ring-spot, which shows no resemblance whatever to an eye.

I long sought in vain for the meaning of this ring-spot, the discovery of which would in this particular case be of the greatest value, because we have here obviously the commencement of the whole development of ring-spots before us—the initial stage from which the marking of all the other species of *Deilephila* has proceeded.

I believe that I have now found the correct answer to this riddle, but unfortunately at a period of the year when I am unable to prove it

easily discovered." This was written in 1843, long before natural selection was thought of.

experimentally. I consider that the ring-spots are crude imitations of the berries of the food-plant. The latter are orange-red, and exactly of the same colour as the spots; the agreement in colour between the latter and the berries is quite as close as that between the leaves and the general colouring of the caterpillar. I know of no species which more closely resembles the colour of the leaves of its food-plant, the dark upper side and light under side corresponding in the leaves and caterpillars. The colour of the *Hippophae* is not an ordinary green, but a grey-green, which shade also occurs, although certainly but rarely, in the larvæ. I may expressly state that I have repeatedly shown to people as many as six to eight of the large caterpillars on one buckthorn branch, without their being able at once to detect them. It is not therefore mere supposition, but a fact, that this species is protected by its general colouring. At first the orange-red spots appear rather to diminish this protection—at least when the insects are placed on young shoots bearing no berries. But since at the same time when the berries become red (end of July and the beginning of August) the caterpillars are in their last stage of development (*i. e.* possess red-spots), it appears extremely probable that these spots are vague representations of the berries. For the same reason that these caterpillars have acquired the habit of feeding only at dusk and during the morning twilight, or at night, and of concealing themselves by day, it must be advantageous for them to have the surface of their large bodies not only divided by white stripes, but also interrupted in yet another manner. How could this be better effected than by two spots which, in colour and position, represent the grouping of the red berries on the branches? When feeding, the insect always rests with the hind segments on a branch, the front segments only being more or less raised and held parallel to the leaves; the red spots thus always appear on the stem, where the berries are likewise situated. It might indeed be almost supposed that the small progress which the formation of secondary ring-spots on the other segments has made up to the present time, is explicable by the fact that such berry-like spots on other portions of the caterpillar would be rather injurious than useful.

It may, however, be asked how an imitation of red berries, which are eaten by birds just as much as other berries, can be advantageous to a caterpillar, since by this means it would rather attract the attention of its enemies?

Two answers can be given to this. In the first place, the berries are so numerous on every plant that there is but a very small chance of the smaller and less conspicuous berry-spots catching the eye of a bird before the true berries; and, secondly, the latter, although beginning to turn red when the caterpillars are feeding, do not completely ripen till the autumn, when the leaves are shed, and the yellowish-red clusters of berries can be seen at a distance. The caterpillar, however, pupates long before this time.

I have considered this case in such detail because it appears to me of special importance. It is the only instance which teaches us that the rows of ring-spots of the *Deilephila* larvæ proceed from one original pair—the only instance which permits of the whole course of development being traced to its origin. Were it possible to arrive at the causes of the formation of these spots, their original or primary significance would thereby be made clear.

I will now briefly summarise the results of the investigation of the biological value of the *Deilephila* ring-spots.

In the known species of the genus now existing these spots have different meanings.

In some species (certainly in *Galii*, and probably in *Euphorbiæ* and *Mauritanica*) the conspicuous ring-spots serve as signals of distastefulness for certain enemies (not for all).

In a second group of species they serve as a means of alarm, like the eye-spots of the *Chærocampa* larvæ (*Nicæa*? light form of *Galii*?).

Finally, in a third group, of which I can at present only cite *Hippophaës*, they act as an adaptive resemblance to a portion of a plant, and enhance the efficacy of the protective colouring.

5. *Subordinate Markings.*—If, from the foregoing considerations, it appears that the three chief elements of the Sphinx-markings—longitudinal and oblique stripes, and spot formations—are not purely morphological characters, but have a very decided significance with respect to their possessors, there should be no difficulty in referring the whole of the markings of the *Sphingidæ* to the action of natural selection, supposing that these three kinds of marking were the only ones which actually occurred.

In various species, however, there appear other patterns, which I have comprised under the term "subordinate markings," some of which I will select, for the purpose of showing the reasons which permit of their being thus designated.

I ascribe to this category, for example, that fine network of dark longitudinal streaks which often extends over the whole upper side of the caterpillar, and which is termed the "reticulation." This character is found chiefly in the adult larvæ of *Chærocampa*, being most strongly pronounced in the brown varieties : it occurs also in *Deilephila Vespertilio*, *Pterogon Œnotheræ*, and *Sphinx Convolvuli*. As far as I know, it is only associated with adaptive colours, and indeed occurs only in those caterpillars which rest periodically at the base of their food-plants among the dead leaves and branches. I do not consider this reticulation to be a distinct imitation, but only as one of the various means of breaking up the large uniform surface of the caterpillar so as to make it present inequalities, and thus render it less conspicuous. There can be no doubt as to the dependence of this character upon natural selection.

There is, however, a second group of markings, which must be referred to another origin. To this group, for instance, belong those light dots in *Chærocampa Porcellus* and *Elpenor* which have been termed "dorsal spots." I know of no other explanation for these than that they are the necessary results of other new formations, and depend on correlation (Darwin), or, as I may express it, they are the result of the action of the law governing the organization of these species.

As long as we are confined to the mere supposition that the character in question may be the outward expression of an innate law of growth, it is permissible to attempt to show that a quite similar formation in another species depends upon such a law.

Many of the dark specimens of *Sphinx Convolvuli* show whitish dots on segments six to eleven, one being situated on the front edge of each of these segments, at the height of the completely vanished subdorsal line (Fig. 52). These spots vary much in size, lightness, and sharpness of definition. Now it might be difficult to attribute any biological significance to this character, but its origin becomes clear on examining light specimens in which the oblique white stripes are distinct on the sides and the subdorsal line is retained at least on the five or six anterior segments. It can then be seen that the spots are located at the points of intersection of the subdorsal and the oblique stripes (Fig. 16, Pl. III.), and they can accordingly be explained by the tendency to the deposition of light pigment being twice as great in these positions as in other portions of the two systems of light lines. Light spots are thus formed when the lines which cross at these points are partially or completely extinct throughout their remaining course.

A marking is therefore produced in this case by a purely innate law of growth—by the superposition of two ancient characters now rudimentary. Many other unimportant details of marking must be regarded as having been produced in a similar manner, although it may not be possible to prove this with respect to every minute spot and stripe. The majority of "subordinate markings" depend on the commingling of inherited, but now meaningless, characters with newly acquired ones.

It would be quite erroneous to attribute to natural selection only those characters which can be demonstrated to still possess a biological value in the species possessing them. They may be equally due to heredity. Thus, it is quite possible that the faint and inconspicuous ring-spots of *Deilephila Vespertilio* are now valueless to the life of the species—they may be derived from an ancestral form, and have not been eliminated by natural selection simply because they are harmless. I only mention this as a hypothetical case.

In the case of markings of the second class, *i.e.* oblique stripes, a transference to later phyletic stages can be demonstrated, although the stripes thereby lose their original biological value. Thus, the *Chærocampa* larvæ, when they were green throughout their whole life and adapted to the leaves, appear to have all possessed light oblique stripes in imitation of the leaf-ribs. All the species of the older type of colouring and marking, such as *Chærocampa Syriaca* (Fig. 29) and *Darapsa Chærilus* (Fig. 34), and also the light green young forms of *C. Elpenor* (Fig. 20), and *Porcellus* (Figs. 25 and 26), show these oblique stripes. In these last species the foliage imitation is abandoned at a later stage, and a dark brown, or blackish brown, ground-colour acquired. Nevertheless the oblique stripes do not disappear, but show themselves—in the fourth stage especially, and sometimes in the fifth—as distinct dirty yellow stripes, although not so sharply defined as in the earlier stages. These persistent stripes, in accordance with their small biological value, are very variable, since they are only useful in so far as they help to break up the large surface presented by the caterpillar, and are of no value as imitations of surrounding objects.

The oblique stripes of *Sphinx Convolvuli* offer a precisely similar case; and it may be safely predicted that the young forms of this species would possess sharply defined light oblique stripes, since more or less distinct remnants of these markings occur in all the adult larvæ, and especially in the green form. The entire pattern of this caterpillar depends essentially on the commingling of characters persisting from an earlier period, *i.e.* of residues of the subdorsal and oblique stripes, both these markings being extraordinarily variable. The black reticulation was added to the ground-colour as a new means of adaptation, this character appearing only in the phyletically younger brown form, and being entirely absent, or only faintly indicated, in the older green variety.

VI.

Objections to a Phyletic Vital Force.

It has been shown in the previous section that the three elements composing the markings of the Sphinx-larvæ originally possessed a distinct significance with respect to the life of the species, and that they were by this means called into existence. It has likewise been shown, that in most of the species which possess these characters at the present time they still have a decided, although sometimes a different use, for their possessors, so that from this point of view no objection can be raised to their being considered as having arisen by natural selection.

On looking at the phenomenon as a whole, however, certain instances occur which appear quite irreconcilable with this view.

The most formidable objection is offered by the genus *Deilephila*. The row of ring-spots which nearly all the existing species have more or less developed, has arisen from a simple subdorsal line. It would not, therefore, be surprising if a species were discovered which possessed this line without any ring-spots as its only marking. If *D. Hippophaës* were thus marked, there would be no objection to the theoretica lassumption that this [1] was the ancestor of the other species. It would then be said that ring-spots were first developed in a later species by natural selection, and that they had been transmitted to all succeeding and younger species.

Certain individuals of *D. Hippophaës*, however, possess small ring-spots, some of which are well developed on several segments. In this species the row of ring-spots is therefore comprised in the development. The remaining species, which are much younger phyletically than *Hippophaës*, could not have inherited their ring-spots from the latter, since this species itself only possesses them occasionally, and, so to speak, in a tentative manner. The spots would therefore appear to have arisen spontaneously in this species, and independently of those in the other species. But if this were the case, how should we be able to prove that in the other species also the ring-spots did not arise independently; and if, moreover, a large number of species showed the same character without its being referable to inheritance from a common ancestor, how could this be otherwise explained than as the result of a force innate in these species and producing similar variations? But this is nothing bnt Askenasy's "fixed direction of variation"—*i.e.*, a phyletic vital force.

[1] Or some other extinct analogously-marked species.

The only escape from this difficulty is perhaps to be found in proving that *D. Hippophaës* formerly possessed ring-spots, and that these have been subsequently either partially or completely lost, so that their occasional appearance in this species would therefore depend upon reversion. The ontogeny, however, teaches us that this is not the case, since the young caterpillar does not possess a greater number of more distinct ring-spots, but wants them altogether with the exception of a red spot on the eleventh segment, which is, however, much fainter than in the last stage.

This last-mentioned fact contains the solution of the problem. The premises from which this reasoning set out were all incorrect—the one red spot on the eleventh segment is likewise a ring-spot, and indeed the most important one of all, being primary, or the first to come into existence. Now all specimens, without exception, possess this first ring-spot, which is useful, and has therefore been called forth by natural selection; it is not inherited, but newly acquired by this species; at least, if the explanation of these spots which I have previously offered is correct.

The primary pair of spots may have been transferred from this to later species by heredity; and since, in all segmented animals there is a tendency for the peculiarities of one segment to be repeated on the others, this repetition must have occurred with greater frequency and more completely in the later species—the more so if the process were favoured by natural selection, *i.e.* if the row of ring-spots which originated in this manner could in any way be turned to the use of the species.

In *Hippophaës* itself there must also be a tendency to the formation of secondary ring-spots, and indeed in a number of specimens we actually see series of such ring-spots, the latter being present in varying numbers, and in very different states of development. The fact that the ring-spots have not become a constant and well-developed character, is simply explained by the circumstance that as such they would have endangered the existence of the species.

In this case there is therefore no necessity for assuming a phyletic vital force. The ring-spots of the genus *Deilephila* rather furnish us with an excellent explanation of a fact which might otherwise have been adduced in support of a phyletic vital force, viz., the strict uniformity in the development of larval markings.

Before I had been led to the discovery, by the study of the marking and development of *Hippophaës*, that the spots of the genus *Deilephila* originated on one segment only, from which they were transferred secondarily to the others, this astonishing regularity appeared to me an incomprehensible problem, which could only be solved by assuming a phyletic vital force. If it be attempted, for the ten species here considered, to construct a genealogical tree based on the supposition that it is the *rows of spots* which have been inherited in cases where they occur, and not the *mere tendency to their production* by the transference of the one originally inherited primary spot to the remaining segments, the attempt will fail. The greater number of the species would have to be arranged in one row, since one species always bears a perfected form of marking, which appears in the young stages of the following species. But it is very improbable that nine different species, derived directly the one from the other, would contemporaneously survive.[2] One species, *D. Vespertilio*, could not be inserted at all in the genealogical tree, since it wants one character which occurs in all the other species, viz., the caudal horn, which is absent even in the third stage, and must therefore have been lost at a very early period of the phyletic development, so that we may consider it to be on this account genetically allied to the oldest known form. But the markings of this larva pass through precisely the same stages of development as do those of the other species. Now if the ring-spots were inherited as such, the existence of a hornless species with ring-spots would be an insoluble riddle, and would favour the admission of parallel developmental series, which

[2] [See Darwin's remarks on the struggle for life being most severe between individuals and varieties of the same species "Origin of Species," 6th ed. p, 59. R.M.]

again could be scarcely otherwise explained than by a "fixed direction of variation." We have here one of that class of cases which the supporters of a phyletic vital force have already so often made use of in support of their view.

The explanation of such a case—*i.e.* its reference to known causes of species transformation—is never easy, and is indeed impossible without a precise knowledge of the ontogeny of many species, as well as of the original significance of the characters in question. In the case of the *Deilephila* larvæ, however, such knowledge is still wanting. It is true that they present us with parallel developmental series, but these do not depend on an unknown phyletic force—the parallelism can be referred to the action of the imperfectly known laws of growth innate in segmented organisms. Because the characters of one segment have a tendency to repeat themselves on the others, from one parent-form possessing ring-spots on one segment only, there may have proceeded several developmental series, all of which developed rows of such spots independently of each other.

From these considerations we may venture to construct the following genealogical tree :—

POSSIBLE GENEALOGICAL-TREE OF THE GENUS DEILEPHILA.

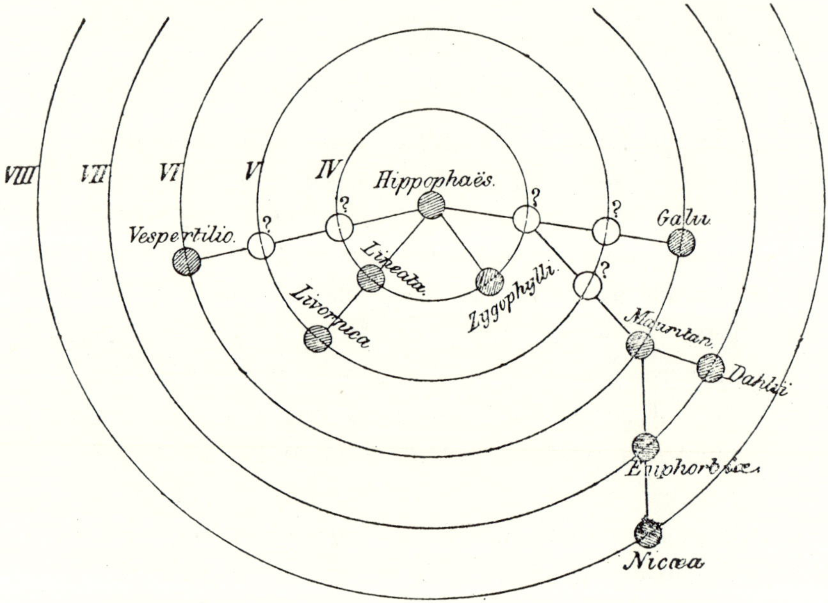

The circles indicate the phyletic stages IV.—VIII.; the eighth is only reached by *Nicæa*, and is distinguished from the seventh chiefly by the ontogeny, in the third stage of which the seventh phyletic stage is reached, whilst in *Euphorbiæ* and *Dahlii* this stage is reached in the fourth ontogenetic stage. The phyletic stages indicated by queries are extinct, and only known through the ontogeny of existing species. It must be understood that this pedigree expresses only the ideal and not the actual relations of the species to one another. Thus, it is possible that *Hippophaës* is not the parent-form, but an unknown or extinct species, which must, however, have possessed the same marking, and so on.

Four parallel series here proceed from the parent-form *Hippophaës;* there may have been five, or possibly only three, but the incomplete state of our knowledge of the ontogeny does not permit of any certain conclusion. For the point under consideration this is, however, quite immaterial. The distance from the central point (the parent-form) indicates the grade of phyletic development which the respective species have at present reached.

There is another case which is no less instructive, because it reveals, although in a somewhat different manner, the action of a law of growth innate in the organism itself, but which can nevertheless by no means be regarded as equivalent to a phyletic vital force. I refer to the coloured edges of the oblique stripes which occur in most of the species of the genus *Sphinx*. It has already been insisted upon in a previous section, that the mode in which this character originates negatives the assumption of a phyletic force, because these coloured edges are gradually built up out of irregularly scattered spots. There is no occasion for a "developmental force" to grope in the dark; if such a power exists, we should expect that it would add new characters to old ones with the precision of a master workman.

If, however, the coloured edges certainly depend on natural selection, this agency causing the scattered spots to coalesce and become linear, we have here the proof that such spots first arose in a precisely similar manner in several species, quite independently of one another—that, in fact, a "fixed direction of variation" in a certain sense exists.

In three species of *Smerinthus*-larvæ, red spots appear towards the end of the ontogeny; in *S. Populi* and *Ocellatus* in only a minority of individuals, and always separate (not coalescent), and in *S. Tiliæ* in a majority of specimens, the spots frequently becoming fused into one large, single, longish marking. These three species cannot have inherited the spots from a common ancestor, since they are absent in the younger ontogenetic stages, or occur only exceptionally, becoming larger and more numerous in the last stage; they obviously form a character which must be considered as a case of "anticipated development."

How is it then that three species vary independently of each other in an analogous manner? I know of no other answer to this question than that similar variations must necessarily arise from similar physical constitutions—or, otherwise expressed, the three species have inherited from an unknown parent species, devoid of spots, not this last character itself, but a physical constitution, having a tendency to the formation of red spots on the skin.[3] The case offers many analogies to

[3] [Compare this with Darwin's remarks on "analogous variations," "Origin of Species," 6th ed., p. 125. R.M.]

that of the colour varieties of *Lacerta Muralis*, to which Eimer[4] briefly calls attention in his interesting communications on the blue lizard of the Faraglioni Rocks at Capri. The South Italian lizards, although having differently formed skulls, show the same brilliantly coloured varieties as those of North Italy; and Eimer believes that these parallel variations in widely separated localities, some of which have long been isolated, must be referred to a tendency towards fixed directions of variation innate in the constitution of the species.

I long ago insisted[5] that it should not be forgotten that natural selection is, in the first place, dependent upon the variations which an organism offers to this agency, and that, although the number of possible variations may be very great for each species, yet this number is by no means to be considered as literally infinite. For every species there may be *impossible* variations. For this reason I am of opinion that the physical nature of each species is of no less importance in the

[4] "Zoologische Studien auf Capri. II. Lacerta muralis cærula, ein Beitrag zur Darwin'schen Lehre." Leipzig, 1874. [The subject of colour-variation in lizards has been much discussed in "Nature" since the publication of the above-mentioned essay; see vol. xix., pp. 4, 53, 97, and 122, and vol. xx., pp. 290 and 480. R.M.]

[5] "Über die Berechtigung der Darwin'schen Theorie." Leipzig, 1868. See also the previous essay "On the Seasonal Dimorphism of Butterflies," pp. 112—116.

production of new characters than natural selection, which must always, in the first place, operate upon the results of this physical nature, *i.e.* upon the variations presented, and can thus call new ones into existence.

It requires but a slight alteration of the definition to make out of this "restricted" or "limited variability," which is the necessary consequence of the physical nature of each species, a "fixed direction of variation" in the sense of a phyletic vital force. Instead of—the *Smerinthus*-larvæ show a tendency to produce red spots on the skin, it is only necessary to say—these larvæ tend to produce red borders to the oblique stripes. The latter statement would, however, be incorrect, since the red borders first arose by the coalescence of red spots through the action of natural selection. It is not even correct to say that *all* the species of *Smerinthus* show this tendency to produce spots, since this character does not seem to occur either in *S. Quercus* or *S. Tremulæ*.

The distinction between the two modes of conception will become clear if we ask, as an example, whether those *Chærocampa*-larvæ which do not at present possess eye-spots will subsequently acquire these markings, supposing that they maintain their existence on the earth for a sufficient period?

The supporters of a "fixed direction of variation" would answer this question in the affirmative. Ocelli constitute a character which occurs in nearly all the species of the group—they are the goal towards which the phyletic force is urging, and which must sooner or later be reached by each member of the group. On the other hand, I cannot express so decidedly my own opinion, viz., that such complicated characters as the many-coloured oblique stripes or eye-spots are never the results of purely internal forces, but always arise by the action of natural selection, *i.e.* by the combination of such minute and simple variations as may present themselves. It may be replied that the formation of eye-spots in those species which are at present devoid of them, cannot indeed be considered impossible, but that they would only appear if the constitution of these species had a tendency to give rise to the production of darker spots on the edge of the subdorsal line, and if, at the same time, the possession of eye-spots would be of use to the caterpillar under its special conditions of life.

The condition of affairs would be quite different if we were simply concerned with the transference of a character from one segment on which it was already present, to the remaining segments. The transference would, in this case, result from causes purely innate in the organism—from the action of laws of equilibration or of growth (correlation), and the external conditions of life would play only a negative part, since they might prevent the complete reproduction of a character, such, for example, as eye-spots, on all the segments, in cases where it was disadvantageous to the species. The fact that our species of *Chærocampa* have only faint indications, and not a completely-developed eye-spot, on the remaining segments, may perhaps be explained in this manner. It is conceivable that the two pairs of ocelli on the front segments are more effective as a means of alarm than if the insects were provided with two long rows of such markings; but nothing can be stated with certainty on this point until experiments have been made with caterpillars having rows of eye-spots.

The question raised above—whether the species of *Chærocampa* at present devoid of eye-spots are to be expected to acquire this character in the course of their further phyletic development—brings with it another point, which cannot be here passed over.

If the *utility* of the four kinds of markings in their perfected form is demonstrated, their origination through natural selection is not, strictly speaking, thereby proved. It must also be shown that the first rudiments of these characters were also of use to their possessors. The question as to the utility of the "initial stages" of useful characters must here be set at rest.

In the case of markings such as longitudinal and oblique stripes, it is quite evident that the initial stages of these simple characters do not differ greatly from the perfected marking, but this is certainly not the case with eye- and ring-spots. The most light is thrown upon this question by the latter, because a species which has remained at the initial stage of the formation of ring-spots here presents itself for examination, viz. *Deilephila Hippophaës*.

I have attempted to show that the orange-red spots, which, as a rule, adorn only the eleventh segment, enhance the adaptive colouring of this caterpillar by their resemblance to the berries of the sea-buckthorn, whilst the general surface resembles the leaves in colour. If this be admitted, the origination of these spots by natural selection offers no difficulty, since a smaller spot, or one of a fainter red, must also be of some use to its possessor.

This case is of importance, as showing that a "change of function" may occur in markings, just as it does in certain organs among the most diverse species of animals, in the course of phyletic development. The spots which in *Hippophaës* are imitations of red berries, in species which have further advanced phyletically play quite another part—they serve as means of alarm, or signals of distastefulness.

It appears to me very improbable, however, that the perfect ocelli of the *Chærocampa*-larvæ have also undergone such a "functional change" (Dohrn). I rather believe that the first rudiments of these markings produced the same effect as that which they now exercise, viz., terror. We are certainly not so favourably circumstanced in this case in knowing a species which shows the initial steps of this character in its last stage of life; but in the initial steps which the second stage of certain species present, we see preserved the form under which the eye-spots first appeared in the phylogeny, and from this we are enabled to judge with some certainty of the effect which they must have produced at the time.

In the ontogeny of *C. Elpenor* and *Porcellus* we see that a small curvature of the subdorsal line first arises, the concavity of which becomes filled with darker green, and soon afterwards with black; the upwardly curved piece of the subdorsal then becomes detached and more completely surrounded by black. The white fragment of the subdorsal which has become separated, in the next place broadens, and a black (dark) pupil appears in its centre.

Now the first rudiments of the eye-spot certainly appear very insignificant in a caterpillar two centimeters long, but we must not forget that in the ancestors of the existing *Chœrocampa*-larvæ this character appeared in the adult state. If we conceive the curvature of the white subdorsal with the underlying dark pigment to be correspondingly magnified, its importance as a means of alarm can scarcely be denied, particularly when we consider that this marking stands on the enlarged fourth segment, which alone invests the caterpillar with a singular, and, to smaller foes, an alarming appearance. We know that in the case of those *Chærocampa*-larvæ which possess no eye-spots, the distension of this segment is employed against hostile attacks. (See the illustration of *Darapsa Chærilus*, Pl. IV., Fig. 34.) Those markings which even only remotely resembled an eye must, in such a position, have increased the terrifying action. On these grounds I believe that it may be safely admitted, that this kind of marking possessed the same significance in its initial stages as it now does when fully perfected. No functional change has here taken place.

Among all the facts brought together in the first section I only know of one group of phenomena which at least permit of an attempt to refer them to a phyletic vital force. This is the occurrence of dark ground-colours in adult larvæ which are of light colours in their young condition. I have already attempted to show that in the *Chærocampa*-larvæ this change of colour depends on a double adaptation, the young caterpillars being adapted to the green colour of the plant and the adults to the soil and dead leaves. This interpretation appears the more correct when we find the same process, viz. the gradual replacement of the original green by brown colours, among species of widely different genera, which, with the dark colouring, possess the necessarily correlated habit of hiding themselves by day when in the adult condition. This is the case with *Sphinx Convolvuli*, *Deilephila Vespertilio*, and *Acherontia Atropos*.

Thus far all has been easily explicable by natural selection; but when we also see a "tendency" to acquire a dark colour in the course of development, in those species which neither conceal themselves nor are adaptively coloured, but are very conspicuously marked—and if, further, it can be shown that these species, such for instance as *Deilephila Galii*, actually possess immunity from the attacks of foes,—how can this tendency to the formation of a dark colour be otherwise explained than by the admission of a phyletic vital force urging the variations in this direction?

Nevertheless I believe that also on this point an appeal to unknown forces can be dispensed with. In the first place, dark ground-colours can be of use to a species otherwise than as means of adaptation. In *D. Galii*, as well as in *D. Euphorbiæ*, the light ring-spots appear rather at their brightest on the pitchy-black ground; and if this caterpillar must (*sit venia verbo!*) become conspicuous, this purpose would be best attained by acquiring a dark ground-colour, such as that of *D. Euphorbiæ*.

The tendency, apparently common to all these *Sphingidæ*, to acquire a dark colour with increasing age, depends therefore on two quite distinct adaptations—first, in species sought by enemies, on an adaptation to the colour of the soil; and secondly, in species rejected by foes, on the endeavour to produce the greatest possible contrast of colour.

Moreover, the supposition from which this last plea for a vital force set out is not universally correct, since there are species, such for instance as *D. Nicæa*, which never acquire a dark colour; and in *D. Galii* also, although all the individuals abandon the protective green of the young stages, they by no means all acquire a dark hue in exchange for this colour; many individuals in their light ochreous-yellow colouring rather strikingly resemble the snake-like caterpillar of *D. Nicæa*.

VII.

Phyletic Development of the Markings of the Sphingidæ: Summary and Conclusion.

If, from the form possessed by many of the caterpillars of the *Sphingidæ* on their emergence from the egg, we may venture to draw a conclusion concerning the oldest phyletic stage, these larvæ were originally completely destitute of marking. The characteristic caudal horn must be older than the existing markings, since it is present in the younger stages (except in cases where it is altogether wanting), and is generally even larger than at a later age.

There is, however, further evidence that there were once Sphinx-larvæ without any markings. Such a species now exists. I do not mean the boring caterpillars of the *Sesiidæ*,[1] which live in the dark, and are therefore colourless, but I refer to a

[1] [Mr. A. G. Butler has recently advanced the view that this family is not allied to the *Sphingidæ*, but is related on the one side to the *Pyrales*, and on the other to the *Gelechiidæ*. See his paper "On the Natural Affinities of the Lepidopterous Family *Ægeriidæ*," Trans. Ent. Soc. 1878, p. 121. R.M.]

large larva (over six centimeters long) preserved in spirit in the Berlin Museum,[2] which, from its form, belongs to the *Smerinthus* group. It possesses a caudal horn, and on the whole upper surface is covered with short and sparsely scattered bristles, such as occur in the *Sesiidæ*. The colour of this unknown insect appears to have been light green, although it now shows only a yellowish shade. Every trace of marking is absent, and it thus corresponds exactly with the youngest stages of the majority of the existing Sphinx-larvæ—even to the short bristles sparsely scattered over the whole upper surface of its body. We have therefore, so to speak, a living fossil before us, and it would be of great interest to ascertain its history.

All the data furnished by the developmental history go to show that of the three kinds of markings which occur in the *Sphingidæ*, viz., longitudinal and oblique stripes and spots, the first is the oldest. Among the species which are ornamented with oblique stripes or spots there are many which are longitudinally striped in their young stages, but the reverse case never occurs—young larvæ never show spots or oblique stripes when the adult is only striped longitudinally.

The first and oldest marking of the caterpillars of the *Sphingidæ* was therefore the longitudinal striping, or, more precisely speaking, the subdorsal,

[2] I am indebted to my esteemed colleague, Prof. Gestäcker, for the knowledge of this specimen.

to which dorsal and spiracular lines may have been added. That this second stage of phyletic development has also been preserved in existing species has already been sufficiently shown ; the greater portion of one group, the *Macroglossinæ*, has indeed remained at this stage of development.

From the biological value which must be attributed to this kind of marking, its origination by natural selection presents no difficulty. The first rudiments of striping must have been useful, since they must have broken up the large surface of the body of the caterpillar into several portions, and would thus have rendered it less conspicuous to its enemies.

Thus it is not difficult to perceive how a whole group of genera could have made shift with this low grade of marking up to the present time. Colour and marking are not the only means of offence and defence possessed by these insects; and it is just such simply-marked larvæ as those of the *Macroglossinæ* which have the protective habit of feeding only at night, and of concealing themselves by day. Moreover, under certain conditions of life the longitudinal stripes may be a better means of protection, even for a Sphinx-larva, than any other marking; and all those species in which this pattern is retained at the present time live either among grasses or on *Coniferæ*.

It cannot be properly said that the second form

of marking—the oblique stripes—has been developed out of the first. If these had arisen by the transformation of the longitudinal stripes, the two forms could not exist side by side. This is the case, however, both in certain species in the adult state (*Calymnia Panopus*[3]), as well as in others during their young stages (most beautifully seen in *Smerinthus Populi*, Fig. 56). Various facts tend to show that the oblique stripes appeared in the phyletic development *later* than the longitudinal lines. In the first place they appear later than the latter in the ontogeny of certain species. This is the case with *Chærocampa Elpenor* and *Porcellus*, in which, however, they certainly do not reach a high state of development. Then again, the longitudinal lines disappear completely in the course of the ontogeny, whilst the oblique stripes alone maintain their ground. Thus, the subdorsal line vanishes at a very early stage, with the exception of a small residue,[4] in all native species of *Smerinthus*. I have already attempted to show that new characters are only acquired *in the last stage*, and that if still newer ones are then added, the former disappear from the last stage, and are transferred back to a younger one. *Characters vanish therefore from a stage in the same order as they were acquired.*

[3] Cat. Lep. East India Co., Pl. VIII.

[4] Such a residue is distinctly visible in *S. Ocellatus*; see Fig. 70, Pl. VII.

Finally, among the genera with longitudinal stripes (*e.g. Macroglossa*) we know certain species which, when at an advanced age, possess oblique stripes (*M. Fuciformis*), although these slant in a direction opposite to those of most of the other larvæ of the *Sphingidæ.* These are, however, always species which differ from their allies in their mode of life, not feeding on grasses or low plants, but on large-leaved shrubs. If we were able to ascertain the ontogeny of these species, we should find that the oblique stripes appeared late in life, as has already been shown in the case of *Pterogon Œnotheræ.*

If it be asked why the longitudinal lines were first formed, and then the oblique stripes, it may be replied that the physical constitution of these caterpillars would be more easily able to give rise to simple longitudinal lines than to complicated oblique stripes crossing their segments.[5] It may

[5] [The question here also suggests itself as to why the *dorsal* line should not have been the primary longitudinal stripe, seeing that such a marking is almost naturally produced in many caterpillars by the food in the alimentary canal; or, in other words, why has not natural selection taken advantage of such an obvious means of producing a stripe in cases where it would have been advantageous? In answer to this I may state, that in large numbers of species the dorsal line has thus become utilized; but in the case of large caterpillars resting among foliage, it can be easily seen that light lateral (*i.e.* sub-dorsal) stripes, are more effective in breaking the homogeneity of the body than a dorsal line only slightly darker than the general ground-colour. Lateral lines are in fact visible from

perhaps also be suggested that the oldest *Sphingidæ* lived entirely on low plants among grasses, and in the course of time gradually took to shrubs and trees. At the present time the majority of the Sphinx-larvæ still live on low plants, and but few on trees, such caterpillars generally belonging to certain special genera.

The character of oblique stripes becomes perfected by the addition of coloured edges, the latter, as is self-evident, having been added subsequently.

The third chief constituent of the Sphinx-markings, *i.e.* the spots—whether perfect ocelli or only ring-spots—in two of the special genera here considered, arise on the subdorsal, where they are either deposited (*Deilephila*), or built up from a fragment of this line (*Chærocampa*). That these markings can, however, also originate independently of the subdorsal, is shown by the ocellus of *Pterogon Œnotheræ*, situated on the segment bearing the caudal horn. In this case, however, the ontogeny teaches us that the spot also succeeds the subdorsal, so that we can state generally that all these spot-markings are of later origin than the longitudinal striping.

two directions of space. If a caterpillar thus marked be placed on a twig, these lines are visible when we look at the creature's back or at either side. That the subdorsal are therefore the primary lines, as shown by Dr. Weismann's observations of the ontogeny of many of the *Sphingidæ*, is quite in harmony with the view of their having been produced by natural selection. R.M.]

The question as to the relative ages of the oblique stripes and the spot-marking does not admit of a general answer. In some cases (*C. Elpenor* and *Porcellus*) the oblique stripes disappear when the ocelli reach complete development, and we may therefore venture to conclude that in these cases the former appeared earlier in the phylogeny. But it is very probable that oblique stripes arose independently at different periods, just as longitudinal lines occur irregularly in quite distinct families. It would be a great error if we were to ascribe the possession of oblique stripes solely to descent from a common ancestor. The oblique markings found on certain species of *Macroglossa* (*M. Corythus* from India) have not been inherited from a remote period, but have been independently acquired by this or by some recent ancestral species. They have nothing to do *genetically* with the oblique stripes which occur in some species of *Chærocampa* (*e.g.* in *C. Nessus*, from India), or with those of the species of *Smerinthus* and *Sphinx*. They depend simply on analogous adaptation (Seidlitz [6]), *i.e.* on adaptation to an analogous environment.

The case is similar with the spot-markings. I have already shown that under certain conditions ring-spots may assume the exact appearance of

[6] "Die Darwin'sche Theorie. Elf Vorlesungen über die Entstehung der Thiere und Pflanzen durch Naturzüchtung." 2nd ed., Leipzig, 1875, p. 195.

eye-spots by the formation of a nucleus in the "mirror," such as occurs occasionally in *Deilephila Euphorbiæ* (Fig. 43), more frequently in *D. Galii*, and as a rule in *D. Vespertilio*. Nevertheless, these markings arise in quite another manner to the eye-spots of the *Chærocampinæ*, with which they conseqently have no genetic relation; the two genera became separated at a time when they neither possessed spot-markings. Further, in *Pterogon Œnotheræ* we find a third kind of spot-marking, which is most closely allied to the ocelli of the *Chærocampa*-larvæ, but is situated in quite another position, and must have originated in another manner, and consequently quite independently of these eye-spots.

It can also be readily understood why the first and second elements of the markings of the *Sphingidæ* should be mutually exclusive, and not the second and third or the first and third.

A light longitudinal line cutting the oblique stripes, considerably diminishes that resemblance to a leaf towards which the latter have a tendency, and it is therefore only found in cases where an adaptive marking can be of no effect on account of the small size of the caterpillar, *i.e.* in quite young stages. (See, for instance, Fig. 56, the first stage of *S. Populi.*) At a later period of life the old marking must give way to the new, and we accordingly find that the subdorsal line vanishes from all the segments on which oblique stripes are situated, and is only retained on the anterior segments where the latter are wanting. In some few cases both elements of marking certainly occur together, such as in *Calymnia Panopus* and *Macroglossa Corythus;* but the oblique stripes are, under these circumstances, shorter, and do not extend above the subdorsal line, and in *Darapsa Chœrilus* even become fused into the latter.[7]

In certain cases there may also be a special leaf structure imitated by the longitudinal lines, but on the whole the latter diminish the effect of the oblique stripes; and we accordingly find that not only has the subdorsal disappeared from those segments with oblique stripes, but that most larvæ with this last character are also without the otherwise broad spiracular and dorsal lines. This is the case with all the species of *Smerinthus*[8] known

[7] [In the following species, already mentioned in previous notes, the oblique stripes are bounded at their upper extremities by a conspicuous subdorsal line:—*Acosmeryx Anceus*, Cram.; *Sphinx Cingulata*, Fabr.; *Pachylia Ficus*, Linn.; *P. Syces*, Hubn. In *Pseudosphinx Cyrtolophia*, Butl., the oblique white stripes, beautifully shaded with pink, run into the white pink-bordered dorsal line, so that when seen from above the markings present the appearance of the midrib and lateral veins of a leaf, and are probably specially adapted for this purpose. R.M.]

[8] [The dorsal line as well as the oblique stripes is present in the caterpillar of *Smerinthus Tartarinovii*, Ménét.; and in *Ambulyx Gannascus*, Stoll., the oblique stripes are bounded above by a subdorsal line, as in the species named in the preceding note. R.M.]

to me, as well as with all the species of the genera *Sphinx*, *Dolba*, and *Acherontia*.

Oblique stripes and spot-markings are not, however, necessarily mutually exclusive in their action, and we also find these in certain cases united in the same larva, although certainly never in an equal state of perfection. Thus, *Chærocampa Nessus*[9] possesses strongly marked oblique stripes, but feebly developed ocelli; and, on the other hand, *Chærocampa Elpenor* shows strongly developed eye-spots, but the earlier oblique stripes are at most only present as faint traces. This is easily explained by the mode of life. These caterpillars—at least such of them as are perfectly known—do not live on plants with large, strongly-ribbed leaves, and are even in the majority of individuals adapted to the colour of the soil; the oblique stripes have therefore in these cases only the significance of rudimentary formations.

That the first and third forms of markings also are not always mutually prejudicial in their action is shown by the case of *Chærocampa Tersa,* in which the eye-spots certainly appear to possess some other significance than as a means of causing terror. In most of the *Chærocampa*-larvæ the subdorsal line disappears in the course of the phylogeny, and it can be understood that the illusive appearance of the eye-spots would be

[9] Cat. Lep. East India Co., Pl. XI.

more perfect if they did not stand upon a white line.

If we consider the small number of facts with which I have here been able to deal, the result of these investigations will not be deemed unsatisfactory. It has been possible to show that each of the three chief elements of the markings of the *Sphingidæ* have a biological significance, and their origin by means of natural selection has thus been made to appear probable. It has further been possible to show that the first rudiments of these markings must also have been of use; and it thus appears to me that their origin by means of natural selection has been proved to demonstration. Moreover, it has not been difficult to understand the displacement of the primary elements of the markings by secondary characters added at a later period, as likewise an essential effect of natural selection. Finally, it has been possible to explain also the subordinate or accessory elements of the markings, partly by the action of natural selection, and partly as the result of markings formerly present acting by correlation.

From the origin and gradual evolution of the markings of the *Sphingidæ* we may accordingly sketch the following picture:—

The oldest Sphinx-larvæ were without markings; they were probably protected only by adaptive colouring, and a large caudal horn, and by being armed with short bristles.

Their successors, through natural selection, became longitudinally striped; they acquired a subdorsal line extending from the horn to the head, as well as a spiracular, and sometimes also a dorsal, line. The caterpillars thus marked must have been best hidden on those plants in which an arrangement of parallel linear parts predominated; and we may venture to suppose that at this period most of the larvæ of the *Sphingidæ* lived on or among such plants (grasses).

At a later period oblique stripes were added to the longitudinal lines, the former (almost always) slanting across the seven hindmost segments from the back towards the feet in the direction of the caudal horn. Whether these stripes all arose simultaneously, or, as is more probable, whether only one at first appeared, which was then transferred to the other segments by correlation assisted by natural selection, cannot, at least from the facts available, at present be determined.

On the whole, as the oblique stripes became lengthened towards the back, the longitudinal lines disappeared, since they injured the deceptive effect of the stripes. In many species also there were formed dark or variegated coloured edges to the oblique stripes, in imitation of the shadow lines cast by the leaf-ribs.

Whilst one group of *Sphingidæ* (*Sphinx*, *Smerinthus*) were thus striving to make their external appearance approximate more and more to that of a ribbed leaf, others of the longitudinally striped species became developed in another manner.

Some of the latter lived indeed on bush-like leaved plants, but no oblique stripes were developed, because these would have been useless among the dense, narrow, and feebly-ribbed leaves of the food-plants. These caterpillars, from the earlier markings, simply retained the longitudinal lines, which, combined with a very close resemblance to the colour of the leaves, afforded them a high degree of protection against discovery. This protection would also have been enhanced if other parts of the food-plant, such as the berries (*Hippophaës*), were imitated in colour and position in such a manner that the large body of the caterpillar contrasted still less with its environment. In this way the first ring-spot probably arose in some species on only one—the penultimate segment.

As soon as this first pair of ring-spots had become an established character of the species, they had a tendency to become repeated on the other segments, advancing from the hind segments towards the front ones. Under certain conditions this repetition of the ring-spots might have been of great disadvantage to the species, and would therefore have been as far as possible prevented by natural selection (*Hippophaës*); in other cases, however, no disadvantage would have resulted—the caterpillar, well adapted to the colour of its food-plant, would not have been made more conspicuous by the small ring-spots, which might thus have become repeated on all the segments (*Zygophylli*). In cases like the two latter, striking colours must have been eliminated when inherited from an immediate ancestor; but on this point nothing can as yet be said with certainty.

In other cases the repetition of the ring-spots with strongly contrasted colours was neither prejudicial nor indifferent, but could be turned to the further advantage of the species. If a caterpillar fed on plants containing acrid juices (*Euphorbiaceæ*) which, by permeating its alimentary system, rendered it repulsive to other animals, the ring-spots commencing to appear (by repetition) would furnish an easy means for natural selection to adorn the species with brilliant colours, which would protect it from attack by acting as signals of distastefulness.

But if the dark spots stood on a light ground (*Nicæa*), they would present the appearance of eyes, and cause their possessors to appear alarming to smaller foes.

From the developmental histories and biological data at present before us, it cannot with certainty be said which of these two functions of the ring-spots was first acquired in the phylogeny, but we may perhaps suppose that their significance as a means of causing alarm was arrived at finally.

It may also be easily conceived that as the ring-spots became more and more complicated, they would occasionally have played other parts, being fashioned once again in these stages into imitations of portions of plants, such as a row of berries or flower-buds. For this, however, there is as yet no positive evidence.

As the ring-spots became detached from the subdorsal line out of which they had arisen, the latter disappeared more and more completely from the last ontogenetic stage, and receded towards the younger stages of life of the caterpillar—it became *historical.* This disappearance of the subdorsal may also be explained by the fact that the original longitudinal stripe imitating the linear arrangement of leaves would become meaningless, even if it did not always diminish the effect of the ring-spots. But characters which have become worthless are known in the course of time to become rudimentary, and finally to disappear altogether. I do not believe that disuse alone causes such characters to vanish, although in the case of active organs it may have a large share in this suppression. With markings it cannot, however, be a question of use or disuse—nevertheless they gradually disappear as soon as they become meaningless. I consider this to be the effect of the arrest of the controlling action of natural selection upon these characters (suspension of the so-called "conservative adaptation," Seidlitz). Any variations may become of value if the character concerned is met with in the necessary state of fluctuation. That this process of extinction does not proceed rapidly, but rather with extreme slowness, is seen in the ontogeny of several species of *Deilephila*, which retain the now meaningless subdorsal line through a whole series of stages of life.

In another group of Sphinx-larvæ with longitudinal stripes, an eye-spot became developed independently of the subdorsal line, in the position of the caudal horn, which has here vanished with the exception of a small knob-like swelling. This character—which we now see perfected in *Pterogon Œnotheræ*—undoubtedly serves as a means of causing terror; but whether the incipient stages possessed the same significance, cannot be decided in the isolated case offered by the one species of the genus *Pterogon* possessing this marking.

In a third group of longitudinally striped caterpillars, the younger genus *Chærocampa*, eye-spots were developed directly from portions of the subdorsal line, at first only on the fourth and fifth segments. It can be here positively asserted that this character served as a means of alarm from its very commencement. It is certainly for this reason that we see the subdorsal line in the immediate neighbourhood of the spots disappear at an early stage, whilst it is retained on the other segments for a longer period. A portion of the younger (tropical) species of this group then developed similar, or nearly similar, ocelli on the remaining segments by correlation; and it may now have occurred that in solitary cases the eye-spots acquired another significance (*C. Tersa* ?), becoming of use as a disguise by resembling berries or flower-buds. It is also conceivable that the eye-spots may in other cases have been converted into a warning sign of distastefulness.

In all those larvæ which possessed purely terrifying markings, however, not only was the original protective colouring preserved, but in most of them this colour gradually became replaced by a better one (adaptation of the adult larva to the soil). The oblique stripes imitating the leaf-ribs also are by no means lost, but are almost always present, although but feebly developed, and often only temporarily.

The pattern formed by the oblique stripes may also be retained, even with perfect adaptation to the soil, and may be converted to a new use by losing its sharpness, and, instead of imitating definite parts of plants, may become transformed into an irregular and confused marking, and thus best serve to represent the complicated lights and shadows, stripes, spots, &c., cast on the ground under low-growing plants from between the stems and dead leaves.

Just as in the case of ocellated species where caterpillars without eye-spots may retain and newly utilize their older markings, so larvæ having oblique stripes with the most diversely coloured edges may show the same markings in allied (younger?) species, both in a rudimentary and in a transformed condition. These markings may thus contribute to the formation of a latticed or reticulated pattern. Even the oldest marking, the subdorsal line, may still play a part, since its remnants cause certain portions of the complicated pattern to appear more strongly marked (*S. Convolvuli*). Finally, when an adaptation to a changing environment intersected by lights and shadows is required, new markings may be here added as in other cases, viz., dark streaks extending over the light surface of the whole caterpillar.

In concluding this essay, I may remark that, with respect to the wide and generally important question which gave rise to these investigations, a clearer and simpler result has been obtained than could have been expected, considering the complexity of the characters requiring to be traced to their causes, as well as our still highly imperfect knowledge of ontogenetic and biological facts.

For a long time I believed that it was not possible to trace all the forms of marking and their combinations to those causes which are known to produce transformation; I expected that there would be an inexplicable residue.

But this is not the case. Although it cannot yet be stated at first sight with certainty in every single instance how far any particular element of marking may have a biological value in the species possessing it, nevertheless it has been established that each of the elements of marking occurring in the larvæ of the *Sphingidæ* originally possessed a decided biological significance, which was produced by natural selection.

In the case of the three chief elements of the markings of the *Sphingidæ*, it can be further shown that not only the initial stages but also their ultimate perfection—the highest stages of their development, are of decided advantage to their possessors, and have a distinct biological value, so that the gradual development and improvement of these characters can be traced to the action of natural selection.

But although natural selection is the factor which has called into existence and perfected the three chief forms and certain of the subsidiary markings, in the repetition of the local character on the other segments, as well as in the formation of new elements of marking at the points of intersection of older characters now rudimentary, we can recognize a second factor which must be entirely innate in the organism, and which governs the uniformity of the bodily structure in such a manner that no part can become changed without exerting a certain action on the other parts—an innate law of growth (Darwin's " correlation ").

Only once during the whole course of the investigations was it for an instant doubtful whether a phyletic vital force did not make itself apparent, viz., in the red spots accompanying the oblique stripes in several *Smerinthus*-larvæ. Closer analysis, however, enabled us to perceive most distinctly the wide gulf that separates "analogous variation" from the mystic phyletic vital force. Nothing further remains therefore for the action of this force in respect to the marking and colouring of the *Sphingidæ*, since several even of the subordinate markings can be traced to their causes, only the "dorsal spots" of our two native species of *Chærocampa* having been referred to correlation without decided proof. From the temporary inability to explain satisfactorily such an insignificant detail, no one will, however, infer the existence of such a cumbrous power as a phyletic vital force.

The final result to which these investigations have led us is therefore the following:—The action of a phyletic vital force cannot be recognized in the marking and colouring of the *Sphingidæ;* the origination and perfection of these characters depend entirely on the known factors of natural selection and correlation.

II.

ON PHYLETIC PARALLELISM IN METAMORPHIC SPECIES.

INTRODUCTION.

In the previous essay I attempted to trace a whole group of apparently "purely morphological" characters to the action of known factors of transformation, to explain them completely by these factors, and in this manner I endeavoured to exclude the operation of an internal power inciting change (phyletic vital force).

In this second study I have attempted to solve the problem as to whether such an innate inciting power can be shown to exist by comparing the forms of the two chief stages of metamorphic species, or whether such a force can be dispensed with.

Nobody has as yet apparently entertained the idea of testing this question by those species which appear in the two forms of larva and imago (insects), or, expressed in more general terms, by those species the individuals of which successively possess quite different forms (metamorphosis), or in which the different forms that occur are distributed among different individuals alternating with and proceeding from one another (alternation of generation). Nevertheless, it is precisely here that quite distinct form-relationships would be expected according as the development of the organic world depended on a phyletic vital force, or was simply the response of the specific organism to the action of the environment.

Assuming the first to be the case, there must have occurred, and must still occur, what I designate "phyletic parallelism," *i. e.* the two stages of metamorphic species must have undergone a precisely parallel development—every change in the butterfly must have been accompanied or followed by a change in the caterpillar, and the systematic groups of the butterflies must be also found in a precisely corresponding manner in a systematic grouping of the caterpillars. If species are able to fashion themselves into new forms by an innate power causing periodic change, this re-moulding cannot possibly affect only one single stage of development—such as the larva only—but would rather extend, either contemporaneously or successively, to all stages—larva, pupa, and imago: each stage would acquire a new form, and it might even be expected that each would change to the same extent. At least, it cannot be perceived why a purely internal force should influence the development of one stage more than that of another. The larvæ and imagines of two species must differ from one another to the same extent, and the same must hold good for the larvæ and imagines of two genera, families, and so forth. In brief, a larval system must completely coincide with the system based entirely on imaginal characters, or, what amounts to the same thing, the form-relationships of the larvæ must correspond exactly with the form-relationships of the imagines.

On the other hand, the condition of affairs must be quite different if an internal power causing phyletic remodelling does not exist, the transformation of species depending entirely on the action of the environment. In this case dissimilarities in the phyletic development of the different stages of life must be expected, since the temporary, and often widely deviating, conditions of life in the two stages can and must frequently influence the one stage whilst leaving the other unacted upon—the former can therefore undergo remodelling while the latter remains unchanged.[1]

[1] [Compare this with Darwin's "Origin of Species" (1st. ed. p. 440), where it is stated that when an animal, during any part of its embryonic career, is active, and has to provide for itself, "the period of activity may come on earlier or later in life; but whenever it comes on, the adaptation of the larva to its conditions of life is just as perfect and beautiful as in the adult animal. From such special adaptations the similarity of the larvæ or active embryos of allied animals is sometimes much obscured." R.M.]

By this means there would arise an unequal difference between the two stages of two species. Thus, the butterflies, supposing these to have become changed, would bear a more remote form-relationship to each other than the caterpillars, and the differences between the former (imagines) would always be greater than that between the larvæ if the butterflies were, at several successive periods, affected by changing influences whilst the larvæ continued under the same conditions and accordingly remained unaltered. The two stages would not coincide in their phyletic development—the latter could not be expressed by parallel lines, and we should accordingly expect to find that there was by no means a complete congruity between the systems founded on the larval and imaginal characters respectively, but rather that the caterpillars frequently formed different systematic groups to the butterflies.[2]

Accordingly, the problem to be investigated was whether in those species which develope by means of metamorphosis, and of which the individual stages exist under very different conditions of life, a complete phyletic parallelism was to be found or not. This cannot be decided *directly* since we cannot see the phyletic development unfolded under our observation, but it can be

[2] [For Fritz Müller's application of this principle to the case of certain groups of Brazilian butterflies see Appendix III. to this Part. R.M.]

established *indirectly* by examining and comparing with each other the form-relationships of the two separate stages—by confronting the larval and imaginal systematic groups. If the phyletic development has been parallel and perfectly equal, so also must its end-results—the forms at present existing—stand at equal distances from one another; larval and imaginal systems must coincide and be *congruent*. If the course of the phyletic development has not been parallel, there must appear inequalities—incongruences between the two systems.

I am certain that systematists of the old school will read these lines wlth dismay. Do we not regard it as a considerable advance in taxonomy that we have generally ceased to classify species simply according to one or to some few characters, and that we now take into consideration not merely the last stage of the development (the imago), but likewise the widely divergent young stages (larva and pupa)? And now shall it not be investigated whether caterpillars and butterflies do not form quite distinct systems? In the case of new species of butterflies of doubtful systematic position was not always the first question:—what is the nature of the caterpillars? and did not this frequently throw light upon the relationships of the imago? Assuredly; and without any doubt we have been quite correct in taking the larval structure into consideration. But in so doing we should always keep in mind that there are two kinds of relationship—form- and blood-relationship—which might possibly not always coincide.

It has hitherto been tacitly assumed that the degree of relationship between the imagines is always the same as that between the larvæ, and if blood-relationship is spoken of this must naturally be the case, since the larva and the imago are the same individual. In all groups of animals we have not always the means of deciding strictly between form- and blood-relationship, and must accordingly frequently content ourselves by taking simply the form-relationship as the basis of our systems, although the latter may not always express the blood-relationship. But it is exactly in the case of metamorphic species that there is no necessity for, nor ought we to remain satisfied with, this mode of procedure, since we have here two kinds of form-relationship, that of the larvæ and that of the imagines, and, as I have just attempted to show, it is by no means self-evident that these always agree; there are indeed already a sufficient number of instances to show that such agreement does not generally exist.

This want of coincidence is strikingly shown in a group of animals widely remote from the Insecta, viz. the Hydromedusæ, the systematic arrangement of which is quite different according as this is based on the polypoid or on the medusoid generation. Thus, the medusoid family of the oceanic

Hydrozoa springs from polypites belonging to quite different families, and in each of these polypoid families there are species which produce *Medusæ* of another family.

Similarly, the larvæ of the Ophiuroidea (*Pluteus*-form) among the Echinodermata are not the most closely related in form to those of the ordinary star-fishes, but rather to the larvæ of quite a distinct order, the sea-urchins.

I will not assert that in these two cases the dissimilarity in the form-relationship, or, as I may designate it, the incongruence of the morphological systems, must depend on an unequal rate of phyletic development in the two stages or generations, or that this incongruence can be completely explained by the admission of such an unequal rate of development: indeed it appears to me probable that, at least in the *Ophiureæ*, quite another factor is concerned—that the form-relationship to the larvæ of the sea-urchins does not depend upon blood-relationship, but on convergence (Oscar Schmidt), *i. e.* on adaptation to similar conditions of life. These two cases, however, show that unequal form-relationship of two stages may occur.

From such instances we certainly cannot infer off-hand that a phyletic force does not exist; it must first be investigated whether and to what extent such dissimilarities can be referred to unequal phyletic development and, should this be the case, whether deviations from a strict congruence of the morphological systems are not compatible with the admission of an internal transforming power. *That a certain amount* of influence is exerted by the environment on the course of the processes of development of the organic world, will however be acceded to by the defenders of the phyletic vital force. It must therefore be demonstrated that deviations from complete congruence occur, which, from their nature or magnitude, are incompatible with the admission of innate powers, and, on the other hand, it must likewise be attempted to show that the departures from this congruence as well as the congruence itself can be explained without admitting a phyletic vital force.

In the following pages I shall attempt to solve this question for the order Lepidoptera, with the occasional assistance of two other orders of insects. Neither the Echinodermata nor the Hydromedusæ are at present adapted to such a critical examination; the number of species in these groups of which the development has been established with certainty is still too small, and their biological conditions are still to a great extent unknown. In both these respects they are far surpassed by the Lepidoptera. In this group we know a large number of species in the two chief stages of their development and likewise more or less exactly the conditions under which they exist during each of these phases. We are thus able to judge, at least to a certain extent, what changes in the conditions of life produce changes of structure. Neither in the number of known species of larvæ, nor in the intimate knowledge of their mode of life, can any of the remaining orders of insects compete with the Lepidoptera. There is no Dipterous or Hymenopterous genus in which ten or more species are so intimately known in the larval stage that they can be employed for the purposes of morphological comparison. Who is able to define the distinctions between the life-conditions of the larvæ of twenty different species of *Culex* or of *Tipula?* The caterpillars of closely allied species of Lepidoptera, on the other hand, frequently live on different plants, from which circumstance alone a certain difference in the life-conditions is brought about.

The chief question which the research had to reply to was the following:—Does there exist a complete phyletic parallelism among Lepidoptera or not? or, more precisely speaking:—Can we infer, from the form-relationships which at present exist between larvæ on the one hand and imagines on the other, an exactly parallel course of phyletic development in both stages; or do incongruences of form-relationship exist which point to unequal development?

Before I proceed to the solution of this question it is indispensable that one point should be cleared up which has not been hitherto touched upon, but which must be settled before the problem can be formally stated in general terms. Before it can be asked whether larvæ and imagines have undergone a precisely parallel development, we must know whether unequal development is possible—whether there does not exist such an intimate structural relationship between the two stages that every change in one of these must bring about a change in the other. Were this the case, every change in the butterfly would cause a correlative change in the caterpillar, and *vice versâ*, so that an inequality of form-relationship between the larvæ on one hand and the imagines on the other would be inconceivable—systems based on the characters of the caterpillars would completely coincide with those based on the characters of the butterflies and we should arrive at a false conclusion if we attributed the phyletically parallel development of the two stages to the existence of an internal phyletic force, whilst it was only the known factor, correlation, which caused the equality of the course of development.

For these reasons it must first be established that the larva and imago are not respectively fixed in form, and the whole of the first section will therefore be devoted to proving that the two stages change independently of one another. Conclusions as to the causes of change will then be drawn, and these will corroborate from another side a subsequent inquiry as to the presence or absence of complete congruence in the two morphological systems. The two questions the answers to which will be successively attempted are by no means identical, although closely related, since it is quite conceivable that the first may be answered by there being no precise correlation of form, or only an extremely small correlation, between the caterpillar and the imago, whilst, at the same time, it would not be thereby decided whether the phyletic development of the two stages had kept pace uniformly or not. A perfect congruence of morphological relationships could only take place if transformations resulted from an internal power instead of external influences. The question:—Does there exist a fixed correlation of form between the two stages? must therefore be followed by another:—Do the form-relationships of the two stages coincide or not—has their phyletic development been uniform or not?

STUDIES IN THE THEORY OF DESCENT

BY

DR. AUGUST WEISMANN
PROFESSOR IN THE UNIVERSITY OF FREIBURG

WITH NOTES AND ADDITIONS BY THE AUTHOR

TRANSLATED AND EDITED, WITH NOTES, BY

RAPHAEL MELDOLA, F.C.S.
LATE VICE-PRESIDENT OF THE ENTOMOLOGICAL SOCIETY OF LONDON

WITH A PREFATORY NOTICE BY

CHARLES DARWIN, LL.D., F.R.S.
Author of "The Origin of Species," &c.

VOL. II.

WITH EIGHT COLOURED PLATES

London
SAMPSON LOW, MARSTON, SEARLE, & RIVINGTON
CROWN BUILDINGS, 188, FLEET STREET
1882

I.

Larva and Imago vary in Structure independently of each other.

It would be meaningless to assert that the two stages above mentioned were *completely* independent of one another. It is obvious that the amount of organic and living matter contained in the caterpillar determines the size of the butterfly, and that the quantity of organic matter in the egg must determine the size of the emergent larva. The assertion in the above heading refers only to the structure; but even for this it cannot be taken as signifying an absolute, but only a relative independence, which, however, certainly obtains in a very high degree. Although it is conceivable that every change of structure in the imago may entail a correlative change of structure in the larva, no such cases have as yet been proved; on the contrary, all facts indicate an almost complete independence of the two stages. It is quite different with cases of *indirect* dependence, such, for example, as are brought about by 'nurse-breeding.' This phenomenon is almost completely absent in Lepidoptera, but is found in Diptera, and especially in Hymenoptera in every degree. The larvæ of ichneumons which live in other insects, require (not always, but in most instances) that the female imago should possess a sharp ovipositor, so that in this case also the structure and mode of life of the larva influences the perfect insect. This does not depend, however, on inherent laws of growth (correlation), but on the action of external influences, to which the organism endeavours to adapt itself by natural selection.

I will now let the facts speak for themselves.

It is shown by those species in which only one stage is di- or polymorphic that not every change in the one stage entails a corresponding change in the other. Thus, in all seasonally dimorphic species we find that the caterpillars of butterflies which are often widely different in the colour and marking of their successive generations are absolutely identical. On the other hand, many species can be adduced of which the larvæ are dimorphic whilst the imagines occur only in one form (compare the first and second essays in this volume).

There are however facts which directly prove that any one stage can change independently of the others; I refer to the circumstance that any one stage may become independently variable—that the property of greater variability or of greater constancy by no means always occurs in an equal degree in all the three stages of larva, pupa, and imago, but that sometimes the caterpillar is very variable and the pupa and imago quite constant. On the other hand, all three stages may be equally variable or equally constant, although this seldom occurs.

If variability is to be understood as indicating the period of re-modelling of a living form, whether in its totality or only in single characters or groups of characters, from the simple fact of the heterochronic variability of the ontogenetic stages, it follows that the latter can be modified individually, and that the re-modelling of one stage by no means necessarily entails that of the others. It cannot however be doubted that variability, from whatever cause it may have arisen, is in all cases competent to produce a new form. From the continued crossing of variable individuals alone, an equalization of differences must at length take place, and with this a new, although not always a widely deviating, constant form must arise.

That the different stages of development of a species may actually be partly variable and partly constant, and that the variable or constant character of one stage has no influence on the other stages, is shown by the following cases, which are, at the same time, well adapted to throw light on the causes of variability, and are thus calculated to contribute towards the solution of the main problem with which this investigation is concerned.

When, in the following pages, I speak of *variability*, I do not refer to the occurrence of local varieties, or to variations which occur in the course of time, but I mean a high degree of individual variability—a considerable fluctuation of characters in the individuals of one and the same district or of the same brood. I consider a species to be constant, on the other hand, when the individuals from a small or large district differ from one another only to a very slight extent. Constant forms are likewise generally, but not invariably, such as are poor in local varieties, whilst variable forms are those which are rich in such variations. Since the terms "variable" and "constant" are but relative, I will confine myself to the most extreme cases, those in which the individual peculiarities fluctuate within very wide or very narrow limits.

As no observations upon the degree of variability shown by a species in the different stages of its development were available, I was obliged to fall back upon my own, at least so far as relates to the larval and pupal stages, whilst for the imaginal stage the wide experience of my esteemed friend Dr. Staudinger has been of essential service to me.

Let us in the first place confine our attention to the three chief forms which every Lepidopteron presents, viz. larva, pupa, and imago. With respect to the constancy or variability of these three forms, we actually find in nature all the combinations which are theoretically conceivable.

(1.) There are species which possess a high degree of constancy in all three stages, such, for example, as *Limenitis Camilla*, *Pieris Brassicæ*,[1] *Sphinx Ligustri*, and *Euchelia Jacobææ*.

(2.) There are species showing a high degree of variability in all three stages. This case must be of rare occurrence, as I am only able to adduce *Araschnia Prorsa-Levana*, a fact which arises from the circumstance that the pupal stage is, as a rule, but seldom variable.

(3.) There are species which are variable in two stages and constant in the third. To this class, for example, belongs *Smerinthus Tiliæ*, of which the larva and imago are very variable, whilst the pupa is quite constant. The same is the case with *Lasiocampa Pini*, the well-known fir moth. Many butterflies show this same phenomenon in other combinations, such, for instance, as *Vanessa Urticæ* and *Polychloros*, in which the larva and pupa are very variable, and the imago very constant. In a less degree the same is also the case with *Vanessa Atalanta*, whilst in *Pieris Napi* the

[1] [The slight variability in the colour of this pupa, opens up the interesting question of the photographic sensitiveness of this and other species, which is stated to cause them to assimilate in colour to the surface on which the larva undergoes its final ecdysis. Some experiments upon this subject have been recorded by Mr. T. W. Wood, Proc. Ent. Soc. 1867, p. xcix, but the field is still almost unexplored. R.M.]

pupa and imago are variable, and the caterpillar remarkably constant, this likewise being the case with the local form *Bryoniæ*, which, according to my theory, is to be regarded as the parent form of *Napi* (See Part I. of the present volume).

(4.) There are species which are constant in two stages, and variable only in the third. Thus, a few species can be found in which the larva and pupa are constant and the imago variable. This is the case with *Saturnia Yamamai*, the imago of which is well known to present numberless shades of colour, varying from light yellow to greyish black, whilst the green caterpillar shows only slight individual differences of marking, and scarcely any differences of colour. The pupa of this species is quite constant. *Arctia Caja* and *Hebe*, and *Chelonia Plantaginis* belong to this same category.

There are a very large number of species which possess very constant imagines and pupæ, but extremely variable larvæ. The following are the cases known to me:—*Macroglossa Stellatarum*, *Fuciformis* and *Bombyliformis*; *Chærocampa Elpenor*, *Celerio*, and *Nerii*; *Deilephila Galii*, *Livornica*, Hübn., *Hippophaës*, *Vespertilio*, and *Zygophylli*; *Sphinx Convolvuli*; *Acherontia Atropos*; *Smerinthus Ocellatus* and *Tiliæ*; *Callimorpha Hera*; *Cucullia Verbasci* and *Scrophulariæ*.

Cases in which the variability depends entirely upon the pupa, while the larva and imago are extremely constant, are of great rarity. *Vanessa Io* is a case in point, the pupa being light or dark brown, or bright golden green, whilst in the two other stages scarcely any light shades of colour or variations in the very complicated marking are to be met with.

The facts thus justify the above view that the individual stages of development change independently—that a change occurring in one stage is without influence on the preceding and succeeding stages. Were this not the case no one stage could possibly become variable without all the other stages becoming so. Did there exist a correlation between larvæ, pupæ, and imagines of such a nature that every change in the larva entailed a corresponding change in the imago, as soon as a large number of larval characters became fluctuating (*i.e.* as soon as this stage became variable), a large number of imaginal characters would necessarily also become fluctuating (*i.e.* this stage would also become correspondingly variable).

There is one other interpretation which might perhaps be attempted from the point of view of the old doctrine of species. It might be said that it is a special property of certain larval or imaginal markings to be variable whilst others are constant, and since the larval and imaginal markings of a species are generally quite distinct, it may easily happen that a butterfly possessing markings having the property of constancy may belong to a caterpillar having variable markings.

There is a soul of truth underlying this objection, since it is true that the various forms of markings which occur in Lepidoptera apparently reach different degrees of constancy. If we speak of the constancy or variability of a species, a different meaning is attached to these expressions according as we are dealing *e.g.* with a species of *Sphinx* or a species of *Arctia*. That which in the latter would be estimated as a high degree of constancy, in the former would be taken as a considerable amount of variability. It is of interest, in connection with the question as to the causes of constancy, to note that the power of any form of marking to attain to a high degree of constancy is by no means inversely proportional to the complication of the marking, as would have been expected *à priori*.

Thus, the species of *Sphinx* and of allied genera possess on their fore-wings, which are mostly coloured with a mixture of dull grey, white and black, an exceedingly complicated arrangement of lines which, in constant species, show a high degree of uniformity : on the other hand, the checquered fore-wings of our *Arctiidæ*, which are far more coarsely marked, always show, even in the most constant species, well-marked individual differences. The different types of marking must therefore be measured by different standards.

But in granting this, we decidedly refute the statement that constancy and variability are inherent properties of certain forms of marking.

This reasoning is based on the simple fact that a given type of marking comprises both species of great constancy and of (relatively) great variability.

Thus, the fore-wings of *Sphinx Ligustri* and *S. Convolvuli* are extremely constant, whilst the very similarly marked *Anceryx* (*Hyloicus*) *Pinastri* is exceedingly variable. Similarly *Deilephila Euphorbiæ* is known by its great variability of colouring and marking, whilst *D. Galii*, which resembles this species so closely as to be sometimes confounded with it, possesses a high degree of constancy, and further, the Corsican and Sardinian *D. Dahlii* is very variable. Among the family *Arctiidæ*, *Callimorpha Hera* and the Alpine *Arctia Flavia* are cases of constancy, whilst *A. Caja*, which is so similar to the last species, is so generally variable that two perfectly identical specimens can scarcely be found together.

The same can be shown to hold good for the markings of caterpillars. Thus, the larva of *D. Dahlii* shows very considerable variability, whilst that of *D. Galii* is very constant in marking (disregarding the ground-colour). So also the larva of *Vanessa Urticæ* is very variable and that of *V. Antiopa* very constant, &c.

The great differences with respect to constancy or variability which are displayed by the different stages of one and the same species, must therefore find their explanation elsewhere than in the type of the marking itself. The explanation must be found in the circumstance that each stage changes independently of the others, and at different periods can enter a new phase of variability.

We are here led in anticipation to the main question :—Are changes produced by internal or external causes? is it the physical nature of the organism which is compelled to become remoulded spontaneously after the lapse of a certain period of time? or does such modification only occur when produced directly or indirectly by the external conditions of life?

In the cases before us the facts undoubtedly indicate a complete dependence of the transformations upon external conditions of life.

The independent appearance of variability in the separate stages of the metamorphosis might, however, be regarded as only apparent. It might still be attempted to attribute the changes to a purely inherent cause, *i.e.*, to a phyletic vital force, by assuming that the latter acts periodically in such a manner that at first one and then the following stage becomes variable, until finally the entire species is transformed.

There is but little to be said in reply to this if we once take refuge in entirely unknown forces, the operation of which can be arbitrarily conceived to be either constant or periodic.

But granting that such a transforming power

exists and acts periodically, the variability must always pass over the different stages in a fixed direction, like a wave over the surface of water—imago, pupa, and larva, or larva, pupa, and imago, must *successively* become variable. Cases like that of *Araschnia Prorsa*, in which all three stages are variable, may certainly be thus explained, but those instances in which the larva and imago are extremely variable, and the pupa quite constant, are entirely inexplicable from this point of view.

The latter can, however, be very simply explained if we suppose the changes to be dependent upon external influences. From this standpoint we not only see how it is possible that an intermediate stage should remain uninfluenced by the changes which affect the two other stages, but we can also understand why it should just be the pupal stage that plays this part so frequently. If we ask why most pupæ are constant and are relatively but very slightly variable, the answer will be found in the facts that all pupæ which remain concealed in the earth or inside plants (*Sesiidæ*), or which are protected by stout cocoons, show complete constancy, whilst any considerable amount of variability occurs only in those pupæ which are suspended or openly exposed. This is closely connected with a fact to which I have called attention on a former occasion,[2] viz., that

[2] "Über den Einfluss der Isolirung auf die Artbildung." Leipzig, 1872, p. 20.

dimorphism occurs in certain pupæ, but only in those which are openly exposed and which are therefore visible to their foes. I am only acquainted with such cases among the pupæ of butterflies, and it is likewise only among these that I have found any considerable amount of variability.

Facts of this kind indicate that Nature does not uselessly sport with forms, but that at any rate changes of this sort result from external influences. The greater frequency of variability among larvæ and its comparative rarity in imagines is also undoubtedly in favour of this view.

It has already been shown that species with variable larvæ and constant imagines are extremely common, but that those with constant larvæ and variable imagines are very rare. This confirms the conclusions, already drawn above, first, that the variability of the imago cannot owe its existence to the variability of the larvæ, and secondly, that the causes which produce variability affect the larval condition more commonly than that of the imago.

Where can these causes be otherwise sought than in the external conditions of life, which are so widely different in the two stages, and which are much more variable for the larva than for the imago?

Let us take the species of one genus, *e.g.* those of *Deilephila*. The imagines of our European species—as far as we know—all live in precisely the same manner; they all fly at twilight,[3] showing a preference for the same flowers and very often frequenting the same spots, so that in the haunts of one species the others are almost always to be met with, supposing them to occur in the same locality. They conceal themselves by day in similar places, and are attacked by similar foes.

It is quite different with the caterpillars. These, even in the case of the most closely allied species, live under different conditions, as appears from the fact that they feed on different plants. The latter can, however, produce changes both directly and indirectly. The larvæ may acquire adaptive colours and markings, and these would vary in accordance with the colour and structure of the food-plant; or they may become brightly coloured as a sign of distastefulness in cases where they are inedible. Then again the colour of the soil on which the larvæ live would act upon their colours making these adaptive. Certain habits of the caterpillars may also be dependent upon the nature of their food-plants. Thus, *e. g. Deilephila Hippophaës* feeds only at night, and conceals itself by day under moss and among the leaves at the base of the food-plant; but *D. Euphorbiæ* could not acquire such a habit, because *Euphorbia*

[3] In some instances *Deïlephila Lineata* has also been seen by day hovering over flowers.

Cyparissias generally grows on arid soil which is poor in vegetation, and which therefore affords no concealment, and furthermore, because a caterpillar, as long as it continues to feed, cannot, and as a matter of fact does not, ever wander far from its food-plant. A habit of concealment by burying in the earth also, such for example as occurs in *Acherontia Atropos*, could not be acquired by *D. Euphorbiæ*, because its food-plant generally grows on hard, dry, and stony ground.

In addition to these considerations, the foes would be different according as the caterpillar lived on plants which formed dense thickets covering large extents of the shore (*Hippophae*), or grew isolated on dry hillocks and declivities where the herbage was scanty or altogether absent; or again, according as the insect, in conjunction with such local differences, fed by day or had acquired the habit of feeding only by night. It must in fact be admitted that new and improved adaptations, or, in more general terms, that inducements to change, when depending on the environment, must be more frequently dissimilar for larvæ than for the imagines. We must accordingly expect to find actual change, or that condition of variability which may be regarded as initiative to change, occurring more commonly in larvæ than in perfect insects.

Since facts are in complete accordance with the results of these *à priori* considerations we may also venture to conclude that the basis of the considerations is likewise correct, viz., the supposition that the changes of colour and marking in caterpillars, pupæ, and imagines result from external influences only.

This must not be taken as signifying that the single stages of the larval development are also only able to change through the action of external influences. The larval stages are correlated with each other, as has already been shown (see the previous essay): new characters arise in the adult caterpillar at the last stage and are then gradually transferred back to the younger stages quite independently of external influences, this recession being entirely brought about by the laws of correlation. Natural selection here only exerts a secondary action, since it can accelerate or retard this transference, according as the new characters are advantageous or disadvantageous to the younger stages.

Now as considerable individual differences appear in the first acquisition of a new character with respect to the rapidity and completeness with which the individuals acquire such a character, the same must obtain for the transference of an improvement acquired in the last stage to the next younger stage. The new character would be acquired by different individuals in different degrees and at different rates—it would have, to a certain extent, to struggle with the older characters of the stage; in brief, the younger stage would become variable.

Variability of this kind might well be designated as *secondary*, in contradistinction to *primary* variability; the latter (primary) depends upon an unequal reaction of the individual organisms to external influences, the former (secondary) results from the unequal strength and rate of the action of the innate laws of growth governing the organism. In both cases alike exceeding variability may occur, but the causes producing this variability are dissimilar.

The different stages of larval development would thus frequently display independent variability in a manner similar to the pupal or imaginal stages, since they can show individual variability while the other stages of development remain constant. This appearance of independent variability in the different stages of the larval development, however, is in truth deceptive—we have here in fact a kind of wave of variability, which passes downwards through the developmental stages, becoming gradually weaker, and finally dying out completely.

In accordance with this, we very frequently find that only the last or two last stages are variable, while the younger stages are constant. Thus in *Macroglossa Stellatarum*, the larvæ are constant in the first, second, and third stages, but become variable in the fourth, and in the fifth stage first show that high degree of variability which has already been described in detail (See. Pl. III., Figs. 3—12).

The larvæ of *Vanessa Cardui* also, according to my notes, are extremely constant in the first four stages in spite of their complicated marking, but become variable in the fifth stage, although to no very great extent.

In *Smerinthus Tiliæ*, *Ocellatus* and *Populi* also, the greatest larval variability is shown only in the last stage, the preceding stages being very constant. These cases by no means depend upon the marking of the young stages being simpler and therefore being less capable of varying. The reverse case also occurs. In a somewhat similar manner as the young of the tapir and wild hog are striped, while the adult animals are plainly coloured, the young caterpillars of *Saturnia Yamamai* possess longitudinal black lines on a yellow ground, while as early as in the second stage a simple green colour appears in the place of this complicated but perfectly constant marking. If the young stages are so frequently constant, this rather depends upon the fact that the transference of a new character to these stages not only takes place gradually, but also with continually diminishing energy, in a manner somewhat similar to physical motion, which continually diminishes in speed by the action of resistance till it is completely arrested. This constancy of the younger stages may further be due to the circumstance that the characters would only be transferred when they had become fixed in the last stage, and were consequently no longer variable. The transferred characters may thus have acquired a greater regularity, *i. e.* a less degree of variability, than they possessed at their first origination. Extensive investigations in this special direction must be made if the precise laws, in accordance with which the backward transference of new characters takes place, are to be discovered. By such researches only should we arrive with certainty at the causes which determine the lesser variability of the young larval stages.

It may also occur that the early stages are variable, whilst the later stages are constant, although this case appears to happen less frequently. Thus, the caterpillars of *Gastropacha Quercifolia* vary considerably in the second stage but are constant at a later period, and the same is the case with *Spilosoma Urticæ*, which in the second stage may be almost considered to be dimorphic, but which subsequently becomes constant.

Cases in which the first stage is variable appear to be of the least frequent occurrence. I know of only one such instance, viz., *Anceryx Pinastri*, of which the newly-hatched larvæ (Pl. VI., Fig. 53) show considerable differences in the brownish-black crescentic spots. The second (Fig. 54), third, and fourth stages are then tolerably constant, while the fifth stage again is very variable.

An instance of this kind can be easily explained by two waves of variation, the first of which now affects only the first stage, while the second has just commenced to affect the fifth stage. Such a supposition is not opposed to any theoretical considerations, but rather has much probability in its favour, since we know that species are from time to time subject to be remodelled; and further, that the coalescence of several stages of phyletic development in the ontogeny of one and the same species (see p. 226, development of the genus *Deilephila*) shows that during the backward transference of one character, new characters may appear in the last stage of the ontogeny, and indeed very frequently at a time when the next youngest character has not been transferred back so far as to the first stage.

That this secondary variability is to a certain extent brought about by the conflict between the old and new characters, the latter striving to suppress the former, is shown by the caterpillar of *Saturnia Carpini* which I have observed for many years from this point of view, and than which I do not know a more beautiful illustration.

When these larvæ leave the egg they are black, but in the adult state are almost bright green —this at least being the case in a local form which, from the district in the vicinity of Genoa where it is found, I will designate as the var. *Ligurica*. Now whilst these two extreme stages of development are relatively constant, the intermediate stages show a variability which becomes greater the nearer the last stage is approached, this variation in the marking depending simply on the struggle between the green colour and the more anciently inherited black. In this manner there arises, especially in the fourth stage of the German local form, an incredible mixture of the most diverse markings, all of which can, however, be very easily explained from the foregoing point of view.

The simpler and, as I am inclined to believe, the older form of the transformation is presented to us in the local variety *Ligurica*. In the last stage, when 7.5 centimeters long, this form is of a beautiful bright green colour without any trace of black marking[4] (Pl. VIII., Fig. 77). The colour of the six orange warts which are situated on each segment is also similar in all specimens, so that this stage is perfectly constant.

Our German *S. Carpini* shows different characters in the fifth stage. It is true that individual specimens occur which are entirely green without any black, but these are rare; the majority possess a more or less broad black ring encircling the middle of each segment (Pl. VIII., Figs. 78 and 79). Those specimens in which the black ring has become broken up into large or small spots

[4] It is true that I only reared one brood, but from this fifty specimens were obtained. It would be interesting to know whether this variety of the caterpillar is distributed over the whole of Southern Europe.

surrounding the base of the warts constitute intermediate forms (Fig. 80). The last stage of the German local form, unlike that of the Genoese local form, is therefore very variable.

The two forms, moreover, do not simply differ in being more or less advanced in phyletic development, but also in several other points. As it is of great theoretical interest to show that a species can develop local differences only in the stage of larva, I will here subjoin the plain facts.

The differences consist in that the Genoese local form goes through five moults whilst the German local form, like most caterpillars, has only four moults. Further, in the Genoese form the light green, which is also possessed by the German form in the fourth stage, when it once appears, is retained to the end of the larval development, whilst in the fifth stage of the German form this colour is replaced by a dull greyish-green (compare Figs. 77 and 78). There is further a very considerable difference in the earlier stages which shows that the phyletic transforming process has taken a quite independent course in the two forms. Since the struggle between the green and black—retaining this idea—appears to be quite finished in the last stage of the Genoese form, we should expect that the new colour, green, would now also have encroached further upon the younger stages than in the German form. Nevertheless, this is not the case, but quite the reverse happens, the black maintaining its ground longer in the Italian than in the German form.

In the Genoese form the two first stages are completely black, and in the third stage an orange-yellow lateral stripe first appears. In the German form this stripe appears in the second stage, and there is not subsequently added, at least on the middle segments, a yellow border surrounding some of the warts of the median series. In the third stage, however, the yellow (which is but the precursor of the later green colour) becomes further extended, so that the caterpillars often appear of an orange colour, some or all of the warts and certain spots and stripes only being black (Figs. 66 and 68). The warts are also often yellow while the ground remains in most part black—in brief, the bright colour is in full struggle with the black, and an endless series of variations is the result of this conflict, whilst in the corresponding stage of the Genoese form almost complete constancy prevails.

This constancy remains also in the following (fourth) stage, the caterpillar still being deep black, only the yellow (sulphur-coloured) lateral stripe, which has now become brighter, indicating the impending change (Fig. 67). This takes place in the fifth stage, in which the ground-colour suddenly becomes bright green, the black remaining at most only in traces on the anterior edges of the segments.

This is the same marking as is shown by the fourth stage of the German form, only in this case individuals quite destitute of black do not occur. In many specimens indeed black forms the ground colour, the green only appearing in certain spots (Figs. 71 to 75); in others the green predominates, and these two extremes are connected by innumerable intermediate forms, so that this stage must be regarded as the most variable of all.

The sixth stage of the Genoese and the fifth of the German form have already been compared together. The results may be thus tabulated:—

A. German form.	*B. Genoese form.*
STAGE I. 9 days.	9 days.
Black; constant.	Black; constant.
STAGE II. 8 days.	11 days.
Black, with orange-yellow lateral stripe; variable.	Black; constant.
STAGE III. 5 days (in some cases as much as 16 days).	12 days.
Black, with yellow; very variable.	Black, with orange-yellow lateral stripes; constant.
STAGE IV. 16 days (in some cases only 5 days).	6 days.
Bright green and black, mixed; very variable.	Black, with bright yellowish lateral stripe; constant.
STAGE V. 6 days (frequently longer).	6 days.
Dark green, with or without black bands; variable.	Bright green, small traces of black; variable.
STAGE VI. Pupation.	18 days.
	Bright green, without any black; constant.
	STAGE VII. Pupation.

From this comparison we perceive that the process of transformation has at least become preliminarily concluded in the Genoese form. Why the backward transference of the newly-acquired character to the young stages has not yet occurred, or, at least, why it is not in progress, does not appear ; neither can it be stated whether this will take place later, although we may venture to suppose that such will be the case. At first sight but a relatively short time appears necessary for the single stage V., which is still in a state of fluctuation (variable), to become constant by continued crossing, like all the other stages.

That the transformation is still in full progress in the German form, is shown by the fact that in this case all the stages are variable with the exception of the first—the second stage being only variable to a small extent, the third to a much greater extent, and the fourth to the highest degree conceivable, whilst the fifth and last stage is again less variable—so that the greatest struggle between the old and new characters takes place in the fourth stage.

Among the innumerable variations presented by this last stage a complete series of transitional forms can be arranged so as to show the gradual conquest of the black by the green, and thus indicating, step by step, the course which the latter colour has taken.

In the blackest specimens there is nothing

green but the lateral (infra-spiracular) line which was yellow in the preceding stage, and a crescent-shaped streak at the base of the middle warts together with a still smaller crescent at the base of the upper warts (Figs. 71 and 81). These spots become extended in lighter specimens and approximate so as to leave only narrow black bridges, a third spot being added at the posterior edge of the warts (Figs. 72 and 82). The three spots then extend on all sides, still leaving for a long period narrow black lines at the boundaries where their growth has caused them to abut. In this manner there frequently arises on the green ground a true hieroglyphic-like marking (Figs. 85 and 86). Finally the black disappears from the anterior edge and diminishes on the middle line of the back where it still partly remains as a T-shaped figure (Figs. 73 and 74), although generally replaced elsewhere by the green with the exception of small residues.

One point remained for a long time inexplicable to me, viz., the change of the light green into dark grey-green which appeared in the last stage in connection with a total change of the black marking.

Supposing that new characters are actually acquired only in the last stage, and that from this they are transferred to the younger stages, we should expect to find completely developed in the last stage the same colouring and markings as are possessed more or less incompletely in the fourth stage. Now since the developmental tendency to the removal of black and to the predominance of green—if we may thus venture to express it—is obvious in the fourth stage, we may expect to find in the fifth stage a bright green ground-colour, either without any mixture of black or with such black spots and streaks as were retained in the fourth stage as residues of the original ground-colour. But instead of this the fifth stage shows a dark green colour, and a more or less developed black marking which cannot in any way be derived from that of the fourth stage.

The Genoese local form observed last year first gave me an explanation to the extent that in this form the last stage is actually only the potential penultimate stage, or, more correctly expressed, that the same characters which at present distinguish the last stage of this form, are already more or less completely transferred to the penultimate stage.

The apparently paradoxical behaviour of the German form can be explained by supposing that before the pure bright green had become completely transferred to the penultimate stage a further change appeared in the last stage, the green ground-colour becoming darker, and black transverse bands being formed. The marking of the last stage would then be regarded as the reverse of that of the preceding stage; the absence of black would be the older, simple black spots at the base of the warts the next in succession, and a connected black transverse band the most advanced state of the development.

Whether this explanation is correct, and if so, what causes have produced the second change, may perhaps be learnt at some future time by a comparison with the ontogeny of other *Saturniidæ;* in the meantime this explanation receives support from another side by the behaviour of the Genoese local form. If the last stage of the German form has actually commenced to be again re-modelled, then this variety is further advanced in phyletic development than the Genoese form ; and this corresponds entirely with the theory that in the former the light colour (the orange considered as preliminary to the transformation into green) has already been carried down into the second stage, whilst in the Genoese variety even in the fourth stage only the first rudiments of the colour-transformation show themselves.

The Genoese form is to a certain extent intermediate between the German form of *Saturnia Carpini* and the nearly related *S. Spini*, a species inhabiting East Germany. In this latter the larvæ, even in the adult state, are completely black with yellow warts. This form of caterpillar must therefore be regarded as phyletically the oldest, and this very well agrees with the character of the moth, which differs essentially from *S.*

Carpini only in not being sexually dimorphic. In *Carpini* the male possesses a far more brilliant colouring than the female, the latter agreeing so completely with the female of *Spini* that it can hardly be distinguished therefrom, especially in the case of the somewhat larger South European specimens of the last species. Now as the more simple colouring of the female must in any case be regarded as the original form, we must consider *Spini*, both sexes of which possess this colouring, to be phyletically the older form, and *Carpini*, the male of which has become differently coloured, must be considered as the younger type. This completely accords with the characters of the larvæ.

I must here mention that I have also asked myself the question whether the variations of the different larval stages are connected together as cause and effect—whether the lightest specimens of the fifth stage may perhaps not also have been the lightest individuals of the third and fourth stages.

Such relationship is only apparent between the third and fourth stages ; the darkest larvæ of the third stage become the darker varieties of the fourth stage, although it is true that the lighter forms of the third sometimes also become dark varieties in the fourth stage. Between the fourth and fifth stages there is scarcely any connection of this kind to be recognized. Thus, the darkest varieties of the fourth stage sometimes become the lightest forms of the fifth stage, whilst in other cases from the lightest individuals of the fourth stage there arise all the possible modifications of the fifth stage. Further details may be omitted: the negative result cannot cause any surprise, as it is a necessary consequence of the continued crossing that must take place.

We thus see that the three chief stages of development (larva, pupa, and imago) actually change in colour independently of each other, the single stages of the larval development being however in greater dependence upon one another, and being connected indeed in such a manner that a new character cannot be added to the last stage without being transferred in the course of time to the preceding stage, and at a later period from this again even to the youngest stage, supposing it not to be previously delayed in the course of its transference by unknown opposing forces. On this last point, however, the facts at present available do not admit of any certain decision.

But why do the individual larval stages behave in this respect so very differently to the chief stages of the whole development? why are the former so exactly correlated whilst the latter are not? If new characters have a general tendency to become transferred to the younger ontogenetic stages, why are not new imaginal characters first transferred to the pupa, and finally to the larva?

The answer to these questions is not far to find. The wider two stages of a species differ in structure, the less does correlation become possible; the nearer the two stages are morphologically related, the more powerful does the action of correlation become. It is readily conceivable that the more widely two succeeding stages deviate in structure and mode of life, the less possible does it become for characters to be transferred from one to the other. How is it possible, for example, that a new character in the proboscis or on the wings of a butterfly can be transferred to the caterpillar? If such correlation existed it could only manifest itself by some other part of the caterpillar changing in correspondence with the change of the proboscis or wings of the butterfly. That this is not the case has, in my opinion, been conclusively shown by all the foregoing considerations respecting the independent variability of the chief stages of the metamorphosis.

There are, moreover, an endless number of facts which prove the independence of the individual stages of development—I refer to the multitudinous phenomena presented by metamorphosis itself. The existence of that form of development which we designate as metamorphosis is alone sufficient to prove incontestibly that the single stages are able to change independently of one another to a most remarkable extent.

If we now ask the question: how has the so-called "complete" metamorphosis of insects arisen? the answer can only be: through the gradual adaptation of the different stages of development to conditions of life which have continually deviated more and more widely from each other.[5]

But if individual stages of the post-embryonic development can finally attain to such complete diversity of structure as that of the larva and imago through gradual adaptations to continually diverging conditions of life, this shows that the characters acquired by the single stages are always only transferred to the same stages of the following generation, whilst the other stages remain uninfluenced thereby. This depends upon that form of heredity designated by Darwin "inheritance at corresponding periods of life," and by Haeckel "homochronic heredity."

[5] In this sense Lubbock says:—"It is evident that creatures which, like the majority of insects, live during the successive periods of their existence in very different circumstances, may undergo considerable changes in their larval organization in consequence of forces acting on them while in that condition; not, indeed, without affecting, *but certainly without affecting to any corresponding extent*, their ultimate form."—"Origin and Metamorphoses of Insects," London, 1874, p. 39.

II.

Does the Form-relationship of the Larva coincide with that of the Imago?

Having thus established the independence in the variability of the individual stages of metamorphosis, I will now turn to the consideration of the question as to how far a parallelism is displayed in the phyletic development of these stages. Is there a complete congruence of form-relationship between larvæ on the one hand and imagines on the other? does the classification founded on the morphology of the imagines agree with that based on the morphology of the larvæ or not?

If, according to Claus,[1] we divide the order Lepidoptera into six great groups of families, it is at once seen that these groups, which were originally founded exclusively on imaginal characters, cannot by any means be so clearly and sharply defined by the larval characters.

This is certainly the case with the *Geometræ*, of which the larvæ possess only ten legs, and on

[1] "Grundzüge der Zoologie," 1875.

this account progress with that peculiar "looping" movement which strikes even the uninitiated. This group, which is very small, is however the only one which can be founded on the morphology of the larvæ; it comprises only two nearly related families (*Phytometridæ* and *Dendrometridæ*), and it is not yet decided whether these should not be united into one group comprising the family characters of the whole of the "loopers."

Neither the group of Micro-lepidoptera, nor those of the *Noctuina*, *Bombycina*, *Sphingina*, and *Rhopalocera*, can be based systematically on larval characters. Several of these groups are indeed but indistinctly defined, and even the imagines present no common characteristics by which the groups can be sharply distinguished.

This is well shown by the *Rhopalocera* or butterflies. These insects, in their large and generally brilliantly coloured wings, which are usually held erect when at rest, and in their clubbed antennæ, possess characters which are nowhere else found associated together, and which thus serve to constitute them a sharply defined group.[2] The caterpillars, however, show a quite different state

[2] [Lepidopterists are of course aware that even these distinctions are not absolute, as no single character can be named which does not also appear in certain moths. The definition in this case, as in that of most other groups of animals and plants, is only a general one. See, for instance, Westwood's "Introduction to the Classification of Insects," vol. ii. pp. 330—332. Also some remarks by C. V. Riley in his "Eighth

of affairs. Although the larval structure is so characteristic in the individual families of butterflies, these "larval-families" cannot be united into a larger group by any common characters, and the "*Rhopalocera*" would never have been established if only the larvæ had been known. It is true that they all have sixteen legs, that they never possess a Sphinx-like horn, and that they are seldom hairy, as is the case with many *Bombycidæ*,[3] but these common *negative* characters occur also in quite distinct groups.

In the butterflies, therefore, a perfect congruence of form-relationship does not exist, inasmuch as the imagines constitute one large group of higher order whilst the larvæ can only be formed into families. If it be admitted that the common characters of butterflies depend on their derivation from a common ancestor, the imagines must have retained certain common characters which enable them to be recognized as allies, whilst the larvæ have preserved no such characters from the period at which the families diverged.

Without going at present into the causes of these phenomena I will pass on to the considera-

Annual Report" on the insects of Missouri, 1876, p. 170. With reference to the antennæ as a distinguishing character, see Mr. A. G. Butler's article in "Science for All," 1880, part xxvii. p. 65. R.M.]

[3] The genus of *Morphinæ*, *Discophora*, possesses hairs very similar to those of the genus *Cnethocampa* belonging to the *Geometræ*.

tion of further facts, and will now proceed to investigate both the form-relationships within the families. Here there can be no doubt that in an overwhelmingly large majority of cases the phyletic development has proceeded with very close parallelism in both stages; larval and imaginal families agree almost completely.

Thus, under the group *Rhopalocera* there is a series of families which equally well permit of their being founded on the structure of the larva or on that of the imago, and in which the larvæ and imagines therefore deviate from one another to the same extent. This is the case, for instance, with the families of the *Pieridæ*, *Papilionidæ*, *Danaidæ*, and *Lycænidæ*.

But there are also families of which the limits would be very different if the larvæ were made the basis of the classification instead of the butterflies as heretofore. To this category belongs the subfamily *Nymphalinæ*. Here also a very characteristic form of caterpillar indeed prevails, but it does not occur in all the genera, being replaced in some by a quite different form of larva.

In the latest catalogue of Diurnal Lepidoptera, that of Kirby (1871), 112 genera are comprised under this family. Of these most of the larvæ possess one or several rows of spines on most or on all the segments, a character which, as thus disposed, is not met with in any other family,

This character is noticeable in genera 1 to 90,

if, from those genera of which the larvæ are known, we may draw a conclusion with reference to their allies. I am acquainted with larvæ of genus 2, *Agraulis*, Boisd. (*Dione*, Hübn.); of genus 3, *Cethosia*, Fabr.; 10, *Atella*, Doubl.; 12, *Argynnis*, Fabr.; 13, *Melitæa*,[4] Fabr.; 19, *Araschnia*, Hübn.; 22, *Vanessa*, Fabr.; 23, *Pyrameis*, Hübn.; 24, *Junonia*, Hübn.; 31, *Ergolis*, Boisd.; 65, *Hypolimnas*, Hübn. (*Diadema*, Boisd.); 77, *Limenitis*, Fabr.; 81, *Neptis*, Fabr.; 82, *Athyma*, Westw.; and finally with those of genus 90, *Euthalia*, Hübn.—which, according to Horsfield's figures, possess only two rows of spines, these being re-

[4] [The larvæ of genera 14, *Phyciodes*, and 35, *Crenis*, are likewise spiny. See Edwards' "Butt. of N. Amer." vol. ii. for figures of the caterpillar of *Phyc. Tharos;* for notes on the larvæ of *Crenis Natalensis* and *C. Boisduvali* see a paper by W. D. Gooch, "Entomologist," vol. xiv. p. 36. The larvæ of genus 55, *Ageronia*, are also spiny. (See Burmeister's figure of *A. Arethusa*, "Lép. Rép. Arg." Pl. V. Fig. 4). The larvæ of genus 98, *Aganisthos*, also appear to be somewhat spiny (see Burmeister's figure of *A. Orion*, *loc. cit.* Pl. V. Fig. 6), and this raises the question as to whether the genus is correctly located in its present position. The larvæ of the following genera figured in Moore's "Lepidoptera of Ceylon," parts i. and ii., are all spiny:—6, *Cirrochroa* (Pl. XXXII); 7, *Cynthia* (Pl. XXVI.); 27, *Kallima* (Pl. XIX.); and 74, *Parthenos* (Pl. XXIV). Many species of caterpillars which are spiny when adult appear to be spineless, or only slighty hairy when young. See Edwards' figures of *Melitæa Phaeton*, *Argynnis Diana*, and *Phyc. Tharos* (*loc. cit.*) and his description of the larva of *Arg. Cybele*, "Canad. Entom." vol. xii. p. 141. The spiny covering thus appears to be a character acquired at a comparatively recent period in the phyletic development. R.M.]

markably long and curved, and fringing both sides. It may be safely assumed that the intermediate genera would agree in possessing this important character of the Nymphalideous larvæ, viz., spines.

After the genus 90 there are 22 more genera, and these are spineless, at least in the case of the two chief genera, 93, *Apatura*, and 104, *Nymphalis*. Of the remainder I know neither figures nor descriptions.[5] In the two genera named the larvæ are provided with two or more spine-like tentacles on the head, and the last segment ends in a fork-like process directed backwards. The body is otherwise smooth, and differs also in form from that of the larvæ of the other *Nymphalinæ*, being thickest in the middle, and tapering anteriorly and posteriorly; neither is the form cylindrical, but somewhat flattened and slug-shaped. If therefore we were to arrange these butterflies by the larvæ instead of by the imagines, these two genera and their allies would form a distinct family, and could not remain associated with the 90 other Nymphalideous genera.

We have here a case of *incongruence;* the

[5] [The larvæ of the 110th genus, *Paphia*, Fabr. (*Anæa*, Hübn.) are also smoothed-skinned. See Edwards' figure (*loc. cit.* vol. i. Pl. XLVI.) of *P. Glycerium.* Also C. V. Riley's "Second Annual Report" on the insects of Missouri, 1870, p. 125. Burmeister figures the larva of a species of *Prepona* (genus 99) which is smooth (*P. Demophon, loc. cit.* Pl. V. Fig. 1). The horns on the head of *Apatura*, &c., may possibly be a survival from a former spiny condition. R.M.]

imagines of the genera 1—90 and 91—112 are more closely allied than their larvæ.

From still another side there arises a similar disagreement. The larvæ of the genera *Apatura* and *Nymphalis* agree very closely in their bodily form and in their forked caudal appendage with the caterpillars of another sub-family of butterflies, the *Satyrinæ*, whilst their imagines differ chiefly from those of the latter sub-family in the absence of an enlargement of certain veins of the fore-wings, an essential character of the *Satyrinæ*.

This double disagreement has also been noticed by those systematists who have taken the form of the caterpillar into consideration. Thus, Morris[6] attempted to incorporate the genera *Apatura* and *Nymphalis* into the family *Libytheidæ*, placing the latter as transitional from the *Nymphalidæ* to the *Satyridæ*. But although the imagines of the genera *Apatura*, *Nymphalis*, and *Libythea* may be most closely related—as I believe they actually are—the larvæ are widely different, being at least as different as are those of *Apatura* and *Nymphalis* from the remaining *Nymphalinæ*.

Now if we could safely raise *Apatura* and *Nymphalis* into a distinct family—an arrangement which in the estimation of Staudinger[7] is correct—

[6] "Synopsis of the described Lepidoptera of North America." Washington, 1862.

[7] "Catalog der Lepidopteren des Europäischen Faunengebietes." Dresden, 1871.

and if this were interpolated between the *Satyridæ* and *Nymphalidæ*, such an arrangement could only be based on the larval structure, and that of the imagines would thus remain unconsidered, since no other common characters can be found for these two genera than those which they possess in common with the other Nymphalideous genera.

The emperor-butterflies (*Apatura*), by the ocelli of their fore-wings certainly put us somewhat in mind of the *Satyrinæ*, in which such spots are always present; but this character does not occur in the genus *Nymphalis*, and is likewise absent in most of the other genera of this group. The genus *Apatura* shows in addition a most striking similarity in the markings of the wings to the purely Nymphalideous genus *Limenitis*, and it is therefore placed, by those systematists who leave this genus in the same family, in the closest proximity to *Limenitis*. This resemblance cannot depend upon mimicry, since not only one or another but *all* the species of the two genera possess a similar marking; and further, because similarity of marking alone does not constitute mimicry, but a resemblance in colour must also be added. The genus *Limenitis* actually contains a case of imitation, but in quite another direction; this will be treated of subsequently.

It cannot therefore be well denied that in this case the larvæ show different relationships to the imagines.

If the "natural" system is the expression of the genetic relationship of living forms, the question arises in this and in similar cases as to whether the more credence is to be attached to the larvæ or to the imagines—or, in more scientific phraseology, which of the two inherited classes of characters have been the most distinctly and completely preserved, and which of these, through its form-relationship, admits of the most distinct recognition of the blood-relationship, or, inversely, which has diverged the most widely from the ancestral form? The decision in single instances cannot but be difficult, and appears indeed at first sight impossible; nevertheless this will be arrived at in most cases as soon as the ontogeny of the larvæ, and therewith a portion of the phylogeny of this stage, can be accurately ascertained.

As in the *Rhopalocera* most of the families show a complete congruence in the form-relationship of the caterpillars and perfect insects, so a similar congruence is also found in the majority of the families belonging to other groups. Thus, the two allied families of the group *Sphingina* can also be very well characterized by their larvæ;[8] both the

[8] This group of moths ("Schwärmer") is regarded as of very different extents by systematists; when I here comprise under it only the *Sphingidæ* proper and the *Sesiidæ*, I by no means ignore the grounds which favour a greater extension of the group; the latter is not rigidly limited. [The affinities of the *Sesiidæ* (*Ægeriidæ*) are by no means clearly made out it

Sphingidæ and the *Sesiidæ* possess throughout a characteristic form of larva.

Of the group *Bombycina* the family of the *Saturniidæ* possess thick cylindrical caterpillars, of which the segments are beset with a certain number of knob-like warts. It is true that two genera of this family (*Endromis* and *Aglia*) are without these characteristic warts, but the imagines of these genera also show extensive and common differences from those of the other genera. A distinct family has in fact already been based on these genera (*Endromidæ*, Boisd.). Thus the congruence is not thereby disturbed.

So also the families *Liparidæ*, *Euprepiidæ*, and *Lithosiidæ* appear sharply defined in both forms; and similar families occur likewise under the *Noctuina*, although in this group the erection of families presents great difficulties owing to the near relationship of the genera, and is always to some extent arbitrary. It is important, however, that it is precisely the transitional families which present intermediate forms both as larvæ and as imagines.

Such an instance is offered by the *Acronyctidæ*, a family belonging to the group *Noctuina*. The imagines here show in certain points an approximation to the group *Bombycina;* and their larvæ, which are thickly covered with hairs, likewise

appears probable that they are not related to the *Sphingidæ*. See note 1, p. 370. R.M.]

possess the characteristics of many of the caterpillars of this group.[9]

A second illustration is furnished by the family *Ophiusidæ*, which is still placed by all systematists under the *Noctuina*, its affinity to the *Geometrina*, however, being represented by its being located at the end of the *Noctuina*. The broad wings and narrow bodies of these moths remind us in fact of the appearance of the "geometers;" and the larvæ, like the imagines, show a striking resemblance to those of the *Geometrina* in the absence of the anterior abdominal legs. For this reason Hübner in his work on caterpillars has termed the species of this family "*Semi-Geometræ.*"

All these cases show a complete congruence in the two kinds of form-relationship; but exceptions are not wanting. Thus, the family *Bombycidæ* would certainly never have been formed if the larval structure only had been taken into consideration, since, whilst the genera *Gastropacha*, *Clisiocampa*, *Lasiocampa*, *Odonestis*, and their allies, are thickly covered with short silky hairs disposed in a very characteristic manner, the caterpillars of the genus *Bombyx*, to which the common silkworm, *B. Mori*, belongs, are quite naked and similar to many Sphinx-caterpillars (*Chærocampa*). Are the imagines of the genera united under this

[9] [For Mr. A. G. Butler's observations on the genus *Acronycta*, see "Trans. Ent. Soc." 1879, p. 313; and note 3, p. 169, of the present volume. R.M.]

family, at any rate morphologically, as unequally related as their larvæ? Whether it is correct to combine them into one family is a question that does not belong here; we are now only concerned with the fact that the two stages are related in form in very different degrees.

An especially striking case of incongruence is offered by the family *Notodontidæ*, under which Boisduval, depending only on imaginal characters, united genera of which the larvæ differed to a very great extent. In O. Wilde's work on caterpillars this family is on this account quite correctly characterized as follows:—"Larvæ of various forms, naked or with thin hairs, sixteen or fourteen legs." [10] In fact in the whole order Lepidoptera there can scarcely be found associated together such diverse larvæ as are here placed in one imago-family; on one side the short cylindrical caterpillars of the genus *Cnethocampa*, Steph. (*C. Processionea*, *Pithyocampa*, &c.), which are covered with fine, brittle, hooked hairs, and are very similar to the

[10] [The following characters are given in Stainton's "Manual of British Butterflies and Moths," vol. i. p. 114:—"Larva of very variable form: at one extreme we find the singular *Cerura* larvæ, with only fourteen legs, and two long projecting tails from the last segment; at the other extreme we have larvæ with sixteen legs and no peculiarity of form, such as *Chaonia* and *Bucephala*; most have, however, the peculiarity of holding the hind segment of the body erect when in repose; generally quite naked, though downy in *Bucephala* and rather hairy in *Curtulu*: very frequently there are projections on the back of the twelfth segment." R.M.]

larvæ of *Gastropacha* with which they were formerly united; and on the other side there are the naked, humped, and flat-headed larvæ of the genus *Harpyia*, Ochs., with their two long forked appendages replacing the hindmost pair of legs, and the grotesquely formed caterpillars of the genera *Stauropus*, Germ., *Hybocampa*, Linn., and *Notodonta*, Ochs.

The morphological congruence between larvæ and imagines declares itself most sharply in genera, where it is the rule almost without exception. In this case we can indeed be sure that a genus or sub-genus founded on the imagines only will, in accordance with correct principles, present a corresponding difference in the larvæ. Had the latter been known first we should have been led to construct the same genera as those which are now established on the structure of the imagines, and these, through other circumstances, would have stood in the same degree of morphological relationship as the genera founded on the imagines. There is therefore a congruence in a double sense; in the first place the differences between the larvæ and imagines of any two genera are equally great, and, in the next place, the common characters possessed by these two stages combined cause them to form precisely the same groups defined with equal sharpness; the genera coincide completely.

So also the butterflies of the sub-family *Nymphalinæ* can well be separated into genera by the characters of the larvæ, and these, as far as I am able to judge, would agree with the genera founded on the imagines.

The genus *Melitæa*, for example, can be characterized by the possession of 7—9 fleshy tubercles bearing hairy spines; the genus *Argynnis* may be distinguished by always having six hairy unbranched spines on each segment, and the genus *Cethosia* by two similar spines on each segment; the genus *Vanessa* shows sometimes as many as seven branched spines; and the genus *Limenitis* never more than two branched blunt spines on each segment, and so forth. If we go further into details it will be seen that the most closely related imagines, as might indeed have been expected, likewise possess the most nearly allied larvæ, whilst very small differences between the imagines are also generally represented by corresponding differences in the larvæ. Thus, for instance, the genus *Vanessa* of Fabricius has been divided into several genera by later authors. Of these subgenera, *Grapta*, Doubl. (containing the European *C.-album*, the American *Fabricii*, *Interrogationis*, *Faunus*, *Comma*, &c.), is distinguished by the fact that the larvæ not only possess branched spines on all the segments with the exception of the prothorax, but these spines are also present on the head; in the genus *Vanessa* (*sensû strictiori*), Doubl., the head and prothorax are spineless (*e.g. V. Urticæ*); in the tropical genus *Junonia*, Hübn., which was also formerly (Godart, 1819 [11]) united with *Vanessa*, the larvæ bear branched spines on all the segments, the head and prothorax included.

It is possible to go still further and to separate two species of *Vanessa* as two new genera, although they have hitherto been preserved from this fate even by the systematists most given to "splitting." This decision is certainly justifiable, simply because these species at present stand quite alone, and the practical necessity of forming a distinct genus does not make itself felt, and this practical necessity moreover frequently comes into conflict with scientific claims : science erects a new genus based on the amount of morphological difference, it being quite immaterial whether one or many species make up this genus ; such an excessive subdivision is, however, a hindrance to practical requirements, as the cumbrous array of names thereby becomes still further augmented.

The two species which I might separate from *Vanessa* on the ground of their greater divergence, are the very common and widely distributed *V. Io* and *Antiopa*, the Peacock Butterfly and the Camberwell Beauty. In the very remarkable pattern of their wings, both show most marked characteristics ; *Io* possesses a large ocellus on each wing, and *Antiopa* has a broad light yellow border which is not found in any other species of

[11] Encycl. Meth. ix. p. 310.

Vanessa. There can be no doubt but that each of these would have been long ago raised into a genus if similarly marked species of *Vanessa* occurred in other parts of the world, as is the case with the other species of the genus. Thus, it is well known that there is a whole series of species resembling our *V. Cardui*, and another series resembling our *V. C.-album*, the two series possessing the same respective types of marking; indeed on these grounds the sub-genera *Pyrameis* and *Grapta* have been erected.[12]

I should not have considered it worth while to have made these remarks if it had not been for the fact that the caterpillars of *V. Io* and *V. Antiopa* differ in small particulars from one another and from the other species of the genus. These differences relate to the number and position of the spines, as can be seen from the following table:—

[12] [The genus *Vanessa* (in the wide sense) appears to be in a remarkable condition of what may be called phyletic preservation. Thus, the group of species allied to *V. C.-album* passes by almost insensible steps into the group of butterflies typified by our "Tortoiseshells." The following is a list of some of the intermediate species in their transitional order:—*I.-album*, *V.-album*, *Faunus*, *Comma*, *Californica*, *Dryas*, *Polychloros*, *Xanthomelas*, *Cashmirensis*, *Urticæ*, *Milberti*, &c. Similarly, our *Atalanta* and *Cardui* are connected by a number of intermediate forms, showing a complete transition from the one to the other. The following is the order of the species so far as I am acquainted with them:—*Atalanta*, *Dejeanii*, *Callirhoë*, *Tammeamea*, *Myrinna*, *Huntera*, *Terpsichore*, *Carye*, *Kershawii*, and *Cardui*. R.M.]

SPECIES OF THE GENUS VANESSA, FABR.

	Number of Spines on the head and segments of the larva.							
	Head.	Segm. I.	Segm. II.	Segm. III.	Segm. IV.	Segm. V.	Segm. VI.—XI.	Segm. XII.
V. Io	0	0	2	2	4	6	6	4
V. Antiopa . .	0	0	4	4	6	6	7	4
V. Urticæ . . .	0	0	4	4	7	7	7	4
V. Polychloros .	0	0	4	4	7	7	7	4
V. Ichnusa . .	0	0	4	4	7	7	7	4
V. Atalanta . .	0	0	4	4	7	7	7	4
V. C.-album . .	2	0	4	4	7	7	7	4
V. Interrogationis	2	0	4	4	7	7	7	4
V. Levana . . .	2	0	4	4	7	7	7	4

This character of the number of spines will not be considered as too unimportant when we observe how perfectly constant it remains in the nearly allied species. This is the case in the three consecutive forms, *Urticæ*, *Polychloros*, and *Ichnusa*. Now when we see that two species which differ in their imaginal characters present correspondingly small differences in their larvæ, this exact systematic congruence indicates a completely parallel phyletic development.

Exceptions are, however, to be met with here. Thus, Hübner has united one group of the species of *Vanessa* into the genus *Pyrameis* just mentioned, on account of certain characteristic distinctions of the butterflies. I do not know, however, how this genus admits of being grounded on the structure of the larvæ; the latter, as appears from the above table, agree exactly in the number and position of the spines with the caterpillars of *Vanessa* (*sensû strictiori*), nor can any common form of marking be detected which would enable them to be separated from *Vanessa*.

Still more striking is the incongruence in the genus *Araschnia*, Hübn. (*A. Prorsa-Levana*), which, like the genus *Pyrameis*, is entirely based on imaginal characters. This is distinguished from all the other sub-genera of the old genus *Vanessa* by a small difference in the venation of the wings (the discoidal cell of the hind-wings is open instead of closed). Now it is well-known that in butterflies the wing-venation, as most correctly shown by Herrich-Schäffer, is the safest criterion of "relationship." It thus happens that this genus, typified by the common *Levana*, is in Kirby's Catalogue separated from *Vanessa* by two genera, and according to Herrich-Schäffer[13] by forty genera! Nevertheless, the larvæ agree so exactly in their spinal formula with *Grapta* that we should have no hesitation in regarding them as a species of this sub-genus. It appears to me

[13] "Prodromus Systematis Lepidopterorum." Regensburg, 1864.

very probable that in this case the form-relationship of the caterpillar gives more correct information as to the blood relationship of the species than that of the imago—in any case the larvæ show a different form-relationship to the imagines.

Just as in the case of butterflies there are many genera of *Sphingidæ* which can be based on the structure of the larvæ, and which agree with those founded on the imagines.

Thus, the genus *Macroglossa* is characterized by a straight anal horn, a spherical head, and by a marking composed of longitudinal stripes, these characters not occurring elsewhere in this combination. The nearly allied genus *Pterogon*, on the other hand, cannot be based on the larvæ only, since not only is the marking of the adult larva very distinct in the different species, but the anal horn is present in two species, whilst in a third (*P. Œnotheræ*) it is replaced by a knob-like eye-spot. The genus *Sphinx* (*sensû strictiori*) is distinguished by the simple, curved caudal horn, the smooth, egg-shaped head and smooth skin, and by a marking mainly composed of seven oblique stripes. The genus *Deilephila* is distinguished from the preceding by a dorsal plate, situated on the prothorax and interrupting the marking, as well as by the pattern, which here consists of a subdorsal line with ring-spots more or less numerous and developed; the skin also is rough, "shagreened," although it must be admitted that there are exceptions (*Vespertilio*). The genus *Chærocampa* admits also of being based on the form-relationship of its caterpillars, although this is certainly only possible by disregarding the marking and taking alone into consideration the peculiar pig-like form of the larvæ. The genus *Acherontia*, so nearly related to *Sphinx*, possesses in the doubly curved caudal horn a character common to the genus (three species known [14]). Finally may be mentioned the genus *Smerinthus*, of which the larvæ, by their anteriorly tapering form, their shagreened skin and almost triangular head with the apex upwards, their simply curved anal horn, and by their seven oblique stripes on each side, constitute a genus as sharply defined as that formed by the moths.

Although in all the systematic divisions hitherto treated of there are cases where the form-relationship of the larva does not completely coincide with that of the imago, such incongruences are of far more frequent occurrence in the smallest systematic group, viz. species.

The larvæ of two species have very frequently a much nearer form-relationship than their imagines. Thus, the caterpillars of *Smerinthus*

[14] [The larva of *Acherontia Morta*, figured by Butler (see note 50, p. 262), possesses the char. cteristically recurved horn; that of *Ach. Medusa* figured by the same author, does not appear to possess this character in any marked degree. R.M.]

Ocellatus and *S. Populi* are closely allied in structure, marking, and colouring, whilst the moths in these two last characters and in the form of the wings are widely separated.[15] Judging from the larvæ we should expect to obtain two very similar moths, but in fact both *Populi* and *Ocellatus* have many near allies, and these closely related species sometimes possess larvæ which differ more widely than those of more distantly related species of imagines.

Thus, in Amur-land and North America there occur species of *Smerinthus* which closely resemble our *Ocellatus* in colour, marking, and form of wing, and which possess the characteristic large blue ocellus on the hind-wings. *S. Excæcatus* is quite correctly regarded as the representative American form of our *Ocellatus*, but its caterpillar, instead of being leaf-green, is of a chrome-yellow, and possesses dark green instead of white oblique stripes, and has moreover a number of red spots, and a red band on the head—in brief, in the very characters (colour and certain of the markings) in which the imagines completely agree it is widely different from *Ocellatus*. It appears also to be covered with short bristles, judging from Abbot and Smith's figure.[16]

Just in the same way that the species having

[15] [See note 28, p. 233. R.M.]

[16] *Loc. cit.* Pl. XXV. [This species is referred by Butler to the genus *Paonias*, Hübn. R.M.]

the nearest conceivable form-relationship to *Ocellatus* possesses a relatively strongly diverging larva, so does the nearest form-relation of *Populi* (imago) offer a parallel case. This species, which is also North American, lives on *Juglans Alba*. The imago of *Smerinthus Juglandis* differs considerably from *S. Populi* in the form of the wings, but it resembles the European species so closely in marking and colouring that no doubt can exist as to the near relationship of the two forms. The caterpillar of *S. Juglandis*,[17] however, differs to a great extent from that of *Populi* in colour—it is not possible to confound these two larvæ; but those of *Populi* and *Ocellatus* are not only easily mistaken for one another, but are distinguished with difficulty even by experts.

In this same family of the *Sphingidæ* cases are not wanting in which, on the other hand, the moths are far more closely allied than the larvæ.

[17] Abbot and Smith, Pl. XXIX. [Placed by Butler in the genus *Cressonia*, Grote and Robinson. Abbot and Smith state that this larva is sometimes green. According to Mr. Herman Strecker (Lepidop. Rhopal. and Hetero, Reading, Pa. 1874, p. 54) it feeds upon black walnut (*Juglans Nigra*), hickory (*Carya Alba*), and ironwood (*Ostrya Virginica*). Of the North American species of *Smerinthus*, the following, in addition to *Excæcatus*, closely resemble our *Ocellatus*:—*S.* (*Calasymbolus*) *Geminatus*, Say; (*C.*) *Cerisii*, Kirby; and *Ophthalmicus*, Boisd. In addition to *S.* (*Cressonia*) *Juglandis*, *S.* (*Triptogon*) *Modesta* much resembles our *Populi*. The larva of *Geminatus*, according to Strecker, is "pale green, lightest above, with yellow lateral granulated stripes; caudal horn violet; stigmata red. It feeds on the willow." R.M.]

This is especially striking in the genus *Deilephila*, eight species of which are allied in the imaginal state in a remarkable degree, whilst the larvæ differ greatly from one another in colour, and to as great an extent in marking. These eight species are *D. Nicæa*, *Euphorbiæ*, *Dahlii*, *Galii*, *Livornica*, *Lineata*, *Zygophylli*, and *Hippophaës*. Of these, *Nicæa*, *Euphorbiæ*, *Dahlii*, *Zygophylli*, and *Hippophaës* are so much alike in their whole structure, in the form of the wings, and in marking, that few entomologists can correctly identify them off-hand without comparison. The larvæ of these four species, however, are of very different appearances. Those of *Euphorbiæ* and *Dahlii* are most alike, both being distinguished by the possession of a double row of large ring-spots. *Zygophylli* (see Fig. 50, Pl. VI.) possesses only faint indications of ring-spots on a white subdorsal line; and in *Hippophaës* there is only an orange-red spot on the eleventh segment, the entire marking consisting of a subdorsal line on which, in some individuals, there are situated more or less developed ring-spots (see Figs. 59 and 60, Pl. VII). If we only compare the larvæ and imagines of *D. Euphorbiæ* and *Hippophaës*, we cannot but be struck with astonishment at the great difference of form-relationship in the two stages of development.

In the case of *D. Euphorbiæ* and *Nicæa* this difference is almost greater. Whilst these larvæ show great differences in colour, marking, and in the roughness or smoothness of the skin (compare Fig. 51, Pl. VI. with Figs. 43 and 44, Pl. V.), the moths cannot be distinguished with certainty. As has already been stated, the imago of the rare *D. Nicæa* is for this reason wanting in most collections; it cannot be detected whether a specimen is genuine, *i. e.* whether it may not perhaps be a somewhat large example of *D. Euphorbiæ*.

An especially striking instance of incongruence is offered by the two species of *Chærocampa* most common with us, viz., *Elpenor* and *Porcellus*, the large and small Elephant Hawk-moths. The larvæ are so similar, even in the smallest details of marking, that they could scarcely be identified with certainty were it not that one species (*Elpenor*) is considerably larger and possesses a less curved caudal horn than the other. The moths of these two species much resemble one another in their dull green and red colours, but differ in the arrangement of these colours, *i. e.* in marking, and also in the form of their wings, to such an extent that *Porcellus* has been referred to the genus *Pergesa* [18] of Walker. If systemy, as is admitted on many sides, has only to indicate the morphological relationship, this author is not to blame—but in this case a special larval classification must likewise be admitted, in a manner somewhat similar to that at present adopted provisionally in text-books of zoology for the

[18] Cat. Brit. Mus.

Hydroid Polypes and inferior Medusæ. This case of *Porcellus*, however, shows that those are correct who maintain that systemy claims to express, although incompletely, the blood-relationship, and that systematists have always unconsciously formed their groups as though they intended to express the genetic connection of the forms. Only on this supposition can it appear incorrect to us to thus separate two species of which the larvæ agree so completely.

I cannot conclude this review of the various systematic groups without taking a glance at the groups comprised within species, viz. varieties. Whilst in species incongruence is of frequent occurrence, in varieties this is the rule, for which reason it admits in this case of being more sharply defined, since we are not concerned with a double difference but only with the question whether in the one stage a difference or an absolute similarity is observable. By far the majority of varieties are either simply imaginal or merely larval varieties—only the one stage diverges, the other is quite constant.

Thus, as has already been shown, in all the seasonally dimorphic butterflies known to me the caterpillars of the two generations of imagines, which are often so widely different, are exactly alike; and the same obtains for the majority of purely climatic varieties of butterflies. Unfortunately there are as yet no connected observations on this point. The only certain instance that I can here mention is that of the Alpine and Polar form of *Pieris Napi.* This variety, *Bryoniæ*, the female of which differs so greatly in marking and colouring, possesses larvæ which cannot be distinguished from those of the ordinary form of *Napi.* (See part I. appendix I. p. 124.)

That caterpillars can also vary locally without thereby affecting the imagines is shown by the frequently mentioned and closly investigated cases of di- and polymorphism in the larvæ of a number of *Sphingidæ* (*M. Stellatarum*, *A. Atropos*, *S. Convolvuli*, *C. Elpenor*, and *Porcellus*, &c.). The same thing is still more clearly shown by those instances in which there are not several but only one distinct larval form occurring in each of two different localities.

To this class belongs the above-mentioned case of *Chærocampa Celerio* (p. 197), supposing our information concerning this species to be correct; likewise the recently-mentioned case of the Ligurian variety of the caterpillar of *Saturnia Carpini;* and finally the case of *Eriogaster Lanestris*, so well known to lepidopterists. This insect inhabits the plains of Germany, and in the Alps extends to an elevation of 7000 feet, where it possesses a larva differently marked and coloured (*E. Arbusculæ*) to those of the lowlands whilst the moths are smaller, but do not differ in other respects from those of the plains.

Among the Alpine species many other such cases may occur, but these could only be discovered by making investigations having special reference to this point. Of the Alpine butterflies, for example, not a single species can have been reared from the caterpillar; for this reason but few observations have on the whole been given by entomologists respecting the Alpine larvæ, which are not known sufficiently well to enable such a question to be decided.

The investigation of the form-relationships existing between larvæ on the one hand and imagines on the other has thus led to the following results:—

We learn on comparison that incongruences or inequalities of form-relationship occur in all systematic groups from varieties to families. These incongruences are of two kinds, in some cases being disclosed by the fact that the larvæ of two systematic groups, *e. g.* two species, are more closely related in form than their imagines (or inversely), whilst in other cases the larvæ form different systematic groups to those formed by the imagines.

The results of the investigation into the occurrence of incongruences among the various systematic groups may be thus briefly summarised:—

Incongruences appear to occur most frequently among varieties, since it very frequently happens that it is only the larva or only the imago which has diverged into a variety, the other stage remaining monomorphic. The systematic division of varieties is thus very often one-sided.

Among species also incongruences are of frequent occurrence. Sometimes the imagines are much more nearly related in form than the larvæ, and at others the reverse happens; whilst again the case appears also to occur in which only the one stage (larva) diverges to the extent of specific difference, the other stage remaining monomorphic (*D. Euphorbiæ* and *Nicæa*).

The agreement in form-relationship appears to be most complete in genera. In the greater number of cases the larval and imaginal genera coincide, not only in the sharpness of their limits, but also—as far as one can judge—in the weight of their distinctive characters, and therefore in the amount of their divergence. Of all the systematic groups, genera show the greatest congruence.

In families there is again an increase of irregularity. Although larval and imaginal families generally agree, there are so many exceptions that the groups would be smaller if they were based exclusively on the larval structure than if founded entirely on the imagines (*Nymphalidæ*, *Bombycidæ*).

If we turn to the groups of families we find a considerably increased incongruence; complete agreement is here again rather the exception, and it further happens in these cases that it is always the larvæ which, to a certain extent, remain at a lower grade, and which form well defined families; but these can seldom be associated into groups of a higher order having a common character, as in the case of the imagines (*Rhopalocera*).

After having thus collected (so far as I am able) the facts, we have now to attempt their interpretation, and from the observed congruence and incongruence of form-relationship of the two stages to endeavour to draw a conclusion as to the underlying causes of the transformations.

It is clear at starting that all cases of incongruence can only be the expression or the consequence of a phyletic development which has not been exactly parallel in the two stages of larva and imago—that one stage must have changed either more rapidly or more slowly than the other. An "unequal phyletic development" is thus the immediate cause of incongruence.

Thus, the occurrence of different larvæ in species of which the imagines have remained alike may be simply understood as cases in which the imago only has experienced a change—has taken a forward step in phyletic development, whilst the larvæ have remained behind. If we conceive this one-sided development to be repeated several times, there would arise two larval forms as widely different as those of *Deilephila Nicæa*, and *Euphorbiæ*, whilst the imagines, as is actually the case in these species, would remain the same.

The more commonly occurring case in which one stage has a greater form-divergence than the other, is explicable by the one stage having changed more frequently or more strongly than the other.

The explanation of the phenomena thus far lies on the surface, and it is scarcely possible to advance any other; but why should one stage become changed more frequently or to a greater extent than the other? why should one portion be induced to change more frequently or more strongly than another? whence come these inducements to change? These questions bring us to the main point of inquiry:—Are the causes which give rise to these changes internal or external? Are the latter the result of a phyletic vital force, or are they only due to the action of the external conditions of life?

Although an answer to this question will be found in the preceding essay, I will not support myself on the results there obtained, but will endeavour to give another solution of the problem on fresh grounds. The answer will indeed be the same as before:—A phyletic force must be discountenanced, since in the first place it does not explain the phenomena, and in the second place the phenomena can be well explained without its assumption.

The admission of a phyletic vital force does not explain the phenomena. The assumption that there is a transforming power innate in the organism indeed agrees quite well with the phenomenon of congruence, but not with that of incongruence. Since a large number of cases of the latter depend upon the fact that the larvæ are more frequently influenced by causes of change than their imagines, or *vice versâ*, how can this be reconciled with such an internal force ? On this assumption would not each stage of a species be compelled to change, if not contemporaneously at least successively, with the same frequency and intensity, by the action of an innate force ? and how by means of the latter can there ever result a greater form-divergence in the larvæ than in the imagines ?

It is delusive to believe that these unequal deviations can be explained by assuming that the phyletic force acts periodically. Granting that it does so, and that the internal power successively compels the imago, pupa, and finally the larva to change, there would then pass a kind of wave of transformation over the different stages of the species, as was actually shown above to be the case in the single larval stages. The only possible way of explaining the unequal distances between larvæ and imagines would therefore be to assume that two allied groups, *e.g.* species, were not contemporaneously affected by the wave, so that at a certain period of time the imago alone of one species had become changed, whilst in the other species the wave of transformation had also reached the larva. In this case the imagines of the two species would thus appear to be more nearly related than their larvæ.

Now this strained explanation is eminently inapplicable to varieties, still less to species, and least of all to higher systematic groups, for the simple reason that every wave of transformation may be assumed to be at the most of such strength as to produce a deviation of form equal to that of a variety. Were the change resulting from a single disturbance greater, we should not only find one-sided varieties, *i. e.* those belonging to *one* stage, but we should also meet as frequently with one-sided species. If, however, a wave of transformation can only produce a variety even in the case of greatest form-divergence, the above hypothetical uncontemporaneous action of such a wave in two species could only give rise to such small differences in the two stages that we could but designate them as varieties. An accumulation of the results of the action of several successive waves passing over the same species could not happen, because the distance from a neighbouring species would always become the same in two stages as soon as *one* wave had ended its course. In this manner there could therefore only arise divergences of the value of varieties, and incongruences in systematic groups of a higher rank could not thus be explained.

All explanations of the second form of incongruence from the point of view of a phyletic force can also be shown to be absurd. How can the fact be explained that larval and imaginal families by no means always coincide; or that the larvæ can only be formed into families whilst the imagines partly form sharply defined groups of a higher order? How can an internal directive force within the same organism urge in two quite distinct directions? If the evolution of a definite system were designed, and the admission of such a continually acting power rendered necessary, why such an incomplete, uncertain, and confused performance?

I must leave others to answer these questions; to me a vital force appears to be inadmissible, not only because we cannot understand the phenomena by its aid, but above all because it is superfluous for their explanation. In accordance with general principles the assumption of an unknown force can, however, only be made when it is indispensable to the comprehension of the phenomena.

I believe that the phenomena can be quite well understood without any such assumption—both the phenomena of congruence and incongruence, in their two forms of unequal divergence and unequal group-formation.

Let us in the first place admit that there is no directive force in the organism inciting periodic change, but that every change is always the consequence of external conditions, being ultimately nothing but the reaction—the response of the organism to some of the influences proceeding from the environment; every living form would in this case remain constant so long as it was not compelled to change by inciting causes. Such transforming factors can act directly or indirectly, *i. e.* they can produce new changes immediately, or can bring about a remodelling by the combination, accumulation, or suppression of individual variations already present (adaptation by natural selection). Both forms of this action of external influences have long been shown to be in actual operation, so that no new assumption will be made, but only an attempt to explain the phenomena in question by the sole action of these known factors of species formation.

If, in the first instance, we fix our attention upon that form of incongruence which manifests itself through unequal divergence of form-relationship, it will appear prominently that this bears precise relations to the different systematic groups. This form of incongruence constitutes the rule in varieties of the order Lepidoptera, it is of very frequent occurrence in species, but disappears almost completely in genera, and entirely in the case of families and the higher groups. On the whole, therefore, as we turn to more and more comprehensive groups, the incongruence diminishes whilst the congruence increases, until finally the latter becomes the rule.

Now if congruence presupposes an equal

number of transforming impulses, we perceive that the number of the impulses which have affected larvæ and imagines agree with one another the more closely the larger the systematic groups which are compared together. How can this be otherwise? The larger the systematic group the longer the period of time which must have been necessary for its formation, and the more numerous the transforming impulses which must have acted upon it before its formation was completed.

But if the supposition that the impulse to change always comes from the environment in no way favours the idea that such impulses always affect both stages contemporaneously, and are equal in number during the same period of time, there is not, on the other hand, the least ground for assuming that throughout long periods the larvæ or the imagines only would have been affected by such transforming influences. This could have been inferred from the fact that varieties frequently depend only upon one stage, whilst specific differences in larvæ only also occur occasionally, the imagines remaining alike; but no single genus is known of which all the species possess similar larvæ. Within the period of time during which genera can be formed the transforming impulses therefore never actually affect the one stage only, but always influence both.

But if this is the case—if within the period of

time which is sufficient for the production of species, the one stage only is but seldom and quite exceptionally influenced by transforming impulses, whilst both stages are as a rule affected, although not with the same frequency, it must necessarily follow that on the whole, as the period of time increases, the difference in the number of these impulses which affect the larva and of those which affect the imago must continually decrease, and with this difference the magnitude of the morphological differences resulting from the transforming influences must at the same time also diminish. With the number of the successively increasing changes the difference in the magnitude of the change in the two stages would always relatively diminish until it had quite vanished from our perception; just in the same manner as we can distinguish a group of three grains of corn from one composed of six, but not a heap of 103 grains from one containing 106 grains.

That the small systematic groups must have required a short period and the large groups a long period of time for their formation requires no special proof, but results immediately from the theory of descent.

All the foregoing considerations would, however, only hold good if the transforming impulses were equal in strength, or, not to speak figuratively, if the changes only occurred in equivalent portions of the body, *i. e.* in such portions as those in which the changes are of the same physiological and morphological importance to the whole organism.

Now in the lower systematic groups this is always the case. Varieties, species, and genera are always distinguished by only relatively small differences; deep-seated distinctions do not here occur, as is implied in the conception of these categories. The true cause of this is, I believe, to be found in the circumstance that all changes take place only by the smallest steps, so that greater differences can only arise in the course of longer periods of time, within which a great number of types (species) can, however, come into existence, and these would be related by blood and in form in different degrees, and would therefore form a systematic group of a higher rank.

The short periods necessary for the production of inferior groups, such as genera, would not result in incongruences if only *un*typical parts of the larvæ, such as marking or spines, underwent change, whilst in the imagines typical parts—wings and legs—became transformed. The changes which could have occurred in the wings, &c., during this period of time would have been much too small to produce any considerable influence on the other parts of the body by correlation; and two species of which the larvæ and imagines, had changed with the same frequency, would show a similar amount of divergence between the larvæ and between the imagines, although on the one side only *un*typical parts—*i. e.* those of no importance to the whole organization—and on the other side typical parts, were affected. The *number* of the changes would here alone determine whether congruence or incongruence occurred between the two stages.

The case would be quite different if, throughout a long period of time, in the one stage only typical and in the other only untypical parts were subjected to change. In the first case a complete transformation of the whole structure would occur, since not only would the typical parts, such as the wings, undergo a much further and increasing transformation in the same direction, but these changes would also lead to secondary alterations.

In this manner, I believe, must be explained the fact that in the higher groups still greater form-divergences of the two stages occur; and if this explanation is correct, the cause of this striking phenomenon, viz., that incongruence diminishes from varieties to genera, in which latter it occurs but exceptionally, whilst in families and in the higher groups it again continually increases, is likewise revealed. Up to genera the incongruence depends entirely upon the one stage having become changed *more frequently* than the other; but in families and groups of families, and in the orders Diptera and Hymenoptera, as will be shown subsequently, in sub-orders and tribes, it depends upon the *importance of the part of the body* affected by the predominant change. In the latter case the number of changes is of no importance, because these are so numerous that the difference vanishes from our perception; but an equal number of changes, even when very great, may now produce a much greater or a much smaller transformation in the entire bodily structure according as they affect typical or untypical portions, or according as they keep in the same direction throughout a long period of time, or change their direction frequently.

Those unequal form-divergences which occur in the higher systematic groups are always associated with a different formation of groups—the larvæ form different systematic groups to the imagines, so that one of these stages constitutes a higher or a lower group; or else the groups are of equal importance in the two stages, but are of unequal magnitude—they do not coincide, but the one overlaps the other.

Incongruences of this last kind appear in certain cases within families (*Nymphalidæ*), but I will not now subject these to closer analysis, because their causes will appear more clearly when subsequently considering the orders Hymenoptera and Diptera. Incongruences of the first kind, however, admit of a clear explanation in the case of butterflies.

They appear most distinctly in the groups composed of families.

Nobody has as yet been able to establish the group *Rhopalocera* by means of any single character common to the larvæ; nevertheless, this group in the imagines is the sharpest and best defined of the whole order. If we inform the merest tyro that clubbed antennæ are the chief character of the butterflies, he will never hesitate in assigning one of these insects to its correct group. Such a typical character, common to all families, is, however, absent in the larvæ; and it might be correctly said that there were no Rhopalocerous larvæ, or rather that there were only larvæ of *Equites*, *Nymphales*, and *Heliconii*. The larvæ of the various families can be readily separated by means of characteristic distinctions, and it would not be difficult for an adept to distinguish to this extent in single cases a Rhopalocerous caterpillar as such; but these larvæ possess only *family* characters, and not those of a higher order.

This incongruence partly depends upon the circumstance that the form-divergence between a Rhopalocerous and a Heterocerous family is much greater on the side of the imagines than on that of the larvæ. Were there but a single family of butterflies in existence, such as the *Equites*, we should be obliged to elevate this to the rank of a sub-order on the side of the imagines, but not on that of the larvæ. Such cases actually occur, and an instance of this kind will be mentioned later in connection with the Diptera. But this alone does not explain why, on the side of the imagines, a whole series of families show the same amount of morphological divergence from the families of other groups. There are two things, therefore, which must here be explained :—First, why is the form-divergence between the imagines of the Rhopalocera and Heterocera greater than that between their larvæ? and, secondly, why can the imagines of the Rhopalocera be formed into one large group by means of common characters whilst the larvæ cannot?

The answers to both these questions can easily be given from our present standpoint. As far as the first question is concerned, this finds its solution in the fact that the form-divergence always corresponds exactly with the divergence of function, *i.e.* with the divergence in the mode of life.

If we compare a butterfly with a moth there can be no doubt that the difference in the conditions of life is far greater on the side of the imagines than on that of the larvæ. The differences in the mode of life of the larvæ are on the whole but very small. They are all vegetable feeders, requiring large quantities of food, and can only cease feeding during a short time, for which reason they never leave their food-plants for long, and it is of more importance for them to remain firmly attached than to be able to run rapidly. It is unnecessary for them to seek long for their food, as they generally find themselves amidst an abundance, and upon this depends the small development of their eyes and other organs of sense. On the whole caterpillars live under very uniform conditions, although these may vary in manifold details.

The greatest difference in the mode of life which occurs amongst Lepidopterous larvæ is shown by wood feeders. But even these, which by their constant exclusion from light, the hardness of their food, their confinement within narrow hard-walled galleries, and by the peculiar kind of movement necessitated by these galleries, are so differently situated in many particulars to those larvæ which live openly on plants, have not experienced any general change in the typical conformation of the body by adaptation to these conditions of life. These larvæ, which, as has already been mentioned, belong to the most diverse families, are more or less colourless and flattened, and have very strong jaws and small feet; but in none of them do we find a smaller number of segments, or any disappearance, or important transformation of the typical limbs; they all without exception possess sixteen legs, like the other larvæ excepting the *Geometræ*.

Now if even under the most widely diverging conditions of life adaptation of form is produced by relatively small, and to a certain extent superficial, changes, we should expect less typical transformations in the great majority of caterpillars which live on the exterior of plants or in their softer parts (most of the Micro-lepidoptera). The great diversity in the forms of caterpillars depends essentially upon a different formation of the skin and its underlying portions. The skin is sometimes naked, and can then acquire the most diverse colours, either protective or conspicuous, or it may develop offensive or defensive markings; in other cases it may be covered with hairs which sting, or with spines which prick; certain of its glands may develop to an enormous size, and acquire brilliant colours and the power of emitting stinking secretions (the tentacles of the *Papilionidæ* and Cuspidate larvæ); by the development of warts, angles, humps, &c., any species of caterpillar may be invested with the most grotesque shape, the significance of which with respect to the life of the insect is as yet in most cases by no means clear: *typical portions* are not, however, essentially influenced by these manifold variations. At most only the form of the individual segments of the body, and with these the shape of the whole insect, become changed (onisciform larvæ of *Lycænidæ*), but a segment is never suppressed, and even any considerable lengthening of the legs occurs but very seldom (*Stauropus Fagi*).[19]

[19] [This lengthening of the true legs is mimetic according

We may therefore fairly assert that the structure of larvæ is on the whole remarkably uniform, in consequence of the uniformity in the conditions of life. Notwithstanding the great variety of external aspects, the general structure of caterpillars does not become changed—it is only their outward garb which varies, sometimes in one direction, and sometimes in another, and which, starting from inherited characters, becomes adapted to the various special conditions of life in the best possible manner.

All this is quite different in the case of the imagines, where we meet with very important differences in the conditions of life. The butterflies, which live under the influence of direct sunlight and a much higher temperature, and which are on the wing for a much longer period during the day, must evidently be differently equipped to the moths in their motor organs (wings), degree of hairiness, and in the development of their eyes and other organs of sense. It is true that we are not at present in a condition to furnish special proofs that the individual organs of butterflies are exactly adapted to a diurnal life, but we may safely draw this general conclusion from the circumstance that no butterfly is of nocturnal habits.[20]

to Hermann Müller, and causes the anterior portion of the caterpillar to resemble a spider. See note 1, p. 290. R.M.]

[20] [Certain butterflies appear to be crepuscular, if not nocturnal in their habits. Thus in his "Notes on the Lepidoptera

It cannot be stated in objection that there are many moths which fly by day. It certainly appears that no great structural change is necessary to confer upon a Lepidopteron organized for nocturnal life the power of also flying by day; but this proves nothing against the view that the structure of the butterflies depends upon adaptation to a diurnal life. Analogous cases are known to occur in many other groups of animals. Thus, the decapodous Crustacea are obviously organized for an aquatic life; but there are some crabs which take long journeys by land. Fish appear no less to be exclusively adapted to live in water; nevertheless the "climbing-perch" (*Anabas*) can live for hours on land.

It is not the circumstance that some of the moths fly by day which is extraordinary and demands a special explanation, but the reverse fact just mentioned, that no known butterfly flies by night. We may conclude from this that the organization of the latter is not adapted to a nocturnal life,

If we assume [21] that the Lepidopterous family

of Natal," Mr. W. D. Gooch states that he never saw *Melanitis, Leda*, or *Gnophodes Parmeno* on the wing by day, but generally during the hour after sunset. He adds :—"My sugar always attracted them freely, even up to 10 or 11 p.m." Many species of *Hesperidæ* are also stated to be of crepuscular habits by this same observer. See "Entomologist," vol. xvi. pp. 38 and 40. R.M.]

[21] I only make this assumption for the sake of simplicity

adapted to a diurnal life gives rise in the course of time to a nocturnal family, there can be no doubt but that the transformation of structure would be far greater on the part of the imagines than on that of the larvæ. The latter would not remain quite unchanged—not because their imagines had taken to a nocturnal life which for the larva would be quite immaterial, but because this change could only occur very gradually in the course of a large number of generations, and during this long period the conditions of life would necessarily often change with respect to the larvæ. It has been shown above that within the period of time necessary for the formation of a new species impulses to change occur on both sides; how much more numerous therefore must these be in the case of a group of much higher rank, for the establishment of which a considerably longer period is required. In the case assumed, therefore, the larvæ would also change, but they would suffer much smaller transformations than the imagines. Whilst in the latter almost all the typical portions of the body would undergo deep changes in consequence of the entirely different conditions of life, the larvæ would perhaps only change in marking, hairs, bristles, or other external characters, the typical parts experiencing only unimportant modifications.

In this manner it can easily be understood why

and not because I am convinced that the existing *Rhopalocera* are actually the oldest Lepidopterous group.

the larvæ of a family of *Noctuæ* do not differ to a greater extent from those of a family of butterflies than do the latter from some other Rhopalocerous family, or why the imagines of a Rhopalocerous and a Heterocerous family present much greater form-divergences than their larvæ. At the same time is therefore explained the unequal value that must be attributed to any single family of butterflies in its larvæ and in its imagines. The unequal form-divergences coincide exactly with the inequalities in the conditions of life.

When whole families of butterflies show the same structure in their typical parts (antennæ, wings, &c.), and, what is of more importance, can be separated as a systematic group of a higher order (*i. e.* as a section or sub-order) from the other Lepidoptera whilst their larval families do not appear to be connected by any common character, the cause of this incongruence lies simply in the circumstance that the imagines live under some peculiar conditions which are common to them all, but which do not recur in other Lepidopterous groups. Their larvæ live in precisely the same manner as those of all the other families of Lepi doptera—they do not differ in their mode of life from those of the Heterocerous families to a greater extent than they do from one another.

We therefore see here a community of form within the same compass as that in which there is community in the conditions of life. In all butterflies such community is found in their diurnal habits, and in accordance with this we find that these only, and not their larvæ, can be formed into a group having common characters.

In the larvæ also we only find agreement in the conditions of life within a much wider compass, viz. within the whole order. Between the limits of the order Lepidoptera the conditions of life in the caterpillars are, as has just been shown, on the whole very uniform, and the structure of the larvæ accordingly agrees almost exactly in all Lepidopterous families in every essential, *i. e.* typical, part.

In this way is explained the hitherto incomprehensible phenomenon that the sub-ordinal group *Rhopalocera* cannot be based on the larvæ, but that Lepidopterous caterpillars can as a whole be associated into a higher group (order) ; they constitute altogether families and an order, but not the intermediate group of a sub-order. By this means we at the same time reply to an objection that may be raised, viz. that larval forms cannot be formed into high systematic groups because of their " low and undeveloped " organization.

To this form of incongruence, viz. to the formation of systematic groups of unequal value and magnitude, I must attach the greatest weight with respect to theoretical considerations. I maintain that this, as I have already briefly indicated above, is wholly incompatible with the admission of a phyletic force. How is it conceivable that such a power could work in the same organism in two entirely different directions—that it should in the same species lead to the constitution of quite different systems for the larvæ and for the imagines, or that it should lead only to the formation of families in the larvæ and to sub-orders in the imagines? If an internal force existed which had a tendency to call into existence certain groups of animal forms of such a nature that these constituted one harmonious whole of which the components bore to one another fixed morphological relationships, it would certainly have been an easy matter for such a power to have given to the larvæ of butterflies some small character which would have distinguished them as such, and which would in some measure have impressed them with the stamp of "*Rhopalocera*." Of such a character we find no trace however; on the contrary, everything goes to show that the transformations of the organic world result entirely from external influences.

III.

Incongruences in other Orders of Insects.

Although the order Lepidoptera is for many reasons especially favourable for an investigation such as that undertaken in the previous section, it will nevertheless be advantageous to inquire into the form-relationships of the two chief stages in some other orders of metamorphic insects, and to investigate whether in these cases the formation of systematic groups also coincides with common conditions of life.

Hymenoptera.

In this order there cannot be the least doubt as to the form-relationship of the imagines. The characteristic combination of the pro- and meso-thorax, the number and venation of the wings, and the mouth-organs formed for biting and licking, are found throughout the whole order, and leave no doubt that the Hymenoptera are well based on their imaginal characters.

But it is quite different with the larvæ. It may be boldy asserted that the order would never have been founded if the larvæ only had been known. Two distinct larval types here occur, the one—caterpillar-like—possessing a distinct horny head provided with the typical masticatory organs of insects, and a body having thirteen segments, to which, in addition to a variable number of abdominal legs, there are always attached three pairs of horny thoracic legs: the other type is maggot-shaped, without the horny head, and is entirely destitute of mouth-organs, or at least of the three pairs of typical insect jaws, and is also without abdominal and thoracic legs. The number of segments is extremely variable; the larvæ of the saw-flies have thirteen besides the head, the maggot-shaped larvæ of bees possess fourteen segments altogether, and the gall-flies and ichneumons only twelve or ten. We should be much mistaken also if we expected to find connecting characters in the internal organs. The intestine is quite different in the two types of larvæ, the posterior opening being absent in the maggot-like grubs; at most only the tracheal and nervous systems show a certain agreement, but this is not complete.

The order Hymenoptera, precisely speaking and conceived only morphologically, exists therefore but in the imagines; in the larvæ there exist only the caterpillar- and maggot-formed groups. The former shows a great resemblance to Lepidopterous larvæ, and in the absence of all knowledge of the further development it might be attempted to unite them with these into one group. The two certainly differ in certain details of structure in the mouth-organs and in the number of segments, abdominal legs, &c., to a sufficient extent to warrant their being considered as two sub-orders of one larval order; but they would in any case be regarded as much more nearly related in form than the caterpillar- and maggot-like types of the Hymenopterous larvæ.

Is it not conceivable, however, that the imagines of the Hymenoptera—that ichneumons and wasps may be only accidentally alike, and that they have in fact arisen from quite distinct ancestral forms, the one having proceeded with the Lepidopterous caterpillars from one root, and the other with the grub-like Dipterous larvæ from another root?

This is certainly not the case; the common characters are too deep-seated to allow the supposition that the resemblance is here only superficial. From the structure of the imagines alone the common origin of all the Hymenoptera may be inferred with great probability. This would be raised into a certainty if we could demonstrate the phyletic development of the maggot-formed out of the caterpillar-formed Hymenopterous larvæ by means of the ontogeny of the former. From the beautiful investigations of Bütschli on the embryonic development of bees[1] we know that the

[1] Zeitschrift für wissenschaftl. Zoologie, vol. xx. p. 519.

embryo of the grub possesses a complete head, consisting of four segments and provided with the three typical pairs of jaws. These head segments do not subsequently become formed into a true horny head, but shrivel up; whilst the jaws disappear with the exception of the first pair, which are retained in the form of soft processes with small horny points. We know also that from the three foremost segments of the embryo the three typical pairs of legs are developed in the form of round buds, just as they first appear in all insects.[2] These rudimentary limbs undergo complete degeneration before the birth of the larva, as also do those of the whole[3] of the remaining segments, which, even in this primitive condition, show a small difference to the three foremost rudimentary legs.

The grub-like larvæ of the Hymenoptera have therefore descended from forms which possessed a horny head with antennæ and three pairs of gnathites and a 13-segmented body, of which the three foremost segments were provided with legs

[2] [See for instance Lubbock's "Origin and Metamorphoses of Insects," chap. iii.; and F. M. Balfour's "Comparative Embryology," vol. i., 1880, pp. 327—356. This last work contains an admirable *résumé* of our knowledge of the embryonic development of insects up to the date of publication. R.M.]

[3] Are not the 4th, 11th, and 12th segments destitute of the rudiments of legs as in the larvæ of all existing saw-flies? I might almost infer this from Bütschli's figures (see for instance Pl. XXV., Fig. 17A).

differing somewhat from those of the other segments; that is to say, they have descended from larvæ which possessed a structure generally similar to that of the existing saw-fly larvæ. The common derivation of all the Hymenoptera from one source is thus established with certainty.[4]

But upon what does this great inequality in the form-relationship of the larvæ and imagines depend? The existing maggot-like grubs are without doubt much further removed from the active caterpillar-like larvæ than are the saw-flies from the Aculeate Hymenoptera. Whilst these two groups differ only through various modifications of the typical parts (limbs, &c.), their larvæ are separable by much deeper-seated distinctions;

[4] [The grub-formed Hymenopterous larvæ, like the larvæ of all other holometabolous insects, thus represent an acquired degenerative stage in the development, *i. e.* an adaptation to the conditions of life at that stage. Bearing in mind the above-quoted observations of Bütschli and the caterpillar-like form of the Terebrantiate group of Hymenopterous larvæ, the following remarks of Balfour's (*loc. cit.* p. 353), appear highly suggestive:—"While in a general way it is clear that the larval forms of insects cannot be expected to throw much light on the nature of insect ancestors, it does nevertheless appear to me probable that such forms as the caterpillars of the Lepidoptera are not without a meaning in this respect. It is easy to conceive that even a secondary larval form may have been produced by the prolongation of one of the embryonic stages; and the general similarity of a caterpillar to *Peripatus*, and the retention by it of post-thoracic appendages, are facts which appear to favour this view of the origin of the caterpillar form." See also Sir John Lubbock, *loc. cit.*, pp. 93 and 95. R.M.]

limbs of typical importance entirely vanish in the one group, but in the other attain to complete development.

In the Hymenoptera there exists therefore a very considerable incongruence in the systems based morphologically, *i. e.* on the pure form-relationships of the larvæ and of the imagines. The reason of this is not difficult to find: *the conditions of life differ much less in the case of the imagines than in that of the larvæ.* In the former the conditions of life are similar in their broad features. Hymenoptera live chiefly in the air and fly by day, and in their mode of obtaining food do not present any considerable differences. Their larvæ, on the other hand, live under almost diametrically opposite conditions. Those of the saw-flies live after the manner of caterpillars upon or in plants, in both cases their peculiar locomotion being adapted for the acquisition and their masticatory organs for the reduction of food. The larvæ of the other Hymenoptera, however, do not as a rule require any means of locomotion for reaching nor any organs of mastication for swallowing their food, since they are fed in cells, like the bees and wasps, or grow up in plant galls of which they suck the juice, or are parasitic on other insects by whose blood they are nourished. We can readily comprehend that in the whole of this last group the legs should disappear, that the jaws should likewise vanish or should become diminished to one pair retained in a much reduced condition, that the horny casing of the head, the surface of attachment of the muscles of the jaws, should consequently be lost, and that even the segments of the head itself should become more or less shrivelled up as the organs of sense therein located became suppressed.

The incongruence manifests itself however in yet another manner than by the relatively greater morphological divergence of the larvæ: a different grouping is possible for the larvæ and for the imagines. If we divide the Hymenoptera simply according to the form-relationships of the imagines, the old division into the two sub-orders *Terebrantia* or *Ditrocha* and *Aculeata* or *Monotrocha* will be the most correct. The distinguishing characters of a sting or ovipositor and a one- or two-jointed trochanter are still of the greatest value. But these two sub-orders do not by any means correspond with the two types of larvæ since, in the *Terebrantia*, there occur families with both caterpillar-formed and maggot-formed larvæ.

The cause is to be found in that a portion of these families possess larvæ which are parasitic in other insects or in galls, their bodily structure having by these means become transformed in a quite different direction. The mode of life of the imagines is, on the other hand, essentially the same.

We have here therefore another case like that which we met with among the Rhopalocerous Lepidoptera, in which the imagines appear to be capable of being formed into a higher group than the larvæ, because the former live under conditions of life which are on the whole similar whilst the latter live under very divergent conditions.

The old division of the Hymenoptera into two sub-orders has certainly been abandoned in the later zoological text-books; they are now divided into three:—saw-flies, parasitic, and aculeate Hymenoptera; but even this arrangement has been adopted with reference to the different structure of the larvæ. Whether this system is better than the older, *i.e.* whether it better expresses the *genealogical* relationship, I will not now stop to investigate.[5]

Diptera.

The imagines of the Diptera (*genuina*), with the exception of the *Aphaniptera* and *Pupipara*, agree in all their chief characters, such as the number

[5] [In the most recent works dealing with this order six groups, based on the character of the imagines are recognized, viz.:—*Tubulifera*, *Terebrantia*, *Pupivora*, *Heterogyna Fossores*, and *Mellifera.* (See, for instance, F. P. Pascoe's "Zoological Classification," 2nd ed. p. 147.) Of these groups the larvæ of the *Terebrantia* as thus restricted are all of the caterpillar type (*Tenthredinidæ* and *Siricidæ*), whilst those of the other groups are maggot-shaped. For a description of the development of the remarkable aberrant larva of *Platygaster*, see Ganin in Zeit. f. wissenschaftl. Zool., vol. xix. 1869. R.M.]

and structure of the wings, the number and joints of the legs, the peculiar formation of the thorax (fusion of the three segments) ;[6] and even the structure of the mouth organs varies only within narrow limits. This is in accordance with their mode of life, which is very uniform in its main features: all the true Diptera live in the light, moving chiefly by means of flight, but having also the power of running; all those which take food in the imago condition feed upon fluids. Their larvæ, on the other hand, are formed on two essentially different types, the one—which I shall designate as the gnat-type—possessing a horny head with eyes, three pairs of jaws, and long or short antennæ, together with a 12- or 13-segmented body, which is never provided with the three typical pairs of thoracic legs, but frequently has the so-called abdominal legs on the first and last segments. The other Dipterous larvæ are maggot-shaped and without a horny head, or in fact without any head, since the first segment, the homologue of the head, can in no case be distinguished through its being larger than the others; it is on the contrary much smaller. The typical insect mouth-parts are entirely absent, being replaced by a variously formed and quite peculiar arrangement of hooks situated on the mouth and capable of protrusion.

[6] [For recent investigations on the structure of the thorax in Diptera, see a paper by Mr. A. Hammond, in Journ. Linn. Soc., Zoology, vol xv. p. 9. R.M.]

Never more than eleven segments are present besides the first, which is destitute of eyes ; neither are abdominal legs ever developed.

The mode of life differs very considerably in the two groups of larvæ. Although the Dipterous maggots are not as a rule quite incapable of locomotion like the grubs of the Hymenoptera (bees, ichneumons), the majority are nevertheless possessed of but little power of movement in the food-substance on which they were deposited as eggs. They do not go in search of food, either because they are parasitic in other insects in the same manner as the ichneumons (*Tachina*), or else they live on decaying animal or vegetable substances or amidst large swarms of their prey, like the larvæ of the *Syrphidæ* amongst *Aphides*. They generally undergo pupation in the same place as that which they inhabit as larvæ and indeed in their larval skin which hardens into an oval pupa-case. Some few leave their feeding place and pupate after traversing a short distance (*Eristalis*).

As in the case of the Hymenoptera the structure of the larvæ can here also be explained by peuliarities in their mode of life. Creatures which live in a mass of food neither require special organs of locomotion nor specially developed organs of sense (eyes). They have no use for the three pairs of jaws since they only feed on liquid substances, and the hooks within the mouth do not serve for the reduction of food but only for fastening the whole body. With the jaws and their muscular system there likewise disappears the necessity for a hard surface of attachment, *i.e.* a corneous head.

The mode of life of the larvæ of the gnat-type is quite different in most points. The majority, and indeed the most typically formed of these, have to go in search of their food, whether they are predaceous, such as the *Culicidæ* and many of the other *Nemocera* (*Corethra, Simulium*), or whether they feed on plants, which they in some cases weave into a protective dwelling tube (certain species of *Chironomus*). Many live in water and move with great rapidity; others bury in the earth or in vegetable substances; and even those species which live on fungi sometimes wander great distances, as in the well-known case of the "army worm" where thousands of the larvæ of *Sciara Thomæ* thus migrate.

Now the two types of larvæ correspond generally with the two large groups into which, as it appears to me correctly, the Diptera (*genuina*) are as a rule divided. In this respect there is therefore an equality of form-relationship—the grouping is the same, and the incongruence depends only upon the form-divergence between the two kinds of larvæ being greater than between the two kinds of imagines.[7]

[7] I am familiar with the fact that the two sub-orders of

That the form-divergence is greater in the larvæ than in the imagines cannot be doubted; that this distant form-relationship cannot, however, be referred to a very remote common origin, *i. e.* to a very remote blood-relationship, not only appears from the existence of transition-forms between the two sub-orders, but can be demonstrated here, as in the case of the Hymenoptera, by the embryonic development of the maggot-like larvæ.

Seventeen years ago I showed[8] that the grub-formed larvæ of the *Muscidæ* in the embryonic state possessed a well-developed head with antennæ and three pairs of jaws, but that later in the course of the embryonic development a marked reduc-

true Diptera, the short-horned (*Brachycera*), and the long-horned (*Nemocera*), are not sharply limited; and I am likewise well acquainted with the circumstance that there are forms which connect the two larval types. The connecting forms of the imagines do not, however, always coincide with the intermediate larval forms, so that there here arises a second and very striking incongruence of morphological relationship which depends only upon the circumstance that the one stage has diverged in form more widely than the other through a greater divergence in the conditions of life. The difficulty is in these cases aggravated because an apparent is added to the true form-relationship through convergence, so that without going into exact details the form and genealogical relationships of the Diptera cannot be distinguished. It would be of great interest for other reasons to make this investigation, and I hope to be able to find leisure for this purpose at some future period.

[8] "Entwicklung der Dipteren." Leipzig, 1864.

tion and transformation of these parts takes place, so that finally the four head segments appear as a single small ring formed from the coalesced pairs of maxillæ, whilst the so-called "fore-head" (the first head segment), together with the mandibles, becomes transformed into a suctorial-head armed with hooks and lying within the body. At the time of writing I drew no conclusion from these facts with reference to the phyletic development of these larval forms; nor did Bütschli, six years later, in the precisely analogous case of the larvæ of the bees. The inference is, however, so obvious that it is astonishing that it should not have been drawn till the present time.[9]

There can be no doubt that the maggot-like larvæ of insects are not by any means ancient forms, but are, on the contrary, quite recent, as first pointed out by Fritz Müller,[10] and afterwards by Packard[11] and Brauer,[12] and as is maintained in the latest work by Paul Mayer[13] on the phylogeny of insects.

[9] Lubbock concludes from the presence of thoracic legs in the embryonic larva of bees that these have been derived from a larva of the *Campodea* type, but he overlooks the fact that the rudiments of the abdominal legs are also present; *loc. cit.*, p. 28.

[10] "Für Darwin," Leipzig, 1864, p. 8.

[11] Mem. Peabody Acad. of Science, vol. i. No. 3.

[12] Verhandl. Wien. Zoolog. Botan. Gesellsch. 1869, p. 310.

[13] Über Ontogenie und Phylogenie der Insekten. Eine akademische Preisschrift. Jen. Zeitschrift. Bd. x. Neue Folge,

The Dipterous maggots have evidently descended from which a larval form possessed a horny head with antennæ and three pairs of jaws, but which had no appendages to the abdominal segments; they are therefore ordinary Dipterous larvæ of the gnat-type which have become modified in a quite peculiar manner and adapted to a new mode of life, just as the grubs of the Hymenoptera are larvæ of the saw-fly type, which have become similarly transformed, although by no means in the same manner. The resemblance between the two types of larva is to a great extent purely external, and depends upon the

iii. Heft 2. 1876. [Some remarks by F. M. Balfour on the origin of certain larval forms have already been quoted in a previous note (p. 485). This author further states:—"The fact that in a majority of instances it is possible to trace an intimate connection between the surroundings of a larva and its organization proves in the clearest way *that the characters of the majority of existing larval forms of insects have owed their origin to secondary adaptations.* A few instances will illustrate this point:—In the simplest types of metamorphosis, *e.g.* those of the Orthoptera genuina, the larva has precisely the same habits as the adult. We find that a caterpillar form is assumed by phytophagous larvæ amongst the Lepidoptera, Hymenoptera, and Coleoptera. Where the larva has not to go in search of its nutriment the grub-like apodous form is assumed. The existence of such an apodous larva is especially striking in the Hymenoptera, in that rudiments of thoracic and abdominal appendages are present in the embryo and disappear again in the larva. . . . It follows from the above that the development of such forms as the Orthoptera genuina is more primitive than that of the holometabolous forms, &c." Comparative Embryology, vol. 1, p. 352. R.M.]

process designated "convergence" by Oscar Schmidt, *i. e.* upon the adaptation of heterogeneous animal forms to similar conditions of life. By adaptation to a life within a mass of fluid nutriment, the caterpillar-formed larvæ of the Hymenoptera and the *Tipula*-like larvæ of the Diptera have acquired a similar external appearance, and many similarities in internal structure, or, in brief, have attained to a considerable degree of form-relationship, which would certainly have tended to conceal the wide divergence in blood-relationship did not the embryological forms on the one side and the imagines on the other provide us with an explanation.

It is certainly of great interest that in another order of insects—the Coleoptera—grub-formed larvæ occur quite irregularly, and their origin can be here traced to precisely the same conditions of life as those which have produced the grubs of bees. I refer to the honey-devouring larvæ of the *Meloïdæ* (*Meloë*, *Sitaris*, *Cantharis*). The case is the more instructive, inasmuch that the six-legged larval form is not yet relegated to the development within the egg, but is retained in the first larval stage. In the *second larval stage* the maggot-form is first assumed, although this is certainly not so well pronounced as in the Diptera or Hymenoptera, as neither the head nor the thoracic legs are so completely suppressed as in these orders. Nevertheless, these parts have made a great advance in the process of transformation.

The grub-like larvæ of the Hymenoptera and Diptera appear to me especially instructive with reference to the main question of the causes of transformation. The reply to the questions: what gives the impetus to change? is this impetus internal or external? can scarcely be given with greater clearness than here. If these larvæ have abandoned their ancestral form and have acquired a widely divergent structure, arising not only from suppression but partly also from an essentially new differentiation (suctorial head of the *Muscidæ*), and if these structural changes show a close adaptation to the existing conditions of life, from these considerations alone it is difficult to conceive how such transformations can depend upon the action of a phyletic force. The latter must have foreseen that at precisely this or that fixed period of time the ancestors of these larvæ would have been placed under conditions of life which would make it desirable for them to be modified into the maggot-type. But if at the same time the imagines are removed in a less degree from those of the caterpillar-like larvæ, this divergence being in exact relation with the deviations in the conditions of life, I at least fail to see how we can escape the consequence that it is the external conditions of life which produce the transformations and induce the organism to change. It is to me incomprehensible how one and the same vital force can in the same individual induce one stage to become transformed feebly and the other stage strongly, these transformations corresponding in extent with the stronger or weaker deviations in the conditions of life to which the organism is exposed in the two stages; to say nothing of the fact that by such unequal divergences the idea of a perfect system (creative thought) is completely upset.

Nor can the objection be raised that we are here only concerned with insignificant changes—with nothing more than the arrested development of single organs and so forth, in brief, only with those changes which can be ascribed to the action of the environment.

We are here as little concerned with a mere suppression of organs through arrested development as in the case of the Cirripedia; the transformation and reconstruction of the whole body goes even much further than in these Crustacea, although not so conspicuous externally. Where do we elsewhere find insects having the head inside a cavity of the body (sectorial head of the *Muscidæ*), and of which the foremost segment—the physiological representative of the head—consists entirely of the coalesced antennæ and pairs of maxillæ?

The incongruences in the form-relationships are, however, exceedingly numerous in the case of the Diptera, and a special treatise would be necessary to discuss them thoroughly. I may here mention only one case, because the inequality shows itself in this instance in a quite opposite sense.

Gerstäcker, who is certainly a competent entomologist, divides the Diptera into three tribes, viz. the *Diptera genuina*, the *Pupipara*, and the *Aphaniptera*. The latter, the fleas, possess in their divided thoracic segments and in their jointed labial appendages characters so widely divergent from those of the true Diptera and of the *Pupipara* that Latreille and the English zoologists have separated them entirely from the Diptera and have raised them into a separate order.[14] Those who do not agree in this arrangement, but with Gerstäcker include the fleas under the Diptera, will nevertheless admit that the morphological divergence between the *Aphaniptera* and the two other tribes is far greater than that which exists between the latter. Now the larvæ of the fleas are completely similar in structure to those of the gnat-type, since they possess a corneous head with the typical mouth parts and antennæ and a 13-segmented body devoid of legs. Were we only acquainted with the larvæ of the fleas we should

[14] [The *Aphaniptera* are now recognized in this country as a sub-order of Diptera. See, for instance, Huxley's "Anatomy of Invertebrated Animals," p. 425, and Pascoe's "Zoological Classification," 2nd ed. p. 122. R.M.]

rank them with the true Diptera under the sub-order *Nemocera*. On first finding such a larva we should expect to see emerge from the pupa a small gnat.

While the imagines of the *Nemocera* and *Aphaniptera* thus show but a very remote form-relationship their larvæ are very closely allied. Can any one doubt that in this case it is not the larva but the imago which has diverged to the greatest extent? Have not the fleas moreover become adapted to conditions of life widely different from those of all other Diptera, whilst their larvæ do not differ in this respect from many other Dipterous larvæ?

We have here, therefore, another case of unequal phyletic development, which manifests itself in the entirely different form-relationship of the larvæ and the imagines. Thus in this case, as in that of the Lepidoptera, it is sometimes the larval and at other times the imaginal stage which has experienced the greatest transformation, and, as in the order mentioned, the objection that a phyletic vital force produces greater and more important differentiations in the higher imaginal stage than in the lower or less developed larval stage, is equally ineffectual.

If, however, it be asked whether the unequal phyletic development depends in this case upon an unequal number of transforming impulses which the two stages may have experienced during an equal period of time, this must be decidedly answered in the negative. The unequal development obviously depends in this case, as in the higher systematic groups of the Lepidoptera, upon the unequal value of the parts affected by the changes. These parts are on the one side of small importance, and on the other side of great importance, to the whole structure of the insect. This is shown in the last-mentioned case of the fleas, where, of the typical parts of the body, only the wings have become rudimentary, whilst the antennæ, mouth-parts, and legs, and even the form and mode of segmentation (free thoracic segments), must have suffered most important modifications; their larvæ, on the other hand, can have experienced only unimportant changes, since they still agree in all typical parts with those of the gnat-type.

Although therefore in this and in similar cases a greater number of transforming impulses may well have occurred on the one side than on the other—and it is indeed highly probable that this number has not been absolutely the same—nevertheless the chief cause of the striking incongruence is not to be found therein, but rather in the *strength* of the transforming impulses, if I may be permitted to employ this figure, or, more precisely expressed, in the importance of the parts which become changed and at the same time in the amount of change.

In this conclusion there is implied as it appears to me an important theoretical result which tells further against the efficacy of a phyletic force.

If the so-called "typical parts" of an animal disappear completely through the action of the environment only, and still further, if these parts can become so entirely modified as to give rise to quite new and again typical structures (suctorial head of the *Muscidæ*) without the typical parts of the other stage of the same individual being thereby modified and transformed into a new type of structure, how can we maintain a distinction between typical and non-typical parts with respect to their origin? But if a difference exists with respect only to the physiological importance of such parts, *i.e.* their importance for the equilibrium of the whole organization, while, with reference to transformation and suppression, exactly the same influences appear to be effective as those which bring about a change in or a disappearance of the so-called adventitious parts, where is there left any scope for the operation of the supposed phyletic force? What right have we to assume that the typical structures arise by the action of a vital force? Nevertheless this is the final refuge of those who are bound to admit that a great number of parts or characters of an animal can become changed, suppressed, or even produced by the action of the environment.

IV.

Summary and Conclusion.

The question heading the second section of this essay must at the conclusion of the investigation be answered in the negative. The form-relationship of the larvæ does not always coincide with that of the imagines, or, in other words, a system based entirely on the morphology of the larvæ does not always coincide with that founded entirely on the morphology of the imagines.

Two kinds of incongruence here present themselves. The first arises from the different amount of divergence between two systematic groups in the larvæ and in the imagines, these groups being of equal extent. The second form of incongruence consists essentially in that the two stages form systematic groups of different extents, either the one stage constituting a group of a higher order than the other and therefore forming a group of unequal value, or else the two stages form groups of equal systematic value, these groups, however, not coinciding in extent, but the one overlapping the other.

This second form of incongruence is very

frequently connected with the first kind, and is mostly the direct consequence of the latter.

The cause of the incongruences is to be found in unequal phyletic development, either the one stage within the same period of time having been influenced by a greater number of transforming-impulses than the other, or else these impulses have been different in strength, *i.e.* have affected parts of greater or less physiological value, or have influenced parts of equal value with unequal strength.

In all these cases in which there are deep-rooted form-differences, it can be shown that these correspond exactly with inequalities in the conditions of life, this correspondence being in two directions, viz. in strength and in extent: the former determines the *degree* of form-difference, the latter its *extent* throughout a larger or smaller group of species.

The different forms of incongruence are manifested in the following manner :—

(1.) Different amount of form-divergence between the larvæ on the one side and the imagines on the other. Among the Lepidoptera this is found most frequently in varieties and species, and there is evidence to show that in this case the one stage has been affected by transforming influences, either alone (varieties), or at any rate to a greater extent (species). In the last case it can be shown in many ways that one stage (the larva) has actually remained at an older phyletic grade (*Deilephila* species). Incongruences of this kind depending entirely upon the more frequent action of transforming impulses can only become observable in the smaller systematic groups, in the larger they elude comparative examination. In the higher groups unequal form-divergence may be produced by the transforming impulses affecting parts of unequal physiological and morphological value, or by their influencing parts of equal value in different degrees. All effects of this kind can, however, only become manifest after a long-continued accumulation of single changes, *i. e.* only in those systematic groups which require a long period of time for their formation. By this means we can completely explain why the incongruences of form-divergence continually diminish from varieties to genera, and then increase again from genera upwards through families, tribes, and sub-orders : the first diminishing incongruence depends upon an *unequal number* of transforming impulses, the latter increasing incongruence depends upon the *unequal power* of these impulses.

Cases of the second kind are found among the Lepidopterous families, and especially in the higher groups (*Rhopalocera* and *Heterocera*), and appear still more striking in the higher groups of the Hymenoptera and Diptera. Thus the caterpillar shaped and maggot-formed larvæ of the

Hymenoptera differ from one another to a much greater extent than their imagines, since the latter have experienced a complete transformation of typical parts; whilst in the caterpillar-formed larvæ these parts vary only within moderate limits. Similarly in the case of the Diptera, of which the gnat-like larvæ diverge more widely from those of the grub type than do the gnats from the true flies. On the other hand the divergence between the imagines of the fleas and gnats is considerably greater than that between their larvæ—indeed the larvæ of the fleas would have to be ranked as a family of the sub-order of the gnat-like larvæ if we wished to carry out a larval classification. By this it is also made evident that these unequal divergences, when they occur in the higher systematic groups, always induce at the same time the second form of incongruence—that of the formation of unequal systematic groups.

In general whenever such unequal divergences occur in the higher groups they run parallel with a strong deviation in the conditions of life. If these differ more strongly on the side of the larvæ, we find that the structure of the latter likewise diverges the more widely, and that their form-relationship is in consequence made more remote (saw-flies and ichneumons, gnats and flies); if, on the other hand, the difference in the conditions of life is greater on the side of the imagines, we find among the latter the greater morphological divergence (butterflies and moths, gnats and fleas).

(2.) The second chief form of incongruence consists in the formation of different systematic groups by the larvæ and the imagines, if the latter are grouped simply according to their form-relationship without reference to their genetic affinities. This incongruence again shows itself in two forms—in the formation of groups of unequal value, and the formation of groups equal in value but unequal in extent, *i. e.* of overlapping instead of coinciding groups.

Of these two forms the first arises as the direct result of a different amount of divergence. Thus the larvæ of the fleas, on account of their small divergence from those of the gnats, could only lay claim to the rank of a *family*, whilst their imagines are separated from the gnats by such a wide form-divergence that they are correctly ranked as a distinct *tribe* or *sub-order*.

The inequalities in the lowest groups, varieties, can be regarded in a precisely similar manner. If the larva of a species has become split up into two local forms, but not the imago, each of the two larval forms possesses only the rank of a *variety*, whilst the imaginal form has the value of a *species*.

Less simple are the causes of the phenomenon that in the one stage the lower groups can be combined into one of higher rank, whilst the other stage does not attain to this high rank. Such a condition appears especially complicated when the two stages can again be formed into groups of a still higher rank.

This is the case in the tribe *Rhopalocera*, which is founded on the imagines alone, the larvæ forming only families of butterflies. Both stages can however be again combined into the highest systematic group of the Lepidoptera.

In this case also the difference in the value of the systematic groups formed by the two stages corresponds precisely with the difference in the conditions of life. This appears very distinctly when there are several sub-groups on each side, and not when, as in the fleas, only one family is present as a tribe on the one side and on the other as a family. Thus in the butterflies, on the one side there are numerous families combined into the higher rank of a sub-order (imagines), whilst on the other side (larvæ) a group of the same extent cannot be formed. In this instance it can be distinctly shown that the combination of the families into a group of a higher order, as is possible on the side of the imagines, corresponds exactly with the limits in which the conditions of life deviate from those of other Lepidopterous families. The group of butterflies corresponds with an equally large circle of uniform conditions of life, whilst a similar uniformity is wanting on the side of the larvæ.

The second kind of unequal group formation arises from the circumstance that groups of equal value can be formed from the two stages, but these groups do not possess the same limits—they overlap, and only coincide in part.

This is most clearly seen in the order Hymenoptera, in which both larvæ and imagines form two well-defined morphological sub-orders, but in such a manner that the one larval form not only prevails throughout the whole of the one sub-order of the imagines, but also extends beyond and spreads over a great portion of the other imaginal sub-order.

Here again the dependence of this phenomenon upon the influence of the environment is very distinct, since it can be demonstrated (by the embryology of bees) that the one form of larva—the maggot-type—although the structure now diverges so widely, has been developed from the other form, and that it must have arisen by adaptation to certain widely divergent conditions of life.

This form of incongruence is always connected with unequal divergence between the two stages of the one systematic group—in this case the *Terebrantia*. The larvæ of this imaginal group partly possess caterpillar-like (*Phytospheces*) and partly maggot-formed (*Entomospheces*) larvæ, and differ from one another to a considerably greater extent than the saw-flies from the ichneumons.[1] The

[1] [This illustration of course only applies to the old arrange-

final cause of the incongruence lies therefore in this case also in the fact that one stage has suffered stronger changes than the other, so that a deeper division of the group has occurred in the former than in the latter.

The analogous incongruences in single families of the Lepidoptera may have arisen in a similar manner, as has already been more clearly shown above; only in these cases we are as yet unable to prove in detail that the larval structure has become more strongly changed through special external conditions of life than that of the imagines.

In the smallest systematic group—varieties, it has been possible to furnish some proof of this. The one-sided change here depends in part upon the *direct action* of external influences (seasonal dimorphism, climatic variation), and it can be shown that these influences (temperature) acted only on the one stage, and accordingly induced change in this alone whilst the other stage remained unaltered.

It has now been shown—not indeed in every individual case, but for each of the different kinds of incongruence of form-relationship—that there is an exact parallelism corresponding throughout with the incongruence in the conditions of life. Wherever the forms diverge more widely in one

ment of the Hymenoptera into *Terebrantia* and *Aculeata*. See also note 5, p. 488. R.M.]

stage than in the other we also find more widely divergent conditions of life; wherever the morphological systemy of one stage fails to coincide with that of the other—whether in the extent or in the value of the groups—the conditions of life in that stage also diverge, either more widely or at the same time within other limits; whenever a morphological group can be constructed from one stage but not from the other, we find that this stage alone is submitted to certain common conditions of life which fail in the other stage.

The law that the divergence in form always corresponds exactly with the divergence in the conditions of life[2] has accordingly received confirmation in all cases where we have been able to pronounce judgment. Unequal form-divergences correspond precisely with unequal divergence in the conditions of life, and community of form appears within exactly the same limits as community in the conditions of life.

These investigations may thus be concluded with the following law:—In types of similar origin, *i.e.* having the same blood-relationship, the degree of morphological relationship corresponds exactly

[2] [Eng. ed. This law is perhaps a little too restricted, inasmuch as it is theoretically conceivable that the organism may be able to adapt itself to similar conditions of life in different ways; differences of form could thus depend sometimes upon differences of adaptation and not upon differences in the conditions of life, or, as I have formerly expressed it, it is not necessary to allow always only *one* best mode of adaptation.]

with the degree of difference in the conditions of life in the two stages.

With respect to the question as to the final cause of transformation this result is certainly of the greatest importance.

The interdependence of structure and function has often been insisted upon, but so long as this has reference only to the agreement of each particular form with some special mode of life, this harmony could still be regarded as the result of a directive power; but when in metamorphic forms we not only see a double agreement between structure and function, but also that the transformation of the form occurs in the two chief developmental stages in successive steps at unequal rates and with unequal strength and rhythm, we must—at least so it appears to me—abandon the idea of an inherent transforming force; and this becomes the more necessary when, by means of the opposite and extremely simple assumption that transformations result entirely from the response of the organism to the actions of the environment, all the phenomena—so far as our knowledge of facts at present extends—can be satisfactorily explained. A power compelling transformation, *i.e.* a phyletic vital force, must be abandoned, on the double ground that it is incapable of explaining the phenomena (incongruence and unequal phyletic development), and further because it is superfluous.

Against the latter half of this argument there can at most be raised but the one objection that the phenomena of transformation are not completely represented by the cases here analysed. In so far as this signifies that the whole organic world, animal and vegetable, has not been comprised within the investigation this objection is quite valid. The question may be raised as to the limit to which we may venture to extend the results obtained from one small group of forms. I shall return to this question in the last essay.

But if by this objection it is meant that the restricted field of the investigation enables us to actually analyse only a portion of the occurring transformations,[3] and indeed only those cases, the dependence of which upon the external conditions of life would be generally admitted, I will not let pass the opportunity of once more pointing out at the conclusion of the present essay that the incongruences shown to exist by no means depend only upon those more superficial characters the remodelling of which in accordance with the external conditions of life may be most easily discerned and is most difficult to deny, but that in certain cases (maggot-like Dipterous larvæ) it is

[3] [It must be understood that the word rendered here and elsewhere throughout this work as "transformation" is not to be taken in the narrow sense of metamorphosis, but as having the much broader meaning of a change of any kind incurred by an organism. Metamorphosis is in fact but one phase of transformation. R.M.]

precisely the "typical" parts which become partly suppressed and partly converted into an entirely new structure. From the ancient typical appendages there have here arisen new structures, which again have every right to be considered as typical. This transformation is not to be compared with that experienced by the swimming appendages of the *Nauplius*-like ancestor of an *Apus* or *Branchipus* which have become mandibulate, nor with the transformation which the anterior limbs must have gone through in the reptilian ancestors of birds. The changes in question (Dipterous larvæ) go still further and are more profound. I lay great emphasis upon this because we have here one of the few cases which show that typical parts are quite as dependent upon the environment as untypical structures, and that the former are not only able to become adapted to external conditions by small modifications—as shown in a most striking manner by the transformations of the appendages in the Crustacea and Vertebrata—but that these parts can become modelled on an entirely new type which, when perfected, gives no means of divining its mode of origin. I may here repeat a former statement:—With reference to the causes of their origination we have no grounds for drawing a distinction between typical and untypical structures.

It may be mentioned in concluding that quite analogous although less sharply defined results are arrived at if, instead of fixing our attention upon the different stages of a systematic group in their phyletic development, we only compare the different functional parts (organs in the wide sense) of the organisms.

A complete parallel can be drawn between the two classes of developmental phenomena. From the very different systematic values attached by taxonomists to this or that organ in a group of animals, it may be concluded that the individual parts of an organism are to a certain extent independent, and that each can vary independently, when affected either entirely alone or in a preponderating degree by transforming impulses, without all the other parts of the organism likewise suffering transformation, or at least without their becoming modified in an equal degree. Did all the parts and organs in two groups of animals diverge from each other to the same extent, the systematic value of such parts would be perfectly equal ; we should, for example, be able to distinguish and characterize two genera of the family of mice by their kidneys, their liver, their salivary glands, or by the histological structure of their hair or muscles, or even by differences in their myology, &c. equally as well as by their teeth, length of toes, &c. It is true that such a diagnosis has yet to be attempted ; but it may safely be predicted that it would not succeed. Judging from all the facts at present before us, the individual parts—and especially those connected in their physiological action, *i.e.* the system of organs—do not keep pace with reference to the modifications which the species undergoes in the course of time; at one period one system and at another period some other system of organs advances while the others remain behind.

This corresponds exactly with the result already deduced from the unparallel development of the independent ontogenetic stages. If the inequality in the phyletic development is more sharply pronounced in this than in the last class of cases, this can be explained by the greater degree of correlation which exists between the individual systems of organs in any single organism as compared with that existing between the ontogenetic stages, which, although developed from one another, are nevertheless almost completely independent. We should have expected *à priori* that a strong correlation would have here existed, but as a matter of fact this is not the case, or is so only in a very small degree.

Just as in the stages of metamorphosis the inequality of phyletic development becomes the more obliterated the more distant and comprehensive, or, in other words, the greater the period of existence of the groups which we compare, so does the unequal divergence of the systems of organs become obliterated as we bring into comparison larger and larger systematic groups.

It is not inconceivable—although a clear proof of this is certainly as yet wanting—that a variety of the ancestral species would differ only in one single character, such as hairiness, colour, or marking, and such instances would thus agree precisely with the foregoing cases in which only the caterpillar or the butterfly formed a variety. All the more profound modifications however—such for instance as those which determine the difference between two species—are never limited to one character, but always affect several, this being explicable by correlation, which, as Darwin has shown in the case of dogs, may cause modifications in the skull of those breeds having hanging ears in consequence of this last character alone. It must be admitted however that one organ only would be originally affected by a modifying influence. Thus, I am acquainted with two species of a genus of Daphniacea which are so closely allied that they can only be distinguished from one another by a close comparison of individual details. But whilst most of the external and internal organs are almost identical in the two species the sperm-cells of the males differ in a most striking manner, in one species resembling an Australian boomerang in form and in the other being spherical! An analogous instance is furnished by *Daphnia Pulex* and *D. Magna,* two species which were for a long time confounded. Nearly all the parts of the body are here exactly alike, but the antennæ of the males differ to a remarkable extent, as was first correctly shown by Leydig.

Similarly in the case of genera there may be observed an incongruence of such a kind that individual parts of the body may deviate to a greater or to a less extent than the corresponding parts in an allied genus. If, for instance, we compare a species of the genus of Daphniacea, *Sida*, with a species of the nearly allied genus *Daphnella*, we find that all the external and internal organs are in some measure dissimilar—nevertheless certain of these parts deviate to an especially large extent, and have without question become far more transformed than the others. This is the case, for example, with the antennæ and the male sexual organs. The latter, in *Daphnella*, open out at the sides of the posterior part of the body as long, boot-shaped generative organs, and in *Sida* as small papillæ on the ventral side of this region of the body. If again we compare *Daphnella* with the nearly allied genus *Latona*, it will be found that no part in the one is exactly similar to the corresponding part in the other genus, whilst certain organs differ more widely than others. This is the case for instance with the oar-like appendages which in *Latona* are triramous, but in *Daphnella*, as in almost all the other Daphniacea, only biramous.

In families the estimation of the form-divergence

of the systems of organs and parts of the body becomes difficult and uncertain: still it may safely be asserted that the two Cladocerous families *Polyphemidæ* and *Daphniidæ* differ much less from one another in the structure of their oar-like appendages than in that of their other parts, such as the head, shell, legs, or abdominal segments. In systematic groups of a still higher order, *i.e.* in orders, and still more in classes, we might be inclined to consider that all the organs had become modified to an equally great extent. Nevertheless it cannot be conclusively said that the kidneys of a bird differ from those of a mammal to the same extent as do the feathers from mammalian hair, since we cannot estimate the differences between quite heterogeneous things —it can only be stated that both differ greatly. Here also the facts are not such as would have been expected if transformation was the result of an internal developmental force; no uniform modification of *all* parts takes place, but first one part varies (variety) and then others (species), and, on the whole, as the systematic divergence increases all parts become more and more affected by the transformation and all tend continually to appear changed to an equal extent. This is precisely what would be expected if the transforming impulses came from the environment. An equalization of the differences caused by transformation must be produced in two ways; first by correlation, since nearly every primary transformation must entail one or more secondary changes, and secondly because, as the period of time increases, more numerous parts of the body must become influenced by primary transforming factors.

A tempting theme is here also offered by attempting to trace the inequality of phyletic development to dissimilar external influences, and by demonstrating that individual organs have as a rule become modified in proportion to the divergence in the conditions of life by which they have been influenced, this action, during a given period of time, having been more frequent in the case of one organ than in that of the others, or, in brief, by showing the connection between the causes and effects of transformation.

It would be quite premature, however, to undertake such a labour at present, since it will be long before physiology is able to account for the fine distinctions shown by morphology, and further because we have as yet no insight into those internal adjustments of the organism which would enable us *à priori* to deduce definite secondary changes from a given primary transformation. But so long as this is impossible we have no means of distinguishing correlative changes from the primary modifications producing them, unless they happen to arise under our observation.

APPENDIX I.[1]

ADDITIONAL NOTES ON THE ONTOGENY, PHYLOGENY, &c., OF CATERPILLARS.

Ontogeny of the Noctua larvæ.—References have already been given in a previous note (2, p. 166) to observations on the number of legs and geometer-like habits of certain *Noctua*-larvæ when newly hatched. This interesting fact in the development of these insects furnishes a most instructive application of the principle of ontogeny to the determination of the true affinities, *i.e.* the blood-relationship of certain groups of Lepidoptera. While the foregoing portions of this work have been in course of preparation for the press, some additional observations on this subject have been published, and I may take the present opportunity of pointing out their systematic bearing—not, indeed, with a view to settling definitively the positions of the groups in question, as our knowledge is still somewhat scanty—but with the object of stimulating further investigation.

Mr. H. T. Stainton has lately recorded the fact that the young larva of *Triphæna Pronuba* is a semi-looper (Ent. Mo. Mag. vol. xvii. p 135); and in a recently published life-history of *Euclidia Glyphica* (*Ibid.* p. 210) Mr. G. T. Porritt states that this caterpillar is a true looper when young, but becomes a semi-looper when adult. To these facts Mr. R. F. Logan adds (*Ibid.* p. 237) that "nearly all the larvæ of the *Trifidæ* are semi-loopers when first hatched." The *Cymatophoræ*

[1] By the Editor.

appear to be an exception, but Mr. Logan points out that this genus is altogether aberrant, and seems to be allied to the *Tortricidæ*. Summing up the results of these and the observations previously referred to, it will be seen that this developmental character has now been established in the case of species belonging to the following families of the section *Genuinæ*:—*Leucaniidæ, Apameidæ, Caradrinidæ, Noctuidæ, Orthosiidæ, Hadenidæ,* and *Xylinidæ*, as well as the other *Trifidæ* (excepting *Cymatophora*).[2] The larvæ of the *Minores* and *Quadrifidæ* are as a rule semi-loopers when adult and may be true loopers when young, although further observations on this point are wanted. These facts point to the conclusion that the *Noctuæ* as a whole are phyletically younger than the *Geometræ*, whilst the *Genuinæ* and *Bombyciformes* have further advanced in phyletic development than the *Minores* and *Quadrifidæ*. The last two sections are therefore the most closely related to the *Geometræ*, as correctly shown by the arrangement given in Stainton's "Manual;" whilst that adopted in Doubleday's "Synonymic List," where the *Geometræ* precede the *Noctuæ*, is most probably erroneous.

Additional descriptions of Sphinx-larvæ.—In the foregoing essay on "The Origin of the Markings of Caterpillars," Dr. Weismann has paid special attention to the larvæ of the *Sphingidæ* and has utilized for this purpose, in addition to his own studies of the ontogeny of many European species, the figures in the chief works dealing with this family published down to the time of appearance of his essay (1876).[3] In order to amplify this part

[2] Mr. C. V. Riley in his excellent "Annual Reports" already quoted in previous notes, states that the larvæ of *Agrotis Inermis, Leucania Unipuncta* (Army-worm), and *L. Albilinea* are all loopers when newly hatched. (See First Report, p. 73; Eighth Report, p. 184; and Ninth Report, p. 53.)

[3] The following species not referred to in the previous part

of the subject I have added references to more recent descriptions and figures of Sphinx-larvæ published by Burmeister and A. G. Butler, and I have endeavoured in these cases to refer the caterpillars as far as possible to their correct position in the respective groups founded on the ontogeny and phylogeny of their allies. It is, however, obvious that for the purposes of this work figures or descriptions of adult larvæ are of but little value, except for the comparative morphology of the markings; and even this branch of the subject only becomes of true biological importance when viewed in the light of ontogeny. As our knowledge of the latter still remains most incomplete in the case of exotic species, it would be at present premature to attempt to draw up any genealogy of the whole family, and I will here only extend the subject by adding some few descriptions of species which are interesting as having been made from the observations of field-naturalists, and which contain remarks on the natural history of the insects.

Mr. C. V. Riley in his "Second Annual Report on the Noxious, Beneficial, and other Insects of the State of Missouri, 1870," gives figures and describes the early

of this work are figured by Semper (Beit. zur Entwicklungsgeschichte einiger ostasiat. Schmet.; Verhandl. d. k.k. zoo. bot. Gesell. in Wien, 1867):—*Panacra Scapularis*, Walk.; *Chœrocampa Clotho*, Drury; and *Diludia* (*Macrosila*) *Discistriga*, Walk. The following are figured by Boisduval and Guenée. (Spéc. Gén. 1874):—*Smerinthus Ophthalmicus*, Boisd.; *Sphinx Jasminearum*, Boisd.; *S.* (*Hyloicus*) *Plebeia*, Fabr.; *S.* (*Hyloicus*) *Cupressi*, Boisd.; *S.* (*Pseudosphinx*) *Catalpæ*, Boisd.; *Philampelus Jussiuæ*, Hübn. (= *Sphinx Vitis*, Linn.?); and *Ceratomia Amyntor*, Hübn. As the works of Abbot and Smith, and Horsfield and Moore have been exhausted by Dr. Weismann, it is quite unnecessary to extend this note by giving a list of the species figured by these authors.

stages and adult forms of certain grape-vine feeding larvæ of the subfamily *Chærocampinæ.* The full-grown larva of *Philampelus Achemon*, Drury, "measures about 3½ inches when crawling, which operation is effected by a series of sudden jerks. The third segment is the largest, the second but half its size, and the first still smaller, and when at rest the two last-mentioned segments are partly withdrawn into the third. . . . The young larva is green, with a long slender reddish horn rising from the eleventh segment and curving over the back." Mr. Riley then states that full grown specimens are sometimes found as green as the younger ones, but "they more generally assume a pale straw or reddish-brown colour, and the long recurved horn is invariably replaced by a highly polished lenticular tubercle." The specimen figured was the pale straw variety, this colour deepening at the sides, and finally merging into a rich brown. The markings appear to consist of an interrupted brown dorsal line, a continuous subdorsal line of the same colour, and six oblique scalloped white bars along the side. Whether the colour and marking is adapted to the vine, as is the case with the two varieties of the dimorphic *Chærocampa Capensis* (*q.v.*), is not stated. The larva of *Philampelus Satellitia*, Linn., when newly hatched, and for some time afterwards is "green with a tinge of pink along the sides, and with an immensely long straight pink horn at the tail. This horn soon begins to shorten, and finally curls round like a dog's tail." The colour of the insect changes to a reddish-brown as it grows older, and the caudal horn is entirely lost at the third moult. The chief markings appear to be five oblique cream-yellow patches with a black annulation on segments 6—10, and a pale subdorsal line. The caterpillar crawls by a series of sudden jerks, and often flings its "head savagely from side to side when alarmed." "When at rest, it draws back the fore part of the body and retracts the head and first two joints into the third." Two points in connection with these species are of interest with respect to the present investigations. The green colour and the possession of a long caudal horn when young shows that these larvæ, like those of *Chærocampa Elpenor* (p. 178), *C. Porcellus* (p. 184), and *Philampelus Labruscæ* (p. 195, note), are descended from ancestors which possessed these characters in the adult state.[4] The next point of interest is the attitude of alarm assumed by these larvæ, and effected by withdrawing the head and two front segments into the third.[5] The importance of this in connection with the similar habit of ocellated species will be seen on reading the remarks on page 367 bearing upon the initial stages of eye-spots. The other species figured by Mr. Riley are *Chærocampa Pampinatrix*, Smith and Abbot, and *Thyreus Abboti*, Swains. The latter has already been referred to (p. 256).

In a paper "On a Collection of Lepidoptera from Candahar" (Proc. Zoo. Soc., May 4th, 1880), Mr. A. G. Butler has described and figured, from materials furnished to him by Major Howland Roberts, the larvæ of three species of *Sphingidæ*. *Chærocampa Cretica*, Boisd., feeds on vine; out of 100 specimens examined, there was not one black variety, while in another closely allied species, found at Jutogh and Kashmir, the larva is stated to be as often black as green. The general colour of the caterpillar harmonizes with that of the under side of the vine leaves; it possesses a thread-like dorsal, and a pale yellow subdorsal line; also "a subdorsal row of eye-spots, each consisting of a green patch in a yellow oval, the first spot on the fifth segment being the largest and most distinct, those on each following segment

[4] The same inference has already been drawn with respect to *Pterogon* (*Proserpinus*) *Œnotheræ*, see pp. 257, 258.

[5] This would of course be the *fourth* segment if the head be considered the first, as on the Continent.

becoming smaller, more flattened, and less distinct, till lost on the twelth segment, sometimes becoming indistinct after the seventh or eighth segment; these spots are only distinct as eye-spots on the fifth and sixth segments, that on the sixth being flatter than that on the fifth, those on the remaining segments appearing like dashes while the larvæ is green, but more like eyes on its changing colour when full fed." The change here alluded to is the dark-brown coloration so generally assumed by green Sphinx-larvæ previous to pupation, and which, as I have stated elsewhere (Proc. Zoo. Soc., 1873, p. 155), is probably an adaptation advantageous to such larvæ when crawling over the ground in search of a suitable place of concealment. Making the necessary correction for the different mode of counting the segments, it will be seen that the primary ocelli of this species are in the same position as those of the other species of this genus as described in a previous part of this essay, and that it belongs to the second phyletic group treated of at p. 193. The interesting fact that this species does not display dimorphism, whilst the closely allied form from Kashmir is dimorphic, shows that in the present species the process of double adaptation has not taken place; and this will probably be found to be connected with the habits of life, *i. e.* the insect being well adapted to the colour of its food-plant may not conceal itself on the ground by day. The caterpillar of *Deilephila Robertsi*, Butl., is found at Candahar on a species of *Euphorbia* growing on the rocky hills, and is so abundant that at the end of May every plant with any leaves left on it had several larvæ feeding upon it. "The larvæ are very beautiful and conspicuous, and are very different in colouring according to their different stages of growth." The general colour is black with white dots and spots; a subdorsal row of large roundish spots, one on each segment, either white, yellow, orange or red; dorsal stripe variable in colour, and sometimes only partially present or altogether absent. "At the end of May most of the larvæ found presented a different appearance; the black disappears more or less, and with it many of the small white spots. In some cases the black only remains as a ring round the larger white spots; the ground-colour therefore becomes yellowish-green or yellow, varying very considerably." The larva does not change colour previous to pupation. This species, according to the outline figure given (*loc. cit.*, Pl. XXXIX., Fig. 9), appears to belong to the first of Dr. Weismann's, groups, comprising *D. Euphorbiæ*, *D. Dahlii* and *D. Nicæa* (see p. 199), and is therefore in the seventh phyletic stage of development (p. 224). From the recorded habits it seems most probable that the colours and markings of this caterpillar are signals of distastefulness. It is much to be regretted that Major Roberts has not increased the value of his description of this species by adding some observations or experiments bearing on this point. *Eusmerinthus Kindermanni*, Lederer, feeds on willow. "General colour green, covered with minute white dots and seven long pale-yellow oblique lateral bands. (The ground-colour is the same as the willow-leaves on which the larva feeds, the yellow stripes the same as the leaf-stalks, and the head and true legs like the younger branches)." As no subdorsal line is mentioned or figured, this species must be regarded as belonging to the third stage of phyletic development (see p. 242).

I have recently had an opportunity of inspecting a large number of drawings of Sphinx-larvæ in the possession of Mr. F. Moore, and of those species not mentioned in the previous portions of this work the following may be noticed:—*Chærocampa Theylia*, Linn., like *Ch. Lewisii* (note 13, p. 194), appears to be another form connecting the second and third phyletic groups of this genus. *Ch. Clotho*, Drury, belongs to the third group (figured by Semper ; see note 3 to this Appendix). The larva of *Ch. Lucasii*, Walk., offers another instance of the retention of the subdorsal line by an ocellated species. The larva of *Ch. Lycetus*, Cram., of which Mr. Moore was so good as to show me descriptions made at the various stages of growth, presents many points of interest. It belongs to the third phyletic group, and all the ocelli appear at a very early stage. The dimorphism appears also in the young larvæ, some being green, and others black, a fact which may be explained by the law of "backward transference" (see p. 274). A most suggestive feature is presented by the caudal horn, which in the young caterpillar is stated to be *freely movable*. It is possible that this horn, which was formerly possessed by the ancestors of the *Sphingidæ*, and which is now retained in many genera, is a remnant of a flagellate organ having a similar function to the head-tentacles of the *Papilio*-larvæ, or to the caudal appendages of *Dicranura* (see p. 289).

Lophostethus Dumolinii, Angas.—The larva of this species differs so remarkably from those of all other *Sphingidæ*, that I have thought it of sufficient interest to publish the following description, kindly furnished by Mr. Roland Trimen, who in answer to my application sent the following notes :—"My knowledge of the very remarkable larva of this large and curious Smerinthine Hawk-moth is derived from a photograph by the late Dr. J. E. Seaman, and from drawings and notes recently furnished by Mr. W. D. Gooch. The colour is greenish-white, inclining to grey, and in the male there is a yellow, but in the female a bluish, tinge in this. All the segments but the second and the head bear strong black spines, having a lustre of steel blue, and springing from a pale yellow tubercular base. The longest of these spines are in two dorsal rows from the fourth to the eleventh segment, the pairs on the fourth and fifth segments being longer than the rest, very erect, and armed with short simple prickles for three-fourths of their upper extremity. The anal horn, which is shorter than the spines, is of the same character as the latter, being covered with prickles, and much inclined backwards. Two lateral rows of similar shorter spines extend from the fourth to the 12th segment, and on each of the segments 6—11 the space between the upper and lower spines is marked with a conspicuous pale yellow spot. Two rows of smaller similar spines extend on each side (below the two rows of larger ones) from the second to the thirteenth segment, one spine of the lowermost row being on the fleshy base of each pro-leg. All the pro-legs are white close to the base, and russet-brown beyond. Head smooth, unarmed in adult, greenish-white with two longitudinal russet-brown stripes on face.

"The young larvæ have proportionally much longer and more erect spines with distinct long prickles on them. There is a short pair besides, either on the back of the head or on the second segment. Moreover, the dorsal spines of the third and fourth segments, and the anal horn (which is quite erect, and the longest of all), are longer than the rest, and distinctly *forked* at their extremity.

"Mr. Gooch notes that these young larvæ might readily be mistaken for those of the *Acrææ*, and suggests that this may protect them. He also states that the yellow lateral spots are only noticed after the last moult before pupation, and that the general resemblance of the larva as regards colour is to the faded leaves of its food-plant, a species of *Dombeia*."

The forked caudal horn in the young larva of this species is of interest in connection with the similar character of this appendage in the young caterpillar of *Hyloicus Pinastri*, p. 265.

Retention of the Subdorsal Line by Ocellated Larvæ.—It has already been shown with reference to the eye-spots of the *Chœrocampa*-larvæ, that these markings have been developed from the subdorsal line, and that, in accordance with their function as a means of causing terror, this line has in most species been eliminated in the course of the phylogeny from those segments bearing the eye-spots in order to give full effect to the latter (see p. 379). In accordance with the law that a character when it has become useless gradually disappears, the subdorsal is more or less absent in all those species in which the ocelli are most perfectly developed; and it can be readily imagined that in cases where adaptation to the foliage exists the suppression of this line would under certain conditions be accelerated by natural selection. On the other hand, it is conceivable that the subdorsal line may under other conditions be of use to a protectively coloured ocellated species by imitating some special part of the food-plant, under which circumstances its retention would be secured by natural selection.

Such an instance is offered by *Chœrocampa Capensis*, Linn.; and as this case is particularly instructive as likewise throwing light upon the retention of the subdorsal by certain species having oblique stripes (see p. 377, and note 7, p. 378), I will here give some details concerning this species which have been communicated to me by Mr. Roland Trimen, the well-known curator of the South African Museum, Cape Town. The caterpillar of *C. Capensis*, like so many other species of the genus, is dimorphic, one form being a bright (rather pale) green, and the other, which is much the rarer of the two, being dull pinkish-red. Both these forms are adapted in colour to the vine on which they feed, the red variety according to some extent with the faded leaves of the cultivated vines, but to a greater extent with the young shoots and underside of the leaves of the South African native vine (*Cissus*

Capensis), on which it also feeds. There are two eye-spots in this species in the usual positions; they are described as being blue-grey in a white ring, and raised so as to project a little. The subdorsal is white, and is bordered beneath by a wide shade of bluish-green irrorated with white dots, and crossed by an indistinct white oblique ray on each segment. These last markings are probably remnants of an oblique striping formerly possessed by the progenitor of this and other species of the genus (see, for instance, Fig. 25, Pl. IV., one of the young stages of *C. Porcellus*). It is possible that these rudimentary oblique stripes are now of service in assisting the adaptation of the larva to its food-plant, but this cannot be decided without seeing the insect *in situ.*

The subdorsal line extends from immediately behind the second eye-spot to the base of the very short and much curved violet anal horn. With reference to the protective colouring Mr. Trimen writes:—"The difficulty of seeing these large and beautifully-coloured larvæ on the vines is quite surprising; six or more may be well within sight, and yet quite unnoticed. The subdorsal stripe greatly aids in their concealment, as it well represents in its artificial light and shade the leaf-stalks of the vine." When this larva withdraws its front segments the eye-spots stand out very menacingly; but in spite of this it is greedily eaten by fowls and shrikes (*Fiscus Collaris*), and Mr. Trimen also found that a tame suricate (*Rhyzœna Suricata*) and a large monitor lizard (*Regenia Albogularis*) did not refuse them. The failure of the eye-spots in causing terror in these particular cases cannot be regarded as disproving their utility in all instances. It must always be borne in mind that no protective character can possibly be of service against *all* foes; natural selection only requires that such characters should be advantageous with respect to the majority of the enemies of any species, and further experiments with this caterpillar may show that in the case of smaller foes the eye-spots are effective as a means of causing alarm. The dimorphism of the larva of *C. Capensis* is of special interest, although we are not yet sufficiently acquainted with the habits of this species to offer a complete explanation. According to Dr. Weismann's conclusions (p. 297), the dimorphism of the *Chærocampa*-larvæ is due to a double adaptation, the insects first having acquired the habit of concealing themselves by day, and the dark form having then been produced by the action of natural selection, in order to adapt such varieties to the colour of the soil, whilst others retained the green colour which adapts them to the foliage of their food-plants. In accordance with this, *C. Capensis* may have a similar habit of concealment, or (should this be found not to be the case) it is possible that this insect at a former period possessed this habit and fed upon some other plant, when it would have become dimorphic in the manner explained, and the *existing* dimorphism may be a survival of the more ancient dimorphism, the red form (corresponding to the older dark form) having been subsequently modified so as to become also adapted to the new food-plant. Much light would be thrown upon this by studying the ontogeny of the species.

Phytophagic Variability.—A number of observations bearing on the phytophagic variability of the Sphinx-larvæ and other caterpillars have been recorded in a previous note (p. 305), and reference has also been made to the food-plants of *Acherontia Atropos* in South Africa (note 50, p. 263). I am now enabled to add some further observations on this species, from notes furnished to me by Mr. Roland Trimen, who states that for many years he has noticed that at the Cape this larva varies greatly in the depth and shade of the green ground-colour, the variability being in strict accordance with the colour of the leaves of the particular plant on which the individual feeds. The phenomenon was particularly noticeable in larvæ feeding on *Buxia Grandiflora,* a shrub in common cultivation in gardens, and of which the foliage is of a very dull pale greyish-green. Another striking instance was noticed in some very fine caterpillars feeding on a large shrubby *Solanum,* which, excepting the bright yellow bands bordering the dorsal violet bars, were generally dull ochreous-yellow, like the leaves and stalks of the *Solanum.* On plants with bright green or deep green leaves, the colour of the larvæ is almost in exact agreement. Mr. Trimen adds :—"These remarks apply principally to the underside and pro-legs and lower lateral regions, the dorsal colours of violet and yellow varying but little. The protection afforded is very considerable, as the larvæ almost always cling to the lower side of the twigs of their food-plants, so that their uniformly-coloured under-surface is upwards, and turned towards the light, and their variegated upper surface turned downwards."

These observations are of the highest importance, not only as adding another instance to the recorded cases of phytophagic variation, but likewise as showing that with this variability a protective habit has been acquired. It is to be hoped that such a promising field for experimental investigation as is offered by this and analogous cases will not long remain unexplored. In attacking the problem two chief questions have in the first place to be settled : (1) Is the variability truly phytophagic, *i. e.* are the colour-variations actually brought about by the chemico-physiological action of the food-plant ? and (2) Are the larvæ at any period of growth susceptible to the action of phytophagic influences ? The first question could be decided by feeding larvæ from the same batch of eggs on different food-plants from the period of their hatching. The second question could be settled by changing the food-plants of a series of selected specimens at various stages of growth, and observing whether any change of colour was produced. In accordance with the principles advocated in a previous note (p. 305), it is conceivable *à priori* that phytophagic variability may occur by direct chemico-physiological action, quite irrespective of any of the changes of colour being of protective use. In the case of brightly-coloured distasteful species phytophagic variability might thus have full play, but in the case of protectively-coloured edible species, phytophagic variability would be under the control of natural selection. These considerations raise a question of the greatest theoretical interest in connection with this phenomenon. If phytophagic variability can have full play uncontrolled by natural selection in brightly-coloured caterpillars, ought not this phenomenon to be of more common occurrence in such species than in those protectively coloured? Although our knowledge of this subject is still very imperfect, as a matter of fact brightly coloured larvæ, so far as I have been able to ascertain, do not appear to be susceptible of phytophagic influences. But this apparent contradiction, instead of opposing actually confirms the foregoing views, as will appear on further consideration. The colours of protected species are as a whole much inferior in brilliancy to those of inedible species, so that any phytophagic effect would be more perceptible in the former than in the latter, in which the highest possible standard of brilliancy appears in most cases to have been attained. Now phytophagic variations of colour appear to be of but small amount, or, in other words, such variations fluctuate within comparatively restricted limits, and as the cases at present known are mostly *adaptive* it is legitimate to conclude that they have been produced and brought to their present standard by natural selection, *i. e.* that they have arisen from phytophagic influences as a cause of variability. The initial stages of phytophagic variations must therefore have been still less perceptible than the now perfected final results; and this leads to the conclusion that minute variations of this character were of sufficient importance to protectively-coloured species to be taken advantage of by natural selection. But minute variations in a dull-coloured larva would, as previously pointed out, produce a comparatively much greater effect than such variations in a brilliantly-coloured species; and as protection is required by the former, the initial phytophagic effects would be accumulated, and the power of adaptability conferred by the continued action of natural selection, whilst in vividly-coloured species where no power of adaptability is required this cause of variation would not only produce a result which, as compared with its effects upon dull species, may be regarded as a "vanishing quantity," but this result would be too insignificant to be taken advantage of by natural selection, which is in these cases dealing only with large "quantities," and striving to make the caterpillars as brilliant as possible. The fact that vividly-coloured distasteful larvæ do not show phytophagic variation is to my mind explained proximately by these considerations; the ultimate cause of phytophagic variability regarded as a chemico-physiological action requires further investigation.

Sexual Variation in Larvæ.—Since most of the markings of caterpillars can be explained by the two factors of adaptation and inheritance, or, in other words, by their present and past relations to the environment, and since sexual selection can have played no direct part in producing these colours and markings, I feel bound to record here some few observations on the sexual differences in larvæ in addition to the cases of *Anapæa* and *Orgyia* already recorded (note i., p. 308) and of *Lophostethus Dumolinii* (p. 527).

Mr. C. V. Riley states[6] with reference to the larva of *Thyreus Abboti* that the ground-colour appears to depend upon the sex, Dr. Morris having described the insect as "reddish-brown with numerous patches of light green," and having expressly stated that "the *female* is of a uniform reddish-brown with an interrupted dark-brown dorsal line and transverse striæ." Mr. W. D. Gooch, who has reared the South African butterflies *Nymphalis Cithæron* and *N. Brutus* from their larvæ, states[7] that these "differed sexually in both instances." Of *Brutus* only a few were bred, but of *Cithæron* many. "The sexual difference of the latter was that the females had a large dorsal sub-cordate cream mark, which was wanting, or only shown by a dot, in the males, and the colour was more vivid in the edgings to the frontal horns."

Although such cases appear to be at present inexplicable, they are of interest as examples of those "residual phenomena" which, as is well known, have in many branches of science so often served as important starting-points for new discoveries and generalizations.[8]

[6] "Second Annual Report," 1870, p. 78.

[7] "Entomologist," vol. xiv. p. 7.

[8] With reference to the habits of *C. Capensis* (p. 531), I have since been informed by Mr. Trimen that this species does not conceal itself by day, so that the dimorphism may be regarded as a character retained from an earlier period and adapted to the present life conditions.

APPENDIX II.

THE following paper by Dr. Fritz Müller [1] forms the third of a series of communications on Brazilian butterflies published in "Kosmos," and as it bears upon the investigations made known in the third essay of the present work, I will here give a translation, by permission of the publisher, Herr Karl Alberts.

"ACRÆA AND THE MARACUJÁ BUTTERFLIES AS LARVÆ, PUPÆ, AND IMAGINES.

"In a thoughtful essay on 'Phyletic Parallelism in Metamorphic Species,' Weismann has shown that in the case of Lepidoptera the developmental stages of larva, pupa, and imago vary independently, and that a change occurring in one stage is without influence upon the preceding and succeeding stages, so that the course which has been followed by the individual stages in their developmental history has not been in all cases identical. This want of agreement may manifest itself both by unequal divergence of form-relationship, and by unequal group formation. With respect to unequal form-divergence the caterpillars are sometimes more closely related in form than their imagines, and at other times the reverse is the case. With respect to unequal group formation again, two cases are possible; the larvæ and imagines may form groups of unequal value, the one stage form-

[1] "Kosmos," Dec. 1877, p. 218. The paper is here introduced chiefly with a view to illustrate an important case of incongruence among Lepidopterous pupæ.

ing higher or lower groups than the other, or they may form groups of unequal size, *i. e.*, groups which do not coincide but which overlap. Form-relationship and blood-relationship do not therefore always agree; the resemblances among the caterpillars would lead to a quite different arrangement to that resulting from the resemblances among the imagines, and it is probable that neither of these arrangements would correspond with the actual relationships.

"Starting from this fact, which he establishes by numerous examples, Weismann proceeds to show most convincingly that an innate power of development or of transformation, such as has been assumed under various names by many adherents of the development theory, has no existence, but that every modification and advancement in species has been called forth by external influences.

"A most beautiful illustration of the want of 'phyletic parallelism,' as Weismann designates the different form-relationships of the larvæ, pupæ, and imagines, is furnished by the five genera *Acræa*, *Heliconius*, *Eueides*, *Colænis*, and *Dione* (= *Agraulis*). This instance seems to me to be of especial value, because it offers the rare case of pupæ showing greater differences than the larvæ and imagines.

"The species of which I observed the larvæ and pupæ are *Acræa Thalia* and *Alalia*, *Heliconius Eucrate*, *Eueides Isabella*, *Colænis Dido* and *Julia*, *Dione Vanillæ* and *Juno*; besides these I noticed the pupa of *Eueides Aliphera*.

"The following remarks apply only to these species, although we may suppose with great probability that the whole of the congeneric forms—excepting perhaps the widely ranging species of *Acræa*—would display similar characters to their Brazilian representatives.

"The imagines of the five genera mentioned form two

sharply defined families, the *Acræidæ* and the butterflies of the Maracujá group.[2] The latter comprises the three genera *Heliconius*, *Eueides*, and *Colænis*, which differ only in very unimportant characters; *Eueides* is distinguished from *Heliconius* by its shorter antennæ, and *Colænis* differs from *Eueides* in having the discoidal cell of the hind-wings open. The genus *Dione* is further removed by the different structure of the legs, and the silvery spots on the underside of the wings. Certain species resemble those of other genera in a most striking manner, and much more closely both in colour and marking, and even in the form of their wings, than they do their own congeners. This is the case with *Acræa Thalia* and *Eueides Pavana*, with *Heliconius Eucrate* and *Eueides Isabella*, and with *Eueides Aliphera* and *Colænis Julia*, which are deceptively alike, and the last two are connected with *Dione Juno*, at least by the upper side of the wings. The difficulty of judging of the relationships of the single species is thus much aggravated; it cannot be said how much of this resemblance is to be attributed to blood-relationship, and how much to deceptive imitation.

"As larvæ all the Brazilian species must be placed *in one genus*, as they agree exactly in the number and arrangement of their spines (4 spines, not in a transverse row, on segments 2 and 3; 6 spines, in a tranverse row, on segments 4—11; 4 spines, not in a transverse row, on the last (12th) segment). They differ from one another much less in this respect than do the German species of *Vanessa*, such, for instance, as *V. Io* or *Antiopa* from *V. Polychloros*, *Urticæ*, and *Atalanta*.[3] The larvæ of *Acræa Thalia* are certainly without the two spines on the head which the others possess, and, on the other hand, they have a well-developed pair of spines on the

[2] [Maracujá, the local name for the Passiflora. R.M.]

[3] See p. 448.

first segment, which, in most of the other species, are completely absent; but this does not justify their separation, since the head spines of *Heliconius*, *Eueides*, and *Colænis Dido*, which are of a considerable length, are shorter than those of the next segment in *Colænis Julia*, and *Dione Vanillæ*, and in *Dione Juno* they dwindle down to two minute points, this last species also bearing a short pair on the first segment. The larva of *Dione Juno* is thus as closely related to that of *Acræa Thalia* as it is to that of its congener *Dione Vanillæ*.

"If it were desired to form two distinct larval groups this could not be effected on the basis of their differences in form, but could only be based on their food-plants. The larvæ of *Heliconius*, *Eueides*, *Colænis*, and *Dione* live on species of Maracujá (Passiflora); those of *Acræa Thalia* and *Alalia* on Compositæ (*Mikania* and *Veronia*). These larval groups would agree with those founded on the form-relationships of the imagines, but unlike the imaginal groups, which can be formed into families, they would scarcely possess a generic value.

"If we arrange the single species of caterpillars according to their resemblances, this arrangement does not agree with that based on the resemblances of the imagines, even if we disregard the different values of the groups. The result is somewhat as follows:—

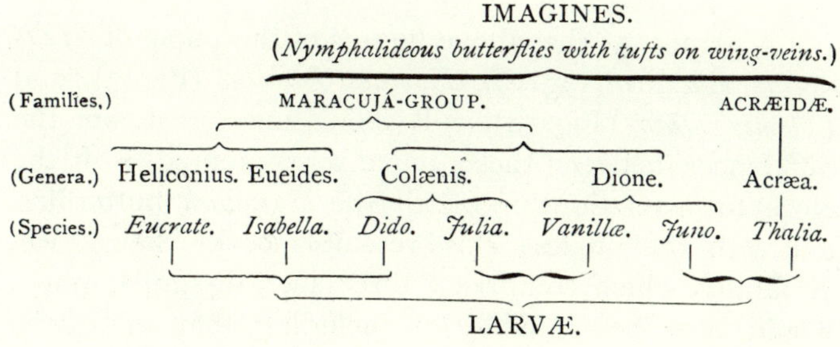

* * * * * *

[Here follow the remarks on the habits of the larvæ in connection with their colours, &c., which have already been quoted in illustration of the use of the spiny protection (note 5, p. 293). From these facts the author draws the conclusion that the form-relationships of the caterpillars depend rather upon their mode of life than upon their blood-relationships, assuming the latter to be correctly expressed by the arrangement of the imagines at present adopted.]

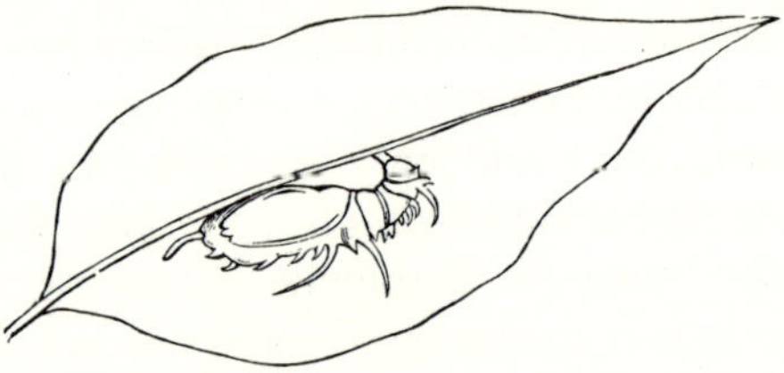

Fig. 1.

Fig. 2.

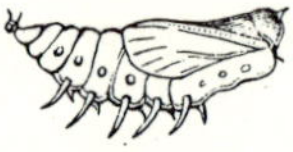

Fig. 3

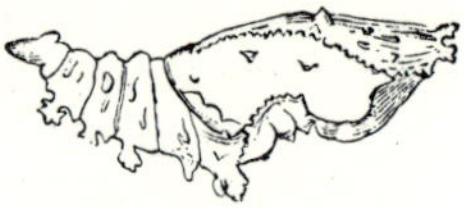

Fig. 4.

Figs. 1—4. Pupæ of *Acræa Thalia ; Heliconius Eucrate ; Eueides Isabella*, and *Colænis Dido ;* life size.

" A glance at the above figures of the pupæ of *Heliconius Eucrate* (Fig. 2), *Eueides Isabella* (Fig. 3), and *Colænis Dido* (Fig. 4), will show how great are the differences between these pupæ as compared with the close form-relationship of all the Maracujá butterflies, and with the no less close resemblance of their larvæ. A family which comprised three such dissimilar pupæ would also be capable of including that of *Acræa Thalia* (Fig. 1).

"The pupa of this last species has nothing peculiar in its general appearance, but possesses the ordinary pupal form; it is tolerably rounded, without any great elevations or depressions; a minute pointed projection is situated on the head over each eye-cover, and a similar process projects from the roots of the wings. Its distinguishing characters are five pairs of spines on the back of the abdominal segments. These spines are found also in *Acræa Alalia*, but appear to be absent in other species, *e.g.* in the Indian *A. Violæ*. Last summer, among some batches of *Thalia* larvæ—each batch being the progeny from one lot of eggs—I found certain individuals which differed from the others in having much shorter spines, and these changed into pupæ in which the five pairs of spines were proportionally shorter than usual, thus being an exception to the rule that changes in one stage of development are without influence on the other stages. I may remark, by the way, that this law, enunciated by Weismann, can only be applied to imagines and pupæ with certain restrictions. The skin of the pupa forms a sheath or cover for the eyes, antennæ, trunk, legs, and wings of the imago, and if these parts undergo any considerable modification in the latter, corresponding changes must appear in the pupa. This is shown, for instance, by many 'Skippers' (*Hesperidæ*), in which the extraordinarily long trunk necessitates a sheath of a corresponding length. The colour of the pupa of *Acræa Thalia* is whitish, the wing-veins with some other markings and the spines are black; metallic spots are absent.

"In the pupa of *Heliconius Eucrate* the laterally compressed region of the wings is raised into a large projection, the antennal sheaths lying on the edges of the wings are serrated and beset with short pointed spines; instead of the minute projections of *Acræa Thalia*, the head bears two large humped processes; the body is raised on each side into a foliaceous border carrying five spines of different lengths, the foremost pair, directed towards the head, being the longest. The pupa is brown, and ornamented with four pairs of brilliant metallic spots, one pair close behind the antennæ, and three pairs, almost coalescent, on the back before the longest pair of spines. A short spine projects from the middle of each of the latter somewhat arched metallic patches.

"In the pupa of *Colænis Dido* (which resembles that of *Colænis Julia*, and to which may be added those of *Dione Vanilla* and *Juno*) the spines are absent, the wing region is but moderately arched, and the antennæ marked only by small elevations; instead of the leaf-like border, there are on each side of the back five knotty or humped processes. The metallic spots are similar in number and position to those of *Heliconius Eucrate*; those on the back have a wart-like process in the middle, instead of a spine.

"The pupæ of *Heliconius* and *Colænis* when moving their posterior segments rapidly, as they do whenever they are disturbed, produce a very perceptible hissing noise by the friction of these segments, this sound, which is especially noticeable in the case of *Heliconius Eucrate*, perhaps serving to terrify small foes. (So loud is the sound produced in this manner by the pupæ of *Epicalia Numilia*, that my children have named them '*Schrei-puppen*.')

"The pupæ of *Heliconius* and *Colænis* thus differ to a much greater extent than the imagines or larvæ, and the same holds good for *Eueides* in a much higher degree as compared with its above-mentioned allies. The larvæ of *Eueides* have no distinctive characters, and even the generic rank of the imagines is doubtful; as pupæ, on the other hand, they are far removed (even by their mode of suspension) not only from the remainder of the Maracujá group and from the whole of the great Nymphalideous group (*Danainæ*, *Satyrinæ*, *Elymniinæ*, *Brassolinæ*, *Morphinæ*, *Acræinæ* and *Nymphalinæ*), but from almost all other butterflies. The larva pupates on the underside of a leaf; the pupa is fastened by the tail, but does not hang down like the pupæ of the other *Nymphalidæ*,—its last segments are so curved that the breast of the chrysalis is in contact with the underside of the leaf. I am not acquainted with any other pupa among those not suspended by a girdle which assumes such a position. Something similar occurs, however, in the pupa of *Stalachtis*, which is without a girdle, and according to Bates, is 'kept in an inclined position by the fastening of the tail.' By this peculiarity Bates distinguishes the *Stalachtinæ* from the *Libytheæ* with pupæ 'freely suspended by the tail.'

"Besides through this peculiar position of the body, the pupa of *Eueides Isabella* is distinguished by short hooked and long narrow sabre-like pairs of processes on the back and head. Its colour is whitish, yellowish, or sordid yellowish-grey; in the last variety both the four long dorsal processes and the surrounding portions, as well as the points of the other processes, remain white or yellowish. The pupa *Eueides Aliphera* is very similar, only all the processes are somewhat shorter, the four longest (dorsal) and some other markings being black.

"Now if, as Weismann has attempted to show for larvæ and imagines, the form-divergence always 'corresponds exactly with the divergence in the mode of life,' the question arises as to what difference in the conditions of life has brought about such a considerable form-divergence between the pupæ of such closely-allied species as the Maracujá butterflies. In pupæ which do not eat or drink, and which have neither to seek in courtship nor to care for progeny, it is only protection from foes that can concern us. But in the pupæ of nearly-allied species of which the larvæ feed on kindred plants in the same districts at the same periods of the year, can the enemies be so different as to produce such a considerable divergence in form? One might answer this question in the negative with some confidence, and affirm that in this case the difference in the pupæ does not result from the 'divergence in the mode of life,' or from the difference in the external conditions, but is accidental, *i.e.* a consequence of some fortunate variation induced by some external cause, which variation afforded protection against common foes—to one species in one way, and to the other species in some other way; this course, once entered upon, having been urged on by natural selection, until at length the wide divergence now shown is attained. How in the case of any of the species the peculiarity in colour or form can actually serve as a protection, I must confess myself at fault in answering. Only in the case of the pupa of *Eueides Isabella* will I venture to offer a supposition. That it is not green like other pupæ which suspend themselves among foliage (*Siderome*, *Epicalia*, *Callidryas*, &c.), but contrasts more or less brightly with the dark green of the leaves, precludes the idea of concealment; on the other hand its colour is too dull to serve as a conspicuous sign of distastefulness. In either case the meaning of the wonderful processes of the pupa would remain unexplained.

"We are thus compelled to seek another possibility in mimicry, by which foes would be deceived by deceptive resemblance. But what is the object imitated? Dead insects overgrown by fungi are often found on leaves, the whitish or yellowish fungi growing from their bodies in various fantastic forms. Such insects of course no longer serve as tempting morsels. The processes of the pupa of *Eueides* suggest such fungoid growths, although I certainly cannot assert that to our eyes in broad daylight the resemblance is very striking. But the pupæ hang among the shadows of the leaves, and a less perfect imitation may deceive foes that are not so sharp-sighted; protective resemblance must commence moreover with an imperfect degree of imitation."

EXPLANATION OF THE PLATES.

PLATE III.

Figs. 1—12 represent larvæ of *Macroglossa Stellatarum*, all bred from one batch of eggs. Most of the figures are enlarged, but sometimes to a very small extent only; the lines show the natural length.

Fig. 1. Stage I.; a caterpillar immediately after hatching. Natural length, 0.2 centim.

Fig. 2. Stage II.; shortly after the first moult. Natural length, 0.7 centim.

Figs. 3—12. Stage V.; the chief colour-varieties.

Fig. 3. The only lilac-coloured specimen in the whole brood. Natural length, 3.8 centim.

Fig. 4. Light-green form (rare) with subdorsal shading off beneath.

Fig. 5. Green form (rare) with strongly-pronounced dark markings (dorsal and subdorsal lines). Natural length, 4.9 centim.

Fig. 6. Dark-brown form (common). Natural length, 4 centim. In this figure the fine shagreening of the skin is indicated by white dots; in the other figures these are partially or entirely omitted, being represented only in Figs. 8 and 10.

Fig. 7. Light-green form (common). Natural length, 4 centim.

Fig. 8. Light-brown form (common). Natural length 3.5 centim.

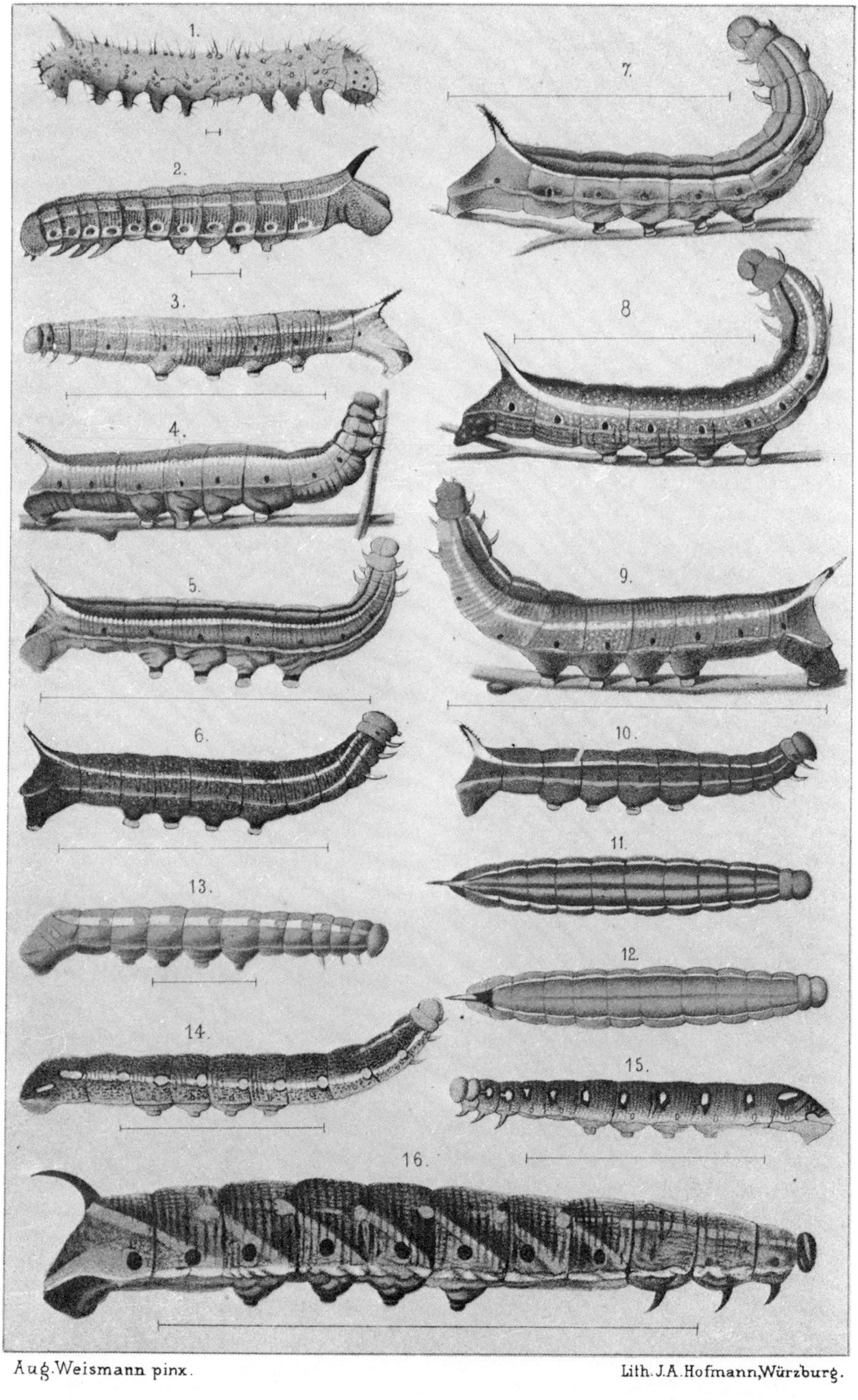

Aug. Weismann pinx. Lith. J. A. Hofmann, Würzburg.

Plate IV.

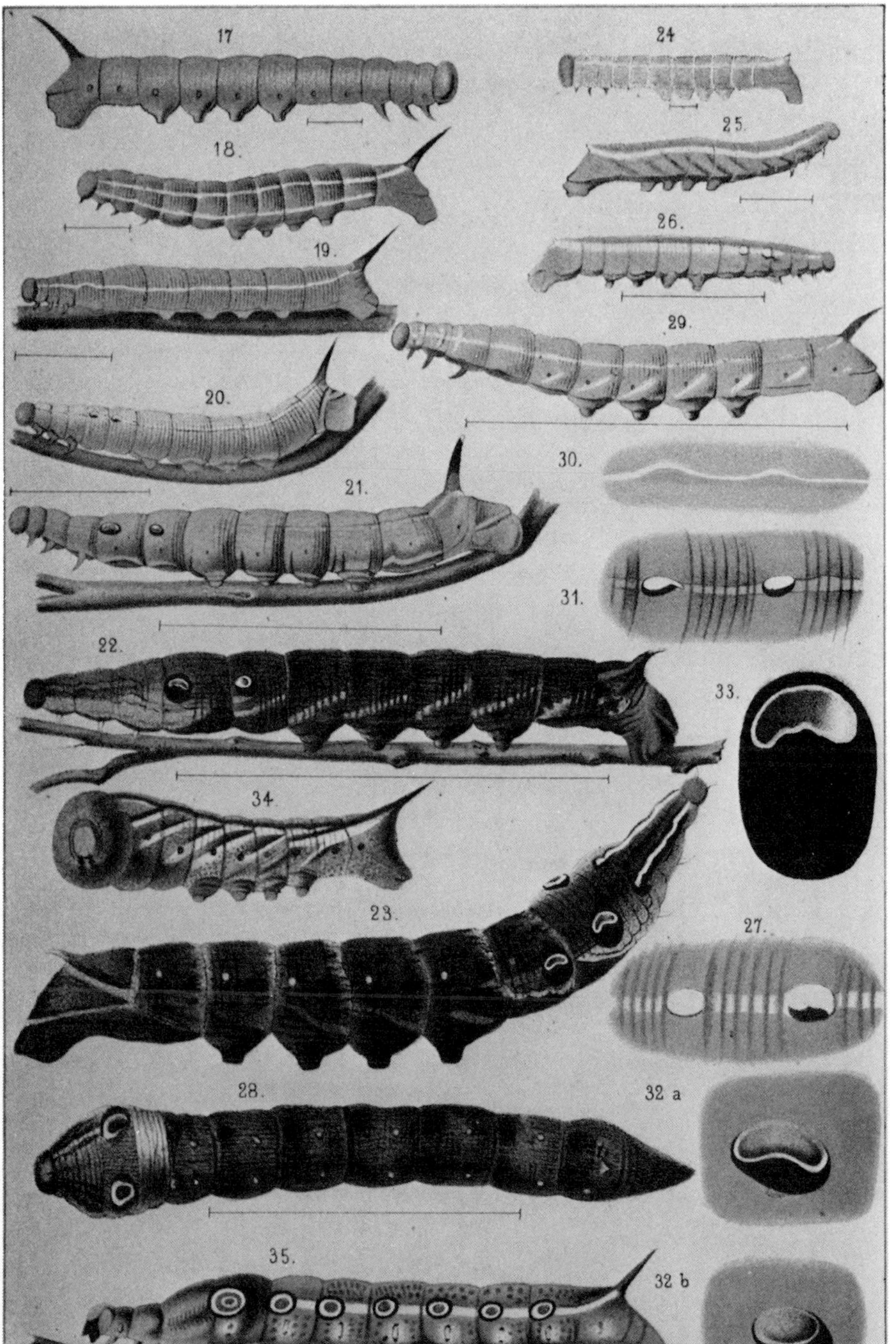

Aug. Weismann pinx.

Lith. J. A. Hofmann, Würzburg

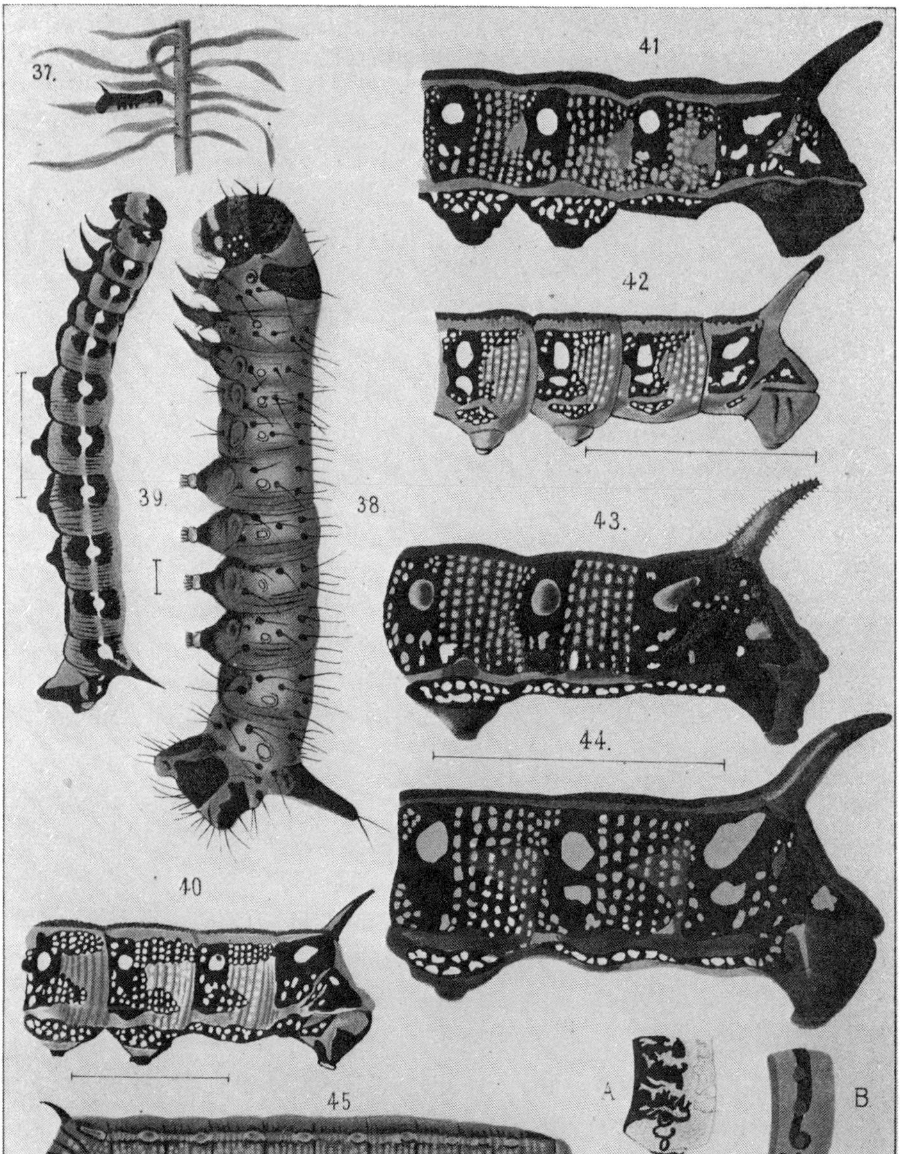

Aug. Weismann pinx.

Lith. J.A. Hofmann, Würzburg.

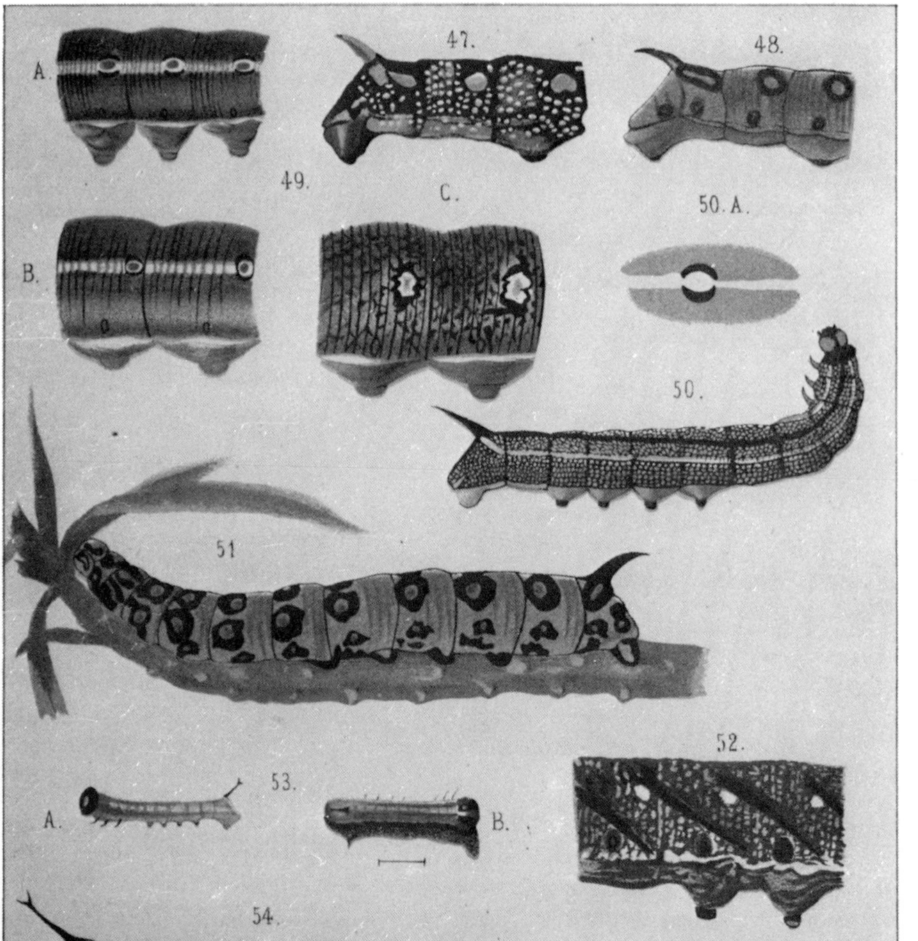

Aug. Weismann pinx.

Lith. J. A. Hofmann, Würzburg.

Fig. 9. Parti-coloured specimen, the only one out of the whole brood. Natural length, 5.5 centim.

Fig. 10. Grey-brown form (rare).

Fig. 11. One of the forms intermediate between the dark-brown and green varieties, dorsal aspect.

Fig. 12. Light-green form with very feeble dorsal line (shown too strongly in the figure), dorsal aspect.

Figs. 13—15. *Deilephila Vespertilio.*

Fig. 13. Stage III. (?) ; the subdorsal bearing yellow spots. Natural length, 1.5 centim.

Fig. 14. Stage IV. ; the subdorsal interrupted throughout by complete ring-spots, the white " mirrors " of which are bordered with black, and contain in their centres a reddish nucleus. Natural length, 3 centim.

Fig. 15. Stage V. ; shortly after the fourth moult. Subdorsal line completely vanished ; ring-spots somewhat irregular, with broad black borders ; natural length, 3.5 centim.

Fig. 16. *Sphinx Convolvuli*, Stage V., brown form. Subdorsal line retained on segments 1—3, on the other segments present only in small remnants ; at the points where the (imaginary) subdorsal crosses the oblique stripes there are large bright spots ; natural length, 7.8 centim.

PLATE IV.

Figs. 17—22. Development of the markings in *Chærocampa Elpenor.*

Fig. 17. Stage I. ; larva one day after hatching. Natural length, 7.5 millim.

Fig. 18. Stage II. ; larva after first moult. Length, 9 millim.

Fig. 19. Stage II. ; immediately before the second moult (Fig. 30 belongs here). Length, 13 millim.

Fig. 20. Stage III. ; after second moult. Length, 20 millim.

Fig. 21. Stage IV.; after third moult (Figs. 32 and 33 belong here). Length, 4 centim.

Fig. 22. Stage V.; after fourth moult. A feeble indication of an eye-spot can be seen on the third segment besides those on the fourth and fifth. Ocelli absent on segments 6—10.

Fig. 23. Stage VI.; after fifth moult. The subdorsal line is feebly present on segments 6—10, and very distinctly on segments 11 and 1—3. Ocelli repeated as irregular black spots above and below the subdorsal line on segments 6—11; a small light spot near the posterior border of segments 5—10 (dorsal spots) and higher than the subdorsal line. Larva adult.

Figs. 24—28. Development of the markings of *Chærocampa Porcellus.*

Fig. 24. Stage I.; immediately after emergence from the egg. Length, 3.5 millim.

Fig. 25. Stage II.; after first moult. Length, 10 millim.

Fig. 26. Stage III.; after second moult. Length, 2.6 centim.

Fig. 27. Eye-spots at this last stage; subdorsal much faded, especially on segment 4. Position the same as in last Fig.; magnified.

Fig. 28. Stage IV.; after third moult; corresponds exactly with Stage VI. of *C. Elpenor.* Dorsal view, with front segments partly retracted (attitude of alarm). Ocelli on segment 5 less developed than in *Elpenor;* repetitions of ocelli as diffused black spots on all the following segments to the 11th; two light spots on each segment from the 5th to the 11th, exactly as in *Elpenor;* subdorsal line visible only on segments 1—3. Length, 4.3 centim.

Fig. 29. *Chærocampa Syriaca.* From a blown specimen in Lederer's collection, now in the possession of Dr. Staudinger. Length, 5.3 centim.

Fig. 30. First rudiments of the eye-spots of *Chærocampa Elpenor*, Stage II. (corresponding also with Fig. 19 in position, the head of the caterpillar being to the left). Subdorsal line slightly curved on segments 4 and 5.

Fig. 31. Eye-spots at Stage III. of the larva Fig. 20 somewhat further developed (larva immediately before third moult). Position as in Fig. 20.

Fig. 32. Eye-spots at Stage IV corresponding to Fig. 21, A being the eye-spot of the fourth and B that of the fifth segment.

Fig. 33. Eye-spot at Stage V. of the larva of *C. Elpenor;* fourth segment.

Figs. 30—33 are free-hand drawings from magnified specimens.

Fig. 34. *Darapsa Chœrilus* from N. America. Adult larva with front segments retracted. Copied from Abbot and Smith.

Fig. 35. *Chœrocampa Tersa*, from N. America. Adult larva copied from Abbot and Smith.

PLATE V.

Fig. 36. Sixth segment of adult *Papilio*-larvæ ; A, *P. Hospiton*, Corsica ; B, *P. Alexanor*, South France ; C, *P. Machaon*, Germany ; D, *P. Zolicaon*, California.

Figs. 37—44. Development of the markings of *Deilephila Euphorbiæ.*

Fig. 37. Stage I. ; young caterpillar shortly after emergence. Natural length, 5 millim.

Fig. 38. Similar to the last, more strongly magnified. Natural length, 4 millim.

Fig. 39. Stage II. ; larva immediately after first moult. The row of spots distinctly connected by a light stripe (residue of the subdorsal line). Natural length, 17 millim.

Fig. 40. Stage III.; after second moult ; magnified

drawing of the last five segments. Only one row of large white spots on a black ground (ring-spots); sub-dorsal completely vanished; the shagreen-dots formerly absent now appear in vertical rows interrupted only by the ring-spots. Below the latter are some enlarged shagreen-dots which subsequently become the second ring-spots. Natural length of the entire caterpillar, 21 millim.

Fig. 41. Stage IV.; the same larva after the third moult. Transformation of the ground-colour from green to black, owing to the spread of the black patches proceeding from the ring-spots in Fig. 40 in such a manner as to leave between them only a narrow green triangle. The shagreen dots below the ring-spots have increased in size, but have not yet coalesced.

Fig. 42. Stage III.; larva, same age as Fig. 40, but with *two rows* of ring-spots. Natural length of the whole caterpillar, 32 millim.

Fig. 43. Stage V.; larva from Kaiserstuhl. Variety with only one row of ring-spots, and with red nuclei in the mirror-spots. Natural length, 5 centim.

Fig. 44. Stage V.; larva from Kaiserstuhl (like the three preceding). The green triangles on the posterior edges of the segments in Fig. 42 have become changed into red. Natural length, 7.5 centim.

Fig. 45. *Deilephila Galii;* Stage IV. Subdorsal with open ring-spots. Natural length, 3.4 centim.

Fig. 46. *D. Galii*; adult larva; Stage V. Brown variety with feeble shagreening; subdorsal completely vanished. Natural length, 6 centim.

PLATE VI.

Fig. 47. The same species at the same stage. Black variety strongly shagreened; similar to *Deil. Euphorbiæ.*

Fig. 48. Similar to the last. Yellow var. without any trace of shagreening.

Fig. 49. *Deilephila Vespertilio.* Three stages in the life of the species, representing three phyletic stages of the genus. A, life-stage III.=phyletic stage 3 subdorsal with open ring-spots); B, life-stage IV.=phyletic stage 4 (subdorsal with closed ring-spots); C, life-stage V.=phyletic stage 5 (subdorsal vanished, only *one* row of ring-spots).

Fig. 50. *Deilephila Zygophylli*, from S. Russia; stage V. From a blown specimen in Staudinger's collection. In this specimen the ring-spots are difficult to distinguish on account of the extremely dark ground-colour; they are nevertheless present, and would probably be more distinct in the living insect. A, open ring-spot from another specimen of this species in the same collection.

Fig. 51. *Deilephila Nicæa*, from South France; Stage V. Copied from Duponchel.

Fig. 52. *Sphinx Convolvuli;* Stage V., segments 10—8. Brown variety, with distinct white spots at the points of intercrossing of the vanished subdorsal with the oblique stripes.

Fig. 53. *Anceyrx Pinastri;* A and B, larvæ immediately after hatching. Natural length, 6 millim.

Fig. 54. Same species; Stage II. Subdorsal, supra- and infra-spiracular lines developed. Natural length, 15 millim.

Fig. 55. *Smerinthus Populi;* Stage I. Immediately after hatching; free from all marking. Length, 6 millim.

Fig. 56. Same species at the end of first stage; lateral aspect. Length, 1.3 centim.

Fig. 57. Same species; Stage II. Subdorsal indistinct; the first and last oblique stripes more pronounced than the others. Length, 1.4 centim.

Fig. 58. *Deilephila Hippophaës* ; Stage III. Subdorsal with open ring-spot on the 11th segment. A, segment 11 somewhat enlarged. Length, 3 centim.

PLATE VII.

Fig. 59. *Deilephila Hippophaës ;* Stage V. Secondary ring-spots on six segments (10—5).

Fig. 60. Same species ; Stage V. One or two red shagreen dots on segments 10—4 in the position of the ring-spots of Fig. 59. Length, 6.5 centim.

Fig. 61. Same species ; Stage V. Segments 9—6 of another specimen, more strongly magnified. A ring-spot on segments 9 and 8 showing its origin from two shagreen-dots ; two red shagreen-dots on segment 7, on segment 6 only one.

Fig. 62. *Deilephila Livornica* (Europe) in the last stage. Green form. Copied from Boisduval.

Fig. 63. *Pterogon Œnotheræ ;* Stage IV. Length, 3.7 centim.

Fig. 64. The same species at the same stage ; dorsal view of the last segment.

Fig. 65. The same segment in Stage V. Eye-spot completely developed.

Fig. 66. *Saturnia Carpini,* larva from Freiburg ; Stage III. Natural length, 15 millim.

Fig. 67. Same species ; larva from Genoa ; Stage IV. Length, 20 millim.

Fig. 68. Same species ; larva from Freiburg ; Stage III. Segments 8 and 9 in dorsal aspect. Length, 15 millim.

Fig. 69. The same caterpillar ; lateral view of segment 8.

Fig. 70. *Smerinthus Ocellatus ;* adult larva with distinct subdorsal on the six foremost segments. The shagreening is only shown in the contour, elsewhere omitted. Length, 7 centim.

Plate VII.

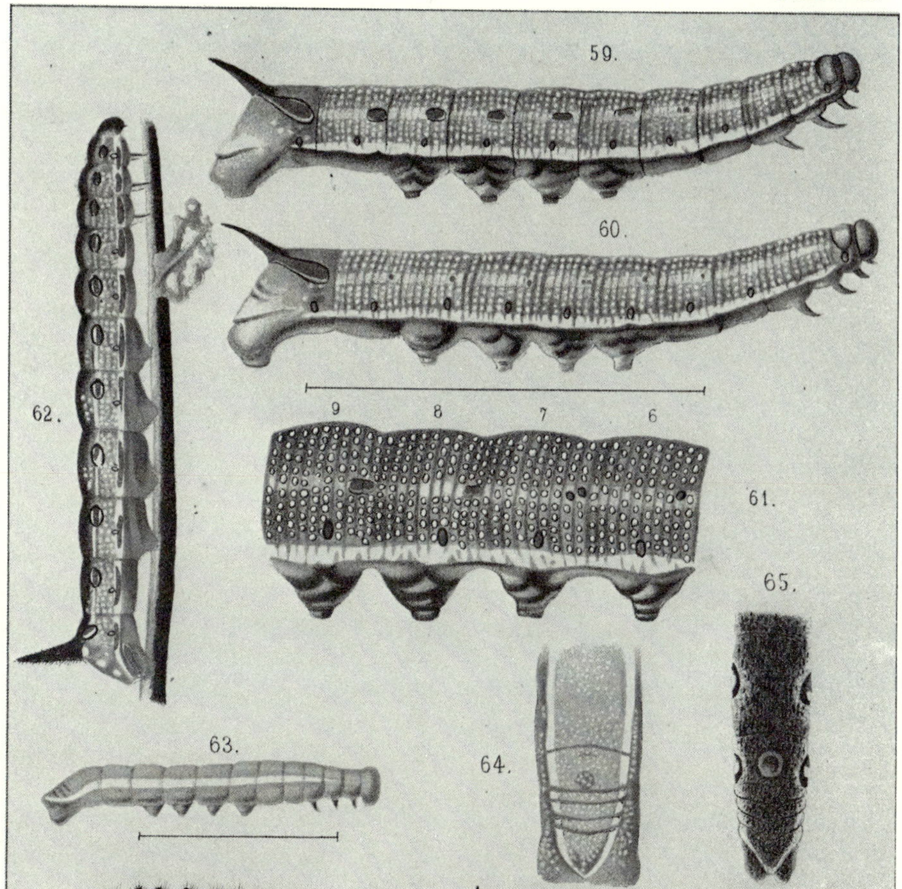

Aug. Weismann pinx.

Lith. J. A. Hofmann, Würzburg.

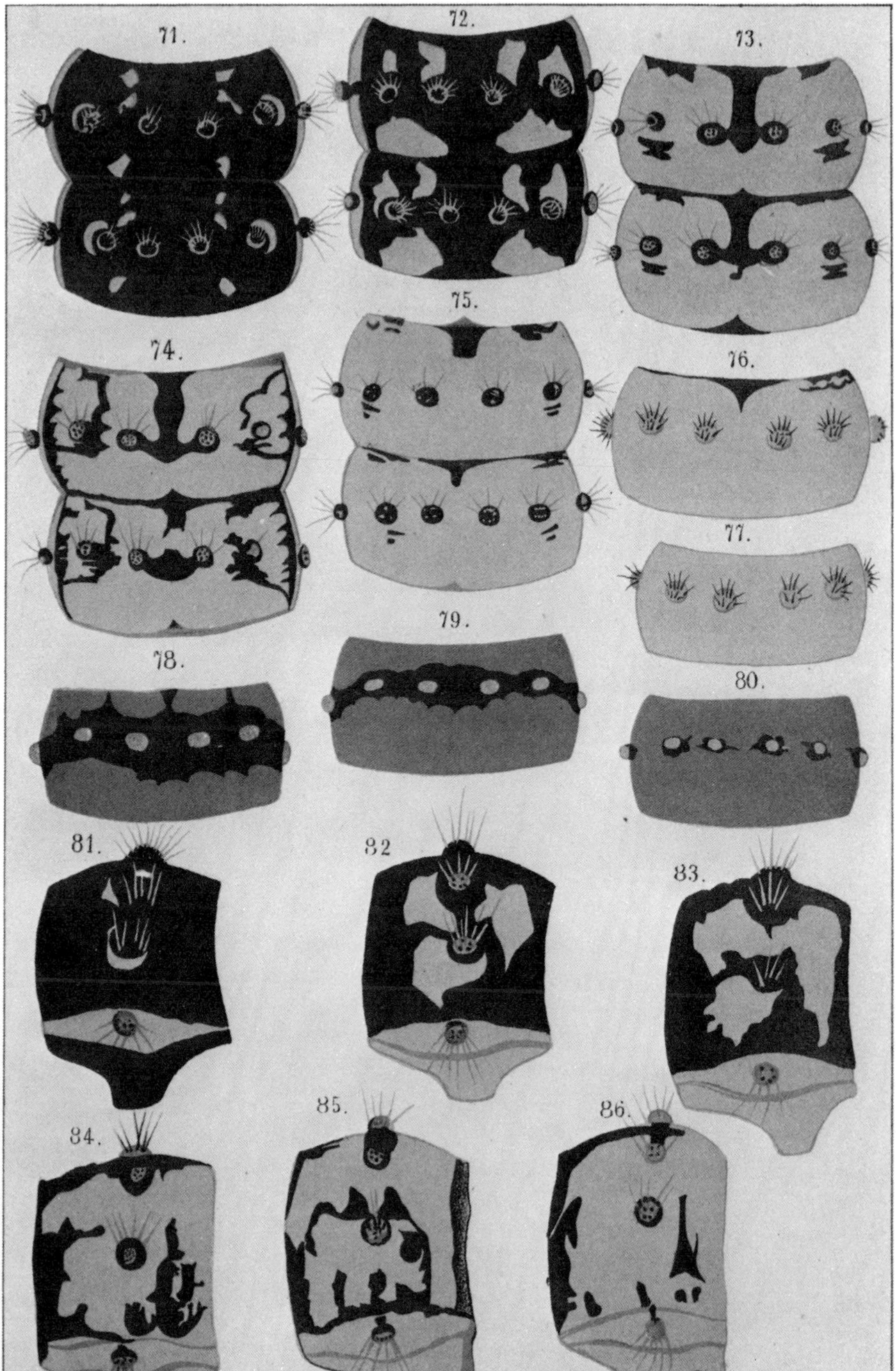

Aug. Weismann pinx.

Lith. J. A. Hofmann, Würzburg.

Plate VIII.

Figs. 71—75 represent segments 8 and 9 of the larva of *Saturnia Carpini* (German form) in dorsal aspect, all at the fourth stage. The head of the caterpillar is supposed to be above, so that the top segment is the eighth.

Fig. 71. *Saturnia Carpini.* Darkest variety.

Fig. 72. Lighter variety.

Fig. 73. Still lighter variety.

Fig. 74. One of the lightest varieties; the black extends further on segments 9 and 10 than on the 8th.

Fig. 75. Lightest variety.

Figs. 76—80 are only represented on a smaller scale than the remaining Figs. in order to save space; were they enlarged to the same scale they would be larger than the other figures.

Fig. 76. *Saturnia Carpini* (Ligurian form); Segment 8; Stage V.

Fig. 77. Same form; same segment in stage VI.

Figs. 78, 79, and 80. *Saturnia Carpini* (German form); dorsal aspect of 8th segment in Stage V. (the last of this form).

Fig. 78. Darkest variety.

Fig. 79. Lighter variety.

Fig. 80. Lightest variety.

Figs. 81—86. *Saturnia Carpini* (German form); Stage IV. Side view of the 8th segment in six different varieties. Fig. 81 shows only two small green spots at the bases of the upper warts besides the green spiracular stripes. Fig. 82 shows the spots enlarged and increased by a third behind the warts; the pro-legs have also become green.

Fig. 83. Two of the three green spots, which have become still more enlarged, are coalescent.

Fig. 84. All three spots coalescent; but here, as also in

Fig. 85, various residues of the original black colour are left as boundary-marks.

Fig. 86. Lightest variety.

END OF PART II.

Part III.

THE TRANSFORMATION OF THE MEXICAN AXOLOTL INTO AMBLYSTOMA

AND

ON THE MECHANICAL CONCEPTION OF NATURE.

STUDIES IN THE THEORY OF DESCENT.

Part III.

ON THE FINAL CAUSES OF TRANSFORMATION.

III.

THE TRANSFORMATION OF THE MEXICAN AXOLOTL INTO AMBLYSTOMA.

INTRODUCTION.

SINCE the time when Duméril made known the transformation of a number of Axolotls into the so-called Amblystoma form, this Mexican Amphibian has been bred in many European aquaria, chiefly with the view to establish the conditions under which this transformation occurred, so as to be enabled to draw further conclusions as to the true causes of this exceptional and enigmatical metamorphosis.

Although the Amphibians propagated freely, the cases in which transformation occurred remained extremely rare, and it was not once possible to reply to the main question, viz. whether this metamorphosis was determined by external conditions or by purely internal causes; to say nothing of the possibility of there perhaps being discoverable certain definite external influences by means of which the metamorphosis could have been induced with certainty. But while these points are undecided all attempted theoretical interpretations of the phenomenon must be devoid of a solid basis.

It appeared to me from the first that the history of this transformation of the Axolotl was of special theoretical value; indeed I believed that it might possibly furnish a special case for deciding the truth of those ground-principles, according to which the origin of this species is represented by the two conflicting schools as a case of transformation or as one of heterogenesis. I therefore determined to make some experiments with the Axolotl myself, in the hopes of being fortunate enough to be able to throw some light upon the subject.

In the year 1872 Prof. v. Kölliker was so good as to leave with me five specimens of his Axolotls, bred in Würzburg, and these furnished a numerous progeny in the following year. With these I carried out the idea, the theoretical bearing of which will be shown subsequently, whether it would not be possible to force all the larvæ, or at any rate, the greater majority, to undergo transformation by exposing them to conditions of life which made the use of gills difficult, and that of lungs more easy; in other words, by compelling them to live partly on land at a certain stage of life.

During that year indeed I obtained no results, most of the larvæ perishing before the time for such an experiment had arrived, and the few survivors did not undergo transformation, but lived on to the following spring and then also died one after the other. Through long absence from Freiburg, necessitated by other labours, I had evidently left them without sufficient care and attention. I was thus led to the conviction, which was more fully confirmed subsequently, that no results can be obtained without the greatest care and attention in rearing, towards which single object all one's interest should be concentrated, and it must not be considered irksome to have to devote daily for many months a large amount of time to this experiment. As it was evident that I could not afford this time without calling in other aid, I hailed with pleasure an opportunity of witnessing the experiment performed by other hands.

A lady living here (Freiburg), Fräulein v. Chauvin, undertook to rear a number of my larvæ of the following year which had just hatched, and in accordance with my idea to make the experiment of forcibly compelling them to adopt the

Amblystoma form. How completely this was accomplished will be seen from the following notes by the lady herself, and it will no less appear that these results were only obtained by that care in treatment and delicacy of observation which she devoted to the experiments.

EXPERIMENTS.

"I began the experiments on June 12th, 1874, with five larvæ about eight days old, these being the only survivors out of twelve. Owing to the extraordinary delicacy of these creatures, the quality and temperature of the water, and the nature and quantity of their food exerts the greatest influence, especially in early life, and one cannot be too cautious in their treatment.

"The specimens were kept in a glass globe of about thirty centimeters in diameter, the temperature of the water being regulated; as food at first Daphnids, and afterwards larger aquatic animals were introduced in large numbers. By this means all the five larvæ throve excellently. At the end of June the rudiments of the front legs appeared in the most vigorous specimens, and on the 9th of July the hind legs also became visible. At the end of November I noticed that one Axolotl remained constantly at the surface of the water, and this led me to suppose that the right period had now arrived for effecting the transformation into Amblystoma. For brevity I shall designate this as No. I., and the succeeding specimens by corresponding Roman numerals.

"In order to bring about this metamorphosis, on December 1st, 1874, No. I. was placed in a large-sized glass vessel containing earth arranged in such a manner that, when the vessel was filled with water, only one portion of the surface of the earth was entirely covered by the liquid, and the creature in the course of its frequent perigrinations was thus more or less exposed to the air. The water was gradually diminished on the following days, during which period the first changes made their appearance in the Amphibian—*the gills commenced to shrivel up*, and at the same time the creature showed a tendency to seek the shallowest spots. On December 4th, it took entirely to the land, and concealed itself among some damp moss which I had placed on a heap of sand on the highest portion of the earth in the glass vessel. At this period the first ecdysis occurred. Within the four days from the 1st to the 4th of December, a striking change took place in the external appearance of No. I., the gill-tufts shrivelled up almost entirely, the dorsal crest completely disappeared, and the tail, which had hitherto been broad, became rounded and similarly formed to that of a land salamander. The grey-brown colour of the body changed gradually into a blackish hue; isolated spots, at first of a dull white, made their appearance and these in time increased in intensity.

"When the Axolotl left the water on December 4th the gill-clefts were still open, but these closed gradually, and after about eight days were overgrown with skin and no longer to be seen.

"Of the other larvæ three appeared at the end of November (*i.e.* at the same time when No. I. came to the surface of the water) to have kept pace in development with No. I., an indication that for these also the right period had arrived for accelerating the developmental processes. They were therefore submitted to the same treatment as No. I. No. II. became transformed at the same time and exactly in the same manner as the latter; its gill-tufts were complete when it was first placed in the shallow water, but after four days these had almost entirely disappeared; in the course of about ten days after it took to the land, the overgrowth of skin on the gill-clefts and the complete assumption of the salamander form occurred. During this last period the creature took food, but only when urged to do so.

"In Nos. III. and IV. the development proceeded more slowly. Neither of these so frequently sought the shallow spots, nor did they as a rule remain so long exposed to the air, so that the greater part of January had expired before they took entirely to the land. Nevertheless the dessication of the gill-tufts did not take a longer time than in Nos. I. and II. as the first ecdysis occurred as soon as they took to the land.

"No. V. showed still more striking deviations in its transformation than Nos. III. and IV., but as this specimen appeared much weaker than the others from the beginning and was retarded in growth to a most notable extent, this is by no means surprising. It took fourteen instead of four days before the transformation had advanced far enough to enable it to leave the water. It was especially interesting to observe the behaviour of this specimen during this period. Its weak and delicate constitution evidently made it much more susceptible to all external influences than the others. If exposed to the air for too long a time it acquired a light colour, and when annoyed or alarmed it emitted a peculiar odour, similar to that of a salamander. As soon as these phenomena were observed it was at once placed in deeper water, into which it immediately plunged and gradually recovered itself, the gills always becoming again expanded. The same experiment was repeated several times and always led to the same result, from which we may venture to conclude that by accelerating the transformation too energetically, the process may come to a standstill, and even by continued compulsion may end in death.

"It yet remains to be mentioned with respect to Axolotl No. V. that this specimen, unlike all the others, did not emerge from the water at the first ecdysis, but at the time of the fourth.

"All the Axolotls are now (July, 1875) living, and are healthy and vigorous, so that with respect to their state of nourishment there is nothing to prevent their propagating. Of the first four the largest is fifteen centim. long; Axolotl No. V. measures twelve centim.

"The preceding statements appear to demonstrate the correctness of the views advanced in the Introduction:—Axolotl larvæ generally but not always complete their metamorphosis if, in the first place, they emerge sound from the egg and are properly fed; and if, in the next place, they are submitted to the necessary treatment for changing aquatic into aërial respiration. It is obvious that this treatment must only be applied very gradually, and in such a manner as not to overtax the vital energy of the Amphibian."

To the foregoing remarks of Fräulein v. Chauvin I may add that in all five cases the transformation was complete, and not to be confounded with that change which occurs more or less in all Axolotls in the course of time when confined in small glass vessels. In this last case there frequently appear changes in the direction of the Amblystoma form without the latter being actually reached. In the five adult Axolotls which I possessed for a short time, and of which two were at least four years old, the gills were much shrivelled, but the aquatic tail and dorsal crest were unchanged. The crest may, however, also disappear, and the tail become shortened without these changes being due to a transformation into Amblystoma, as will be shown further on.

With respect to the duration of the transformation, this amounted in Axolotls Nos. I. to IV. altogether to twelve or fourteen days. Of these, four days were taken by the first changes which occurred while the creature was still in the water; the remaining time, to the completion of the metamorphosis, was passed on land. Duméril gives the duration of the metamorphosis as sixteen days.

The following results of the experiments just described appear to me to be especially noteworthy:—The five Axolotl larvæ which can alone be taken into consideration, the others having soon perished, all experienced metamorphosis, and without an exception became Amblystomas. Only one of them, No. I., by persistently swimming at the surface, as was observed at the end of six months, showed a decided tendency to undergo metamorphosis and to adopt aërial respiration. With respect to this specimen it may therefore be confidently assumed that it would have taken to the land, and that metamorphosis would have occurred without artificial aid, just as was the case in the thirty specimens which Duméril altogether observed.

Respecting Nos. II., III., and IV., on the other hand, such a supposition is but little probable. These three larvæ endeavoured to keep in deep water and avoided as long as possible the shallow places which would have enforced them to take entirely to lung breathing. Metamorphosis thus occurred more than a month later in these individuals.

Finally, there can scarcely be any doubt that No. V. would not have become transformed without forcible adaptation to an aërial life.

From these results we may venture to conclude that most Axolotl larvæ change into the Amblystoma form when, at the age of six to nine months, they are placed in such shallow water that they are compelled to respire chiefly by their lungs. The experiments before us are certainly at present but very few in number, but such a conclusion cannot be termed premature if we consider that out of several hundred Axolotls (the exact number is not given) Duméril obtained only about thirty Amblystomas, while v. Kölliker bred only one Amblystoma out of a hundred Axolotls.

It now only remains questionable whether *each larva* could have been forced to undergo metamorphosis, but this could only be decided by new experiments. It was originally my intention to have delayed the publication of the experiments till Fräulein v. Chauvin had repeated them in larger numbers, but as my Axolotls have not bred this year (1875) I must abandon my scheme, and this can be done the more readily because, for the theoretical consideration of the facts, it is immaterial whether *all* or only *nearly all* the Axolotls could have been compelled to undergo transformation. I must not, however, omit to mention that Herr Gehrig, the curator of our Zoological Museum, bred a considerable number of larvæ from the same brood as that with which Fräulein v. Chauvin experimented, and that of these larvæ six lived over the winter *without undergoing metamorphosis.* They were always kept in deep water and thus furnished the converse experiment to those recorded above; they further prove that this whole brood did not have a previous tendency to undergo metamorphosis.

If these new facts are to be made use of to explain the nature of this extraordinary process of transformation in accordance with our present conception, the data already known must in the first place be called to our aid.

It has first to be established that *Siredon Mexicanus* never, as far as we know, undergoes metamorphosis in its native country. This Amphibian is there only known in the *Siredon* form, a statement which I have taken from De Saussure,[1] who has himself observed the Axolotl in the Mexican lakes. This naturalist never found a single

[1] Verhandl. Schweiz. Naturforsch. Gesellschaft. Einsiedeln, 1868.

Amblystoma in the neighbourhood of the lakes, "nevertheless the larva (Axolotl) is so common there that it is brought into the market by thousands." De Saussure believes that in Mexico the Axolotl does not undergo transformation.[2] The same statement is distinctly made by Cope,[3] whose specimens of *Siredon Mexicanus* bred in America, even in captivity showed "no tendency to become metamorphosed." On the other hand Tegetmeier observed [4] that one out of five specimens obtained from the Lake of Mexico underwent metamorphosis, and this accordingly establishes the second fact, viz. that the true Axolotl becomes transformed under certain conditions into an Amblystoma when in captivity.

This last remark would be superfluous if, as was for a long time believed, the Paris Axolotls, of which the metamorphosis was first observed and which at the time made such a sensation, were actually *Siredon Mexicanus, i.e.* the *Siredon* which alone in its native country bears the name of Axolotl. In his first communication Duméril was himself of this opinion; he then termed the animal

[2] [Eng. ed. In 1878 Señor José M. Velasco published a paper entitled "Descripcion, metamorfosis. y costumbres de una especie nueva del genero *Siredon*." Memor. Sociedad Mexicana de Historia Natural, December 26th. See Addendum to this essay].

[3] Dana and Silliman's Amer. Journ., 3rd series, i. p. 89. Annals Nat. Hist. vii. p. 246.

[4] Proc. Zoo. Soc. 1870, p. 160.

"*Siredon Mexicanus vel Humboldtii,*" [5] but subsequently, in his amplified work [6] on the transformation of the Axolotl observed in the Jardin des Plantes, he retracted this view, and after a critical comparison of the five described species of *Siredon*, he came to the conclusion that the species in the possession of the Paris Museum was probably *Siredon Lichenoides* (Baird). All the transformations of Axolotls observed in Europe must consequently be referred to this species, since they were—at least as far as I know—all derived from the Paris colony. My own experimental specimens were also indirectly descended from these.

Now it must be admitted that this does not coincide with the fact that the Amblystoma form which Duméril first obtained from his Axolotls agreed with Cope's species, *A. Tigrinum,* while on the other hand we learn from Marsh [7] that *Siredon Lichenoides* (Baird), when it does undergo metamorphosis, becomes transformed into *Amblystoma Mavortium* (Baird).

Marsh found *Siredon Lichenoides* in mountain lakes (7000 feet above the sea) in the south-

[5] Compt. Rend., vol. lx. p. 765 (1865).

[6] Nouvelles Archives du Muséum d'Histoire Nat. Paris, 1866, vol. ii. p. 268.

[7] Proc. Boston Soc., vol. xii. p. 97; Silliman's Amer. Journ., vol. xlvi. p. 364; reference given in "Troschel's Jahresbericht" for 1868, p. 37.

west of the United States (Wyoming Territory), and obtained from them, by breeding in aquaria, *Amblystoma Mavortium* (Baird). He considers it indeed doubtful whether the Amphibian undergoes this transformation in its native habitat, although he certainly states this opinion without rigorous proof on purely theoretical considerations, because, according to his view, "the low temperature is there less favourable."[8]

If I throw doubt upon this last statement it is simply because *Amblystoma Mavortium* is found native in many parts of the United States, viz :—in California, New Mexico, Texas, Kansas, Nebraska, and Minnesota. It is indeed by no means inconceivable that in the mountain lakes where Marsh obtained this species, it may behave differently with respect to metamorphosis than in other habitats, and this appears probable from certain observations upon *Triton* which will be subsequently referred to.

Meanwhile, in the absence of further observations, we must admit that the Paris Axolotls were not *Siredon Lichenoides*, but some nearly allied and probably new species. But little information is furnished by observing the course of the transformation, although it is at least established that this

[8] Proc. Boston Soc., vol. xii. p. 97; Silliman's Amer. Journ., vol. xlvi. p. 364. I have not been able to get a copy of this paper, and quote from a reference in "Troschel's Jahresbericht." See preceding note.

Axolotl in its native habitat does not undergo metamorphosis or does so as exceptionally as in Europe. Unfortunately in his papers Duméril gives no precise statement respecting the locality of this species imported from "Mexico"—it is probable that he was himself unacquainted with it, so that I can only state on the authority of Cope that Amblystoma has never been brought from south of the provinces of Tamaulipas and Chilhuahua, *i. e.* south of the Tropic of Cancer.[9]

This last statement, however, gives no certainty to the matter. Of much more importance is the above-mentioned fact, that the true Axolotl of the lakes about the city of Mexico does not, as a rule, become transformed into an Amblystoma in that locality, although this species in certain cases undergoes metamorphosis when in confinement. From this circumstance and from the fact that the Paris Axolotl likewise experienced but a very small percentage of metamorphosis in captivity, we may venture to conclude that this species also, in its native habitat, either does not become transformed at all or does so only exceptionally.

But there is yet another body of facts which come prominently into view on considering the history of the transformations. I refer to the existence of numerous species of *Amblystoma* in a natural state. In the "Revision of the genera of

[9] Dana and Silliman's Amer. Journ. See note 3.

Salamandridæ," published some years ago by Strauch,[10] this author, following Cope,[11] gives twenty species of *Amblystoma* as inhabiting North America. Although some of these species are based on single examples, and consequently, as Strauch justly remarks, "may well have to be reduced in the course of time," there must nevertheless always remain a large number of species which live and propagate as true Amblystomas, and of which the habitat extends from the latitude of New York to that of New Mexico. There are therefore true species of *Siredon* which regularly assume the Amblystoma form under their natural conditions of life, and which propagate in this form, while, on the other hand, there are at least two species which, under their existing natural conditions of life, always propagate as *Siredon*. It is but another mode of expression for the same facts if we say that the Mexican Axolotl and the Paris *Siredon*—whether this is *Lichenoides* or some other species—stand at a lower grade of phyletic development than those species of *Amblystoma* which propagate in the salamander form. No one can raise any objection to this statement, while the alternative view maintained by all authors contains a theory either expressed or implied which is, as I believe, incorrect, viz. that the Mexican Axolotl

[10] Proc. Acad. Philadelph. xix. 1867, pp. 166—209.

[11] Mém. Acad. Petersb. vol. xvi.

has remained at an inferior stage of phyletic development.

All zoologists[12] who have expressed an opinion upon the transformation of the Axolotl, and who are not, like the first observer of this fact, embarrassed by Cuvier's views as to the immutability of species, regard the phenomenon as though a species, which owing to some special conditions had hitherto remained at a low stage of development, had now through some other influences been compelled to advance to a higher stage.

I believed for a long time that the phenomenon could not otherwise be comprehended, so little was I then in a position to bring all the facts into harmony with this view. Thus in the year 1872 I expressed myself as follows[13]:—" Why should not a sudden change in all the conditions of life (transference from Mexico to Paris) have a direct action on the organization of the Axolotl, causing it suddenly to reach a higher stage of development, such as many of its allies have already attained, and which obviously lies in the nature of its organization—a stage which it would perhaps itself have

12 [Eng. ed. Seidlitz is an exception, since in his work on Parthenogenesis (Leipzig, 1872, p. 13) he states that " In the Axolotl, Pædogenesis, which is not in this case . . . . monogamous, but sexual, and indeed gynækogenetic, has already become so far constant that it has perhaps entirely superseded the orthogenetic reproduction."]

13 Über den Einfluss der Isolirung auf die Artbildung. Leipzig, 1872, p. 33.

reached, although later, in its native country? Or is it inconceivable that the sudden removal from 8000 feet above the sea (Mexican plateau) to the altitude of Paris, may have given the respiratory organs an impetus in the direction of the transformation imminent? In all probability we have here to do with the direct action of changed conditions of life."

That the substance of this last statement must still hold good is obvious from the experiments previously described, which show that by the application of definite external influences, we have it to a certain extent in our power to produce the transformation. It is precisely in this last point that there lies the new feature furnished by these experiments.

But are we also compelled to interpret the phenomenon in the above manner? *i. e.* as a sudden advance in the phyletic development of the species occurring, so to speak, at one stroke? I believe not.

What first made this view appear to me erroneous, was the appearance of the living Amblystomas bred from my Axolotl larvæ. These creatures by no means differed from the Axolotls merely in single characters, but they were distinct from the latter in their entire aspect; they differed in some measure in all their parts, in some but slightly and in other parts strongly—in brief, they had become quite different animals. In accordance with this, their mode of life had become completely modified; they no longer lived in the water, but remained concealed by day among the damp moss of the vivarium, coming forth only by night in search of food in dry places.

I had been able to perceive the great difference between the two stages of development from the anatomical data with which I had long been familiar, and which Duméril had made known with respect to the structure of his Amblystomas. But the collecting of numerous details gives no very vivid picture of the metamorphosis; it was the appearance of the living animal that first made me conscious how deep-seated was the transformation which we have here before us, and that this change not merely affected those parts which would be directly influenced by the change in the conditions of life, such as the gills, but that most if not all the parts of the animal underwent a transformation, which could in part be well explained as morphological adaptation to new conditions of life, and partly as a consequence of this adaptation (correlative changes), but could not possibly be regarded as the sudden action of these changed conditions.

Such at least is my view of the case, according to which a *per saltum* development of the species of such a kind as must here have taken place, is quite inconceivable.

I may venture to assume that most observers

of the metamorphosis of Axolotl have, like myself, not been hitherto aware of the extent of the transformation, and it may thus be explained why the theoretical bearings of the case have on all sides been incorrectly conceived. We have here obviously a quite extraordinary case of the first order of importance. I believe that it can easily be shown that the explanation of the history of the metamorphosis of the Paris Axolotl which has hitherto been pretty generally accepted, necessarily comprises a very far-reaching principle. If this interpretation is correct, then in my opinion must also hold good the ideas of those who, like Kölliker, Askenasy, Nägeli, and, among the philosophers, Hartmann and Huber, would refer the transformation of species in the first instance to a power innate in the organism, to an active, *i.e.* a self-urging "law of development"—a phyletic vital force.

Thus, if the Axolotls transformed into Amblystomas are regarded as individuals which, impelled by external influences, have anticipated the phyletic development of the others, then this advance can only be ascribed to a phyletic vital force, since the transformation is sudden, and leaves no time for gradual adaptation in the course of generations. The *indirect* influence of the external conditions of life, *i. e.* natural selection, is thus excluded from the beginning. But the *direct* action of the changed conditions of life by no means furnishes an explanation of the complete transformation of the whole structure, such as I have already alluded to, and which I will now enter into more closely.

The differences between the Paris Axolotl and its Amblystoma according to Duméril, Kölliker, and my own observations are as follow:—

1. The gills disappear; the gill-clefts close up, and of the branchial arches only the foremost remain, the posterior ones disappearing. At the same time the *os hyoideum* becomes changed (Duméril).

2. The dorsal crest completely disappears (Duméril).

3. The aquatic tail becomes transformed into one like that of the salamanders (Duméril), which, however, is not as in the salamander cylindrical, but somewhat compressed laterally (Weismann).

4. The skin becomes yellowish white, irregularly spotted on the sides and back (Duméril), whilst at the same time its former grey-black ground-colour changes into a shining greenish black (Weismann); it loses, moreover, the slimy secretion of the skin, and the cutaneous glands become insignificant (Kölliker).

5. The eyes become prominent and the pupils narrow (Kölliker), and eye-lids capable of completely covering the eyes are formed; in Axolotl only a narrow annular fold surrounds the eyes, so that these cannot be closed (Weismann).

6. The toes become narrowed and lose their

skin-like appendages (Kölliker), or more precisely, the half webs which connect the proximal extremities of the toes on all the feet (Weismann).

7. The teeth are disposed in this species, as in all other *Amblystomæ*, in transverse series; whilst in Axolotl, as in *Triton* larvæ, they are arranged at the sides of the jaw in the form of a curved arch-like band beset with several rows of teeth.[14] (Duméril. See his fig., *loc. cit.* p. 279).

8. In Axolotl the lower jaw, in addition to the teeth on the upper edge of the bone, also bears "de très petites dents disposées sur plusieurs rangs;" these last disappear after metamorphosis (Duméril). I will add that the persistent teeth belong to the *os dentale* of the lower jaw, and those that are shed to the *os operculare*.[15]

9. The surface of the posterior moveable part of the body is slightly concave both before and after transformation; the anterior part is, how-

[14] Duméril represents the teeth of the *vomer* as separated from those of the *os palatinum* by a gap. This is probably accidental, since Gegenbaur (Friedrich u. Gegenbaur, the skull of Axolotl, Würzburg, 1849) figures the rows of teeth as passing over from the one bone to the other without interruption. This was the case with the Axolotls which I have been able to examine on this point; but this small discrepancy is, however, quite immaterial to the question here under consideration.

[15] See O. Hertwig "Über das Zahnsystem der Amphibien und seine Bedeutung für die Genese des Skelets der Mundhöhle." Archiv. für microsc. Anat., vol. xi. Supplement, 1874.

ever, less concave in *Amblystoma* than in *Siredon* (Duméril).

I have not yet been able to verify Duméril's 7th and 9th statements, as I did not want to kill any of my living Amblystomas,[16] simply in order to confirm the observations of a naturalist in whom one may certainly place complete confidence. Neither have I as yet observed the transformation of the branchial arches, but all the other statements of Kölliker and Duméril I can entirely corroborate.

The structural differences between Axolotl and Amblystoma are considerably greater and of more importance than those between allied genera, or indeed than between the families of the Urodela. The genus *Siredon* undoubtedly belongs to a different sub-order to the genus *Amblystoma* into which it occasionally becomes transformed. Strauch, the most recent systematic worker at this group, distinguishes the sub-order *Salamandrida* from that of the *Ichthyodea* by the possession of eyelids, and by the situation of the palatine teeth in single rows on the posterior edge of the palatal bone: in *Ichthyodea* the eyelids are wanting and the palatine teeth are either "situated on the anterior edge of the palatal bone," or

[16] [Eng. ed. These Amblystomas have since died and have been minutely described by Dr. Wiedersheim. See his memoir, "Zur Anatomie des *Amblystoma Weismanni*," in Zeit. für wiss. Zool., vol. xxxii. p. 216.]

"cover the whole surface of the palatal plates in brush-like tufts."

How is it possible to regard such widely divergent anatomical characters as changes suddenly produced by the action (but once exerted) of deviating conditions of life? Hand in hand with the shedding of the old and the appearance of new palatine teeth, there occurs a change in the anatomical structure of the vertebral column, and also —as we may fairly conclude from Kölliker's correct observation of the cessation of the slimy secretion—in the histological structure of the skin. Who would undertake to explain all these profound modifications as the direct and sudden action of certain external influences? And if any one were inclined to explain such changes as a consequence of the disappearance of the gills, *i.e.* as correlative changes, what else is such a correlation than the phyletic vital force under another name?

If from one change arising from the direct action of external agencies, the whole body can in two days become transformed in all its parts, in the precise manner which appears best adapted for the new conditions of life under which it is henceforward to exist, then the word "correlation" is only a phrase which explains nothing, but which prevents any attempt at a better explanation, and it would be preferable to profess simply the belief in a phyletic vital force.

Moreover, it is hardly permissible to seek such an explanation, since Urodela are known which have no gills in the adult state, and which nevertheless possess all the other characters of the *Ichthyodea*, viz. want of eyelids, characteristic palatine teeth, and the tongue bone. This is the case with the genera *Amphiuma* (Linn.), *Menopoma* (Harl.), and *Cryptobranchus* (v.d. Hoev.). The two first genera, as is known, still possess gill-clefts, but *Cryptobranchus* has even lost these clefts, which, as in *Amblystoma*, are overgrown by skin; nevertheless *Cryptobranchus* is, according to the concurrent testimony of all systematists, a true salamander in habits, tongue bone, palatine teeth,[17] &c. It must further be added that the Axolotl itself can lose the gills without thereby becoming transformed into an Amblystoma. I have previously mentioned that in Axolotls which were kept in shallow water the gills frequently became diminutive, and it also sometimes happens that they completely shrivel up. I possess an Axolotl preserved in alcohol in which the gills have shrivelled up into small irregular bunches, and the dorsal crest is also so completely absent that its place is occupied by a long furrow, and even on the tail the crest has entirely disappeared from the lower edge and about half from the upper edge. Notwithstanding this, the creature is widely removed from Amblystoma in structure; it possesses

[17] See Strauch, *loc. cit.* p. 10.

the arched branchial apparatus, the palatine teeth, the skin, &c., of the Axolotl.

These facts prove, therefore, that the shedding of the gills by no means always entails all the other modifications which we observe in the metamorphosis of Axolotl, so that these modifications are thus not by any means the necessary and immediate consequence of such gill shedding.

Whether these modifications will occur after a long series of generations—whether the successors of *Cryptobranchus* will also one day acquire the salamandriform structure is another question, and one which I could not exactly answer in the negative. But this question does not here come into consideration, as we are now only concerned with the *immediate result* of the shedding of the gills.

The problem appears therefore to be as follows:—Either the hitherto received interpretation of the transformational history of the Axolotl as a further development of the species is incorrect, or else the case of Axolotl incontestably proves the existence of a phyletic vital force.

We have now to ask whether the facts of this transformational history are not capable of another explanation.

I believe that this is certainly possible, and that another interpretation can be shown to be correct with some degree of probability.

I am of opinion that those Amblystomas which

have been developed in captivity in certain instances from *Siredon Mexicanus* (*S. Pisciformis*), as well as from the Paris Axolotls, are not progressive, but reversion forms; I believe that the Axolotls which now inhabit the Mexican lakes were Amblystomas at a former geological (or better, zoological) epoch, but that owing to changes in their conditions of life, they have reverted to the earlier perennibranchiate stage.

I was undoubtedly first led to this conception by the results which arose from my studies on the seasonal dimorphism of butterflies.[18] In this case we were also concerned with the two different forms under which one and the same species appears, and of which it was shown to be probable that the one is phyletically older than the other. The younger summer form, according to my view, has arisen, through the gradual amelioration of the climate, from the winter form, which at an earlier zoological epoch was the only one in existence; but the latter, the primary form, has not for this reason ceased to exist, but now alternates in each year as a winter form with the secondary summer form.

Now with seasonally dimorphic butterflies, it was easily possible to induce the summer brood to assume the winter form by exposing their pupæ for a long time to a low temperature; and it was shown to be highly probable that this abrupt and

[18] See Part I. of this volume.

often very extensive change or transformation, only apparently takes place suddenly, and is but the apparent result of the action of cold upon this generation, whilst in fact it depends upon reversion to the primary form of the species, so that the low temperature, which is only once applied, gives but the impetus to reversion, and is not the true cause of the transformation. This cause must rather be sought in the long continued action of the cold to which the ancestors of our existing butterflies were subjected for thousands of generations, and of which the final result is the winter form.

If we assume for an instant that my interpretation of the transformation of Axolotl as just offered is correct, we should have conditions in many respects analagous to those of seasonal dimorphism. It is true that in this case the two forms no longer alternate regularly with each other, but the primary form may occasionally appear instead of the secondary form, owing to the action of external conditions.

Just as in the case of seasonal dimorphism it is possible to compel the summer generation to abandon the summer form, and to assume the winter guise by the action of cold; so in the present case we are able to induce the Axolotl to adopt the Amblystoma form by making aërial respiration compulsory at a certain stage of life; and further, just as in seasonal dimorphism it can be shown that this artificially produced change is only apparently an abrupt transformation, and is actually a reversion to the much older winter form; so here we have not an actual, but only an apparent remodelling of the species—a reversion to the phyletically older form.

This certainly appears a paradox, inasmuch as a form here arises by reversion which must yet undoubtedly rank as the more highly developed. I believe, however, that much which seems paradoxical in this statement will disappear on further examination.

It must in the first place be taken into consideration that the phyletic development of species need not by any means always take place by advancement. We have indeed many cases of retrogressive development, although in a somewhat different sense, as with parasites and those forms which have degenerated from free locomotion to a sedentary mode of life.[19] I do not confuse this kind of retrogressive development, arising from the arrest of certain organs and

[19] [This is the principle of "Degeneration" recognized by Darwin (see "Origin of Species," 6th ed. p. 389, and "Descent of Man," vol. i. p. 206), and given fuller expression to by Dr. Anton Dohrn (see his work entitled "Der Ursprung der Wirbelthiere und das Princip des Functionswechsels." Leipzig, 1875). A large number of cases have been brought together by Prof. E. R. Lankester, in his recent interesting work on "Degeneration, a Chapter in Darwinism." Nature series, 1880. R.M.]

systems of organs, with true reversion. The latter is a return to a form which has already been once in existence; but in the former case, in spite of all simplification of the organization, some entirely new feature always comes into existence. But I am not able to see any absurdity in the assumption that even true reversion, whether of a whole species or of the individuals of a certain district, may be regarded as possible, and I require no further concession. Why, for example, should it be inconceivable that at a very remote period the Axolotl was adapted to a life on land; that through the direct and indirect action of changed conditions of life it gradually acquired the salamander form, but that subsequently, through new and unfavourable changes in the conditions of life, it again relapsed to the older form, or at least to one nearly related thereto?

At any rate such an assumption contains nothing opposed to known facts, but can be supported in many ways, and finally it commends itself, at least in my opinion, as offering the only admissible explanation of the facts before us.

The existence of a whole series of species of *Amblystoma*, as already mentioned, at once shows that species of *Siredon* can become elevated into the salamander form, and can propagate regularly in this state, and further, that this phyletic advance has already actually taken place in many species.

That degeneration may also occur from this high stage to a lower stage of development, is shown by many observations on our water-salamanders. It is known that under certain circumstances Tritons, as it is generally expressed, become "sexually mature in the larval condition."

In the year 1864 De Filippi[20] found fifty Tritons in a pool at Andermatten, in the neighbourhood of Puneigen, and of these only two showed the structure of the adult water-salamander; all the others still possessed gills, but notwithstanding this, they agreed in both sexes, in size and in the development of the sexual organs, with mature animals. De Filippi established that these "sexually mature larvæ" not only resembled larvæ externally through the possession of gills, but that they also possessed all the other anatomical characters of the larvæ, *i. e.* the characteristic bunches of palatine teeth situated on both sides in the position of the subsequent single rows, and a vertebral column represented throughout its whole length by the *chorda dorsalis.*

According to my view this would be a case of the reversion of the Triton to the immediately anterior phyletic stage, *i. e.* to the perennibranchiate stage, and in the present instance the majority of zoologists who take their stand by the theory of descent, would certainly concur in this view. I

[20] "Sulla Larva del *Triton Alpestris.*" Archivio per la Zoologia. Genova e Torino, 1861, vol. i. pp. 206—211.

should at least consider it to be a useless play upon words did we here speak of larval reproduction, and thereby believe that we had explained something. The animal certainly becomes sexually mature in the same condition as that in which it first appears as a larva, but we first get an insight into the nature of this process by considering that this so-called "sexually mature larva" has the precise structure which must have been possessed by the preceding phyletic stage of the species, and that an individual reversion to the older phyletic stage of the species is consequently before us. I maintain that Duméril is in error in regarding this case of the Triton as parallel with the true larval reproduction of Wagner's *Cecidomyia* larva. In this last case it is certainly not reversion to an older phyletic stage that confers the power of reproduction upon the larvæ, since the latter do not represent an older phyletic stage of the species, but must have arisen contemporaneously with this last stage. The enormous structural difference between the larvæ and the imagines is not explained by the latter having arisen from the former supplementarily as a finished production, but by both having been contemporaneously adapted to continually diverging conditions of life.[21] Considered phyletically, these larvæ are by no means necessarily transitional to the origination

[21] See also Lubbock "On the Origin and Metamorphoses of Insects," London, 1874.

of the flies. They could have been quite different without the form of the imagines having been thereby modified, since the stages of insect metamorphosis vary independently of each other in accordance with the conditions of life to which they are subjected, and exert scarcely any, or only a very small form-determining influence upon each other, as has been amply proved in the preceding essay. In any case the power of these larvæ (the *Cecidomyiæ*) to propagate themselves asexually was first acquired as a secondary character, as appears from the fact that there exist numerous species of the same genus which do not "nurse." In the form which they now possess they could never have played the part of the final stage of the ontogeny, nor could they formerly have possessed the power of sexual reproduction.[22] In brief, we are here concerned with true larval reproduction, whilst in Triton we have reversion to an older phyletic stage.[23]

[22] See the first essay "On the Seasonal Dimorphism of Butterflies," p. 82.

[23] [Eng. ed. It has frequently been objected to me that the existing Axolotl is not a form resulting from atavism, but a case of "arrested growth." The expression "atavism" is certainly to be here taken in a somewhat different sense than, for example, in the case of the reversion of the existing Axolotl to the Amblystoma form. Further on, I have myself insisted that in the first case the phyletic stage in which the reversion occurred is still completely preserved in the ontogeny of each individual, whilst the Amblystoma stage has become lost in the ontogeny of the Axolotl. If, therefore, we apply the term "atavism" only to

I cannot agree with my friend Professor Haeckel when he occasionally designates the reversion of

such characters or stages (*i. e.* complexes of characters) as are no longer preserved in the ontogeny, we cannot thus designate the present arrest of the Axolotl at the perennibranchiate stage. Such a restriction of the word, however, appears to me but little desirable, since the process is identical in both cases, *i. e.* it depends upon the same law of heredity, in accordance with which a condition formerly occurring as a phyletic stage suddenly reappears through purely internal processes. It is true that the reversion is not *complete*, *i. e.* the present sexually mature Axolotl does not correspond in all details with its perennibranchiate ancestors. Since Wiedersheim has shown that the existing Axolotl possesses an intermaxillary gland, this can be safely asserted. This gland occurs only in *land* Amphibians, and therefore originated with the Amblystoma form, afterwards becoming transferred secondarily to the larval stage. Nevertheless, the present Axolotl must resemble its perennibranchiate ancestors in most other characters, and we should be the more entitled to speak of a reversion to the perennibranchiate stage as we speak also of the reversion of single characters. To this must be added that the Axolotl does not correspond exactly with an Amblystoma larva, since Wiedersheim has shown that the space for the intermaxillary gland is present, but that the gland itself is confined to a few tubes which do not by any means fill up this space. ("Das Kopfskelet der Urodelen." Morph. Jahrbuch, vol. iii. p. 149). By the expression "arrested growth" not much is said, if at the same time the cause of the arrest is left unstated. But what can be the cause why the whole organization remains stationary at the perennibranchiate stage, the sexual organs only undergoing further development? Surely only that law or force of heredity known by its effects, but obscure with respect to its causes, through which old phyletic stages sometimes suddenly reappear, or in other words, that power through which reversion takes place. It must not be forgotten that all these cases of "larval reproduction" in the Tritons as an "adaptation" to a purely aqueous existence.[24] We could here only speak of "adaptation" if we took the word in a quite different sense to that in which it was first introduced into science by Darwin and Wallace. These naturalists thereby designate a gradual bodily transformation appearing in the course of generations in correspondence with the new requirements of altered conditions of life or, in other words, the action of natural selection, and not the result of a suddenly and direct acting transforming cause exerted but once on a generation.

Just because the word "adaptation" can be used in ordinary language in many senses, it is desirable that it should have only *one* precise signification, and above all that we should not speak of adaptation where scarcely any morphological

Amphibians appear suddenly. The present sexually mature form of the Axolotl has not arisen by the sexual maturity gradually receding in the ontogeny from generation to generation, but by the occurrence of single individuals which were sexually mature in the perennibranchiate stage, these having the advantage over the *Amblystomæ* in the struggle for existence under changed climatic conditions.

By admitting a reversion, we perfectly well explain why arrest at the perennibranchiate stage can be associated with complete development of the sexual organs; the assumption of an "arrested growth" leaves this combination of characters completely unexplained. Moreover, I am of opinion that the expressions "arrested growth" or "reversion" are of but little importance so long as the matter itself is clear.]

[24] See Haeckel's "Anthropogenie," p. 449.

change occurs, but only a kind of functional change in the sense used by Dohrn.[25] This is the case for example, when Forel[26] shows that fresh water Pulmonifera, the organization of which is attributed to the direct respiration of air, can nevertheless become settled in the greatest depths of mountain lakes through their lungs being again employed as gills. That not the least change in the lungs hereby takes place is shown by the observations of Von Siebold,[27] who saw the shallow water Pulmonifera using their lungs alternately for direct aërial and aquatic respiration, according to the amount of air contained in the water. If with Von Siebold we merely apply the word "adaptation" to such cases, this expression would lose the special sense which it originally conveyed, and the word would have to be abandoned as a *terminus technicus;* still, such cases may perhaps be spoken of as *physiological* adaptation.

In any case the reproductive "larvæ" of the Tritons as little present a case of true adaptation as the Axolotl, which occasionally becomes transformed into an Amblystoma. In both cases the transformation referred to is by no means indispensable to the life of the individual. Mature Tritons

25 "Der Ursprung der Wirbelthiere und das Princip des Functionswechsels," Liepzig, 1875.

26 Bull. Soc. Neuchâtel. vol. viii. p. 192. Reference given in "Troschel's Jahresbericht" for 1869.

27 Sitzungsberichte d. math. phys. Klasse der Akad. d. Wiss. zu München, 1875. Heft i.

(devoid of gills) can exist, as I have myself seen, for many months, and probably also for a year in deep water, although adapted for purely pulmonary respiration; whilst Axolotls, as I have already mentioned, can live well for a year in shallow water poor in air. If their gills by this means become shrivelled up or completely disappear, even this is not adaptation in the Darwinian sense, but the effect of directly acting external influences, and chiefly of diminished use.

A case entirely analagous to that of Filippi's was observed by Jullien in 1869. Four female larvæ of *Lissotriton Punctatus* (Bell)—(synonymous with *Triton Tæniatus*, Schnd.), taken from a pool, proved to be sexually mature. They contained mature eggs in their ovaria ready for laying, and two of them actually deposited eggs. Four male larvæ found in the same pool, appeared to be equally developed with respect to size, but their testicles contained no free spermatozoa, but only sperm-cells.[28]

I have met with a third case of a similar kind mentioned by Leydig in his memoir, rich in interesting details, "on the tailed Amphibians of the Wurtemburg fauna."[29] Schreibers, the former director of the Vienna Museum, also found "larvæ" of Tritons with well-developed gills, but of the size of the "adult male individuals," and,

[28] Compt. Rend. vol. lxviii. pp. 938 and 939.
[29] Archiv f. Naturgeschichte, 1867.

as shown by anatomical investigation, with well "developed sexual organs," the ovaria especially being distended with eggs.

It is thus established that species which long ago reached the salamander stage in phyletic development, may occasionally degenerate to the perennibranchiate stage. This fact obviously makes my conception of the Axolotl as a reversion form appear much less paradoxical—indeed, the cases of reversion in Triton are precisely analagous to the process which I suppose to have taken place in the Axolotl. We have only to substitute Amblystomas for Tritons, to imagine the pool in which De Filippi found his "sexually mature Triton larvæ" enlarged to the size of the Lake of Mexico, and to conceive the unknown, and perhaps here transitory, causes of the reversion to be permanent, and we have all that is necessary, so far as we at present know, for the restoration of the Axolotl; we obtain a perennibranchiate population of the lake.

It has not yet been determined whether the perennibranchiate form of the Triton actually prevailed permanently in De Filippi's pool, since, so far as I know, this has not since been examined.

Let us, however, assume for an instant that this is really the case, and that there exists at that spot a colony of sexually reproductive perennibranchiate Tritons: should we wonder if a true Triton occasionally appeared among their progeny, or if we were able to induce the majority of the individuals of this brood to become metamorphosed into Tritons by keeping them in shallow water? According to my view this is precisely the case of the Mexican Axolotl.

I need not, however, restrict myself to this in order to support my hypothesis, but must also directly combat the view hitherto received, since the latter is in contradiction with facts.

Did there really exist in the Axolotl a tendency to sudden phyletic advancement, then one fact would remain quite incomprehensible, viz. the sterility of the Amblystomas.

Out of about thirty Amblystomas obtained by Duméril down to the year 1870, there was not one in a state of sexual maturity; neither copulation nor deposition of eggs took place, and the anatomical investigation of single specimens showed that the eggs were immature, and that the spermatozoa, although present, were without the undulating membrane characteristic of the salamanders, but were not devoid of all power of movement, only, as established by Quatrefages, were "incompletely motile."[30]

So also the five Amblystomas about which I have been writing, show up to the present time no appearance of reproduction.

The objection raised by Sacc,[31] that the sterility

[30] Compt. Rend. vol. v. 1870, p. 70.

[31] Bull. Soc. Neuchâtel. vol. viii. p. 192. Referenc given in "Troschel's Jahresbericht" for 1869.

of the Amblystomas bred from Axolotls is attributable to "bad nourishment," is obviously of but little avail. How is it that the Axolotls, which are fed in a precisely similar manner, propagate so readily? Moreover, I am able to expressly assert that my Amblystomas were very well fed. It is true that they have as yet scarcely reached the age of two years, but the Axolotl propagates freely in the second year, and some of Duméril's Amblystomas were five years old in 1870.

This fact of the sterility is strongly opposed to the idea that these Amblystomas are the regular precursors of the phyletically advancing genus *Siredon*.[32] I will by no means assert that my

[32] [Eng. ed. It was mentioned in the German edition of this work that in the spring of 1876 a female Amblystoma of the Jardin des Plantes in Paris had laid eggs (see Blanchard in the Compt. Rend. 1876, No. 13, p. 716). Whether these eggs were fertile, or whether they developed was not then made known. Thus much was however at the time clear, that even if this had been the case, the reproduction of this Amblystoma would have been only an *exceptional* occurrence. At that time there were in the Jardin des Plantes Amblystomas which had been kept for more than ten years, and only on one occasion was there a deposition of eggs, and this by only one specimen. That I was correct in speaking of the "sterility" of these Amblystomas in spite of this one exception, is proved by the latest communication from the Jardin des Plantes. We learn from this (Compt. Rend. No. 14, July, 1879, p. 108) that in the years 1877 and 1878 none of the Amblystomas laid any more eggs, although all means were exerted to bring about propagation. In April, 1879, eggs were again laid by one female, and by a second in May. These eggs certainly developed, as did those of 1876, and produced tadpoles. These

theory of reversion actually explains the sterility, but it is at least not directly opposed to it. Mere reversion forms may die off without propagating themselves; but a *new form* called forth by the action of a phyletic vital force should not be sterile, because this is the precise "aim" which the vital force had in view. The conception of a vital force comprises that of teleology.

The sterility of Amblystoma moreover, although not completely explicable from our standpoint, can be shown to be a phenomenon not entirely isolated. In the above mentioned case of *Lissotriton Punctatus*, the female "larvæ" were certainly sexually mature and laid eggs, but the males of the same period contained in their testicles no fully developed spermatozoa.

Other cases of this kind are unknown to me; at the time when I made the experiments with butterflies already recorded (see the first essay), this point of view was remote, and I therefore neglected to examine the artificially bred reversion forms with respect to their organs of reproduction. But general considerations lead to the supposition that atavistic forms may easily remain sterile.

Amblystomas are therefore not absolutely, but indeed relatively sterile. Whilst the Axolotl propagates regularly and freely every year, this occurs with the Amblystoma but rarely and sparsely. The degree of their sterility can only be approximately established when we know the number of Amblystomas that have since been kept in the Jardin des Plantes. Unfortunately nothing has been said with respect to this.]

Darwin[33] finds the proximate causes of sterility in the first place in the action of widely diverging conditions of life, and in the next place in the crossing of individuals widely different in constitution. Now it is certainly deviating conditions of life which lead to the metamorphosis of the Axolotl, and from this point of view it cannot be surprising if we find those individuals sterile which show themselves so especially affected by these changed conditions as to revert to the salamander form.

By this it is not in any way meant to be asserted that reversion is invariably accompanied by sterility, and one cannot raise as an objection to my interpretation of the metamorphosis of the Axolotl, that a reproductive colony of Axolotls could never have arisen by reversion. On the contrary, Jullien's egg-depositing female Triton-larvæ show that also with reversion the power of reproduction may be completely preserved.[34] From the above-mentioned

[33] Origin of Species, 6th ed. p. 252.

[34] In plants also reversion forms show sterility in different degrees. Mr. Darwin has called my attention to the fact that the peloric (symmetrical) flowers which occasionally appear as atavistic forms in *Corydalis solida* are partly sterile and partly fertile. That in other causes of sterility, and above all by bastardizing, the reproductive power is lost in the most varying degrees, has been known since the celebrated observations of Kölreuter and Gärtner. [Eng. ed. An Orchid (*Catasetum tridentatum*) has the sexes separate, and the male flowers (*Myanthus barbatus*) differ considerably from the female (*Monachanthus viridis*); besides these, there occurs a form with bisexual flowers which must be considered as a reversion

general causes of sterility, it may even be inferred that fertility can be lost in different degrees, and it can be further understood to a certain extent why this fertility is more completely lost by reversion to the Amblystoma, than by the reversion of the Triton to the perennibranchiate form.

If in these cases the reversion is brought about by a change in the conditions of life, we may perhaps suppose that the magnitude of this change would determine the degree of fertility, and the preservation of the reversion form. Still more, however, would the fertility be influenced by the extent of the morphological difference resulting from the reversion. We know that the blending of very different constitutions (*e. g.* the crossing of different species) produces sterility. Something similar results from the sudden reversion to a stage of development widely different in its whole structure. Here also we have in a certain sense the union of two very different constitutions in one individual—a kind of crossing.

From this point of view it can in some measure be comprehended why sterility may be a result of reversion; on the other hand, we thereby obtain no explanation why, with the same amount of morphological difference, in one case complete sterility, and in another relative fertility occurs. The morphological difference between Axolotl and Amblystoma

(*Cat. tridentatum*) and *this is always sterile.* Darwin, "Fertilization of Orchids," 2nd ed. p. 199.]

is exactly the same as between Triton and its "sexually mature larva;" the difference between the two cases of reversion depends entirely upon the direction of the leap, that taken in the former case being precisely opposite in direction to that taken in the latter.

Herein might be sought the explanation of the different strength with which the reproductive power is affected; not indeed in the direction of the leap itself, but in the differences in the ontogeny which are determined by the differences in the direction of the leap. The reversion of the Triton to an older phyletic stage coincides with the arrest at a younger ontogenetic stage; or, in other words, the older stage of the phylogeny to which reversion takes place is still entirely comprised in the ontogeny of each individual. Each Triton is perennibranchiate throughout a long period of its life; the reverting individual simply reverts to the older phyletic stage by remaining at the larval stage of its individual development.

But it is quite different with the reversion of the Axolotl to the formerly acquired, but long since abandoned Amblystoma form. This is not retained in the ontogeny of Axolotl, but has been completely lost; for a long series of generations—so must we suppose—the ontogeny has always only attained to the perennibranchiate form. Now if at the present time certain individuals were compelled to revert to the Amblystoma form, certainly no greater leap would have been made from a morphological point of view, than in the reversion of Triton to the perennibranchiate form, but at the same time the leap would be in another direction, viz. over a long series of generations back to a form which the species had not produced for a long period, and which had to a certain extent become foreign to it. We should thus have here also the grafting of a widely different constitution upon that of the Axolotl, or, if one prefers it, the commingling of two widely different constitutions.

Of course I am far from wishing to pretend that this "explanation" is exact; it is nothing more than an attempt to point out the direction in which the causes affecting the reproductive powers in different degrees are to be looked for. A deeper penetration into and special demonstration of the manner in which these causes bring about such results, must be reserved for a future period. For the present it must suffice to have indicated that there is an essential distinction between the two kinds of reversion, and to have made it to some extent comprehensible that this distinction may be the determining impulse with respect to the question of sterility. Perhaps the law here concealed from us may one day be thus formulated :—Atavistic individuals lose the power of reproduction the more completely, the greater the number of generations of their ancestors whose ontogeny no longer comprises the phyletically older stage to which the reversion takes place.

The hypothesis which interprets the transformation of the Axolotl as a case of reversion, thus holds out the possibility of our being able to comprehend the sterility of the Amblystomas arising in this manner, whilst, on the other hand, for the adherents of a phyletic vital force, not only is this observed sterility as Duméril expresses it "un véritable énigme scientifique," but an absolute paradox. We should expect such a directive and inciting principle to call into existence new forms having vitality and not destined to perish, the more so when it is concerned with a combination of structural characters which, when originating in another manner (viz. from other species of *Siredon*), have long since shown themselves to have vitality and reproductive power. We are indeed acquainted with species of *Amblystoma* which propagate as such, and each of which arises from an Axolotl-like larva. Thus we cannot regard the sterile Amblystomas produced by the Paris Axolotls as abortive attempts of a vital force—an interpretation which is certainly in itself already sufficiently rash.

Now if it be asked what change in the conditions of life could have led to the reversion in the Lake of Mexico[35] of the Amblystoma to the Siredon

[35] As we do not know the origin of the "Paris Axolotl" I must restrict myself in the following remarks to *Siredon Mexicanus* (Shaw).

form, I must admit that I can only offer a conjectural reply, having but a conditional value so long as it is not supported by a precise knowledge of the conditions there obtaining, and of the habits both of the Axototl and of the Amblystoma.

It may be supposed generally that reversion is brought about by the same external conditions as those which formerly produced the perennibranchiate stage. This supposition is in the first place supported by the experiments here recorded, since it is evidently the inducement to aërial respiration which causes the young Axolotl to revert to the Amblystoma form, *i.e.* the inciting cause under whose domineering influence the Amblystoma form must have arisen.

Here again the case is quite similar to that of seasonally dimorphic butterflies. Reversion of the summer brood to the winter form is there most easily caused by the action of cold, *i.e.* by the same influence as that under whose sway the winter form was developed.

We know indeed that reversion may also arise by the crossing of races and species, and I have attempted to show that reversion in butterflies may also be brought about by other influences than cold; but still the most probable supposition obviously is, that reversion would be caused by the persistent action of the same influences as those which in a certain sense created the perennibranchiate form. That the latter was produced under the influence of an aquatic life there can be no doubt, and thus, in accordance with my supposition, the hypothetical *Amblystoma Mexicanum*, the supposed ancestral form of the Axolotl of the Mexican Lake, might have been caused to revert to the perennibranchiate form by a reduction in the possibilities of its living upon land, and by its being compelled to frequent the water.

I will not here return to the consideration of every other opinion *ab initio*. It is very advisable to distinguish between the mere impulses which are able to produce sudden reversion, and between actual transforming causes which result directly or indirectly in the remodelling of a species. Thus, it is conceivable *à priori* that reversion may occur by the action of an inciting cause having nothing to do with the origin of the phyletically older form. Temperature can certainly have played no part, or only a very small part, in the formation of the perennibranchiate form; nevertheless cold may well have been one of the inciting causes which induced the Amblystoma at one time to revert to the *Siredon* form, and we cannot at present consider De Saussure to be incorrect when he maintains that the low temperature of the Mexican winter might prevent that transformation (of the Axolotl into the Amblystoma) which would occur "in the warm reptile-house" of the Jardin des Plantes. He supports this view by stating that "Tschudi has found the Amblystoma" (of course another species) "in the hottest parts of the United States." "On the Mexican plateau, however, it snows every winter, and if the lake does not actually freeze, its temperature must fall very considerably in the shallowest parts."

But although this view is not opposed by any theoretical considerations, I still hold it to be incorrect. I doubt whether it is temperature that has brought about the reverse transformation of the Amblystoma into the Axolotl, or which, according to De Saussure's conception, at the present time prevents the transformation of the Axolotl in the Lake of Mexico. I doubt this because Amblystomas are now known from all parts of the United States as far north as New York, a proof that a winter cold considerably greater than that of the Mexican plateau is no hindrance to the metamorphosis of the Axolotl, and that the genus does not show itself to be in this respect more sensitive than our native genera of *Salamandridæ*.

The following observations of De Saussure, in which he calls attention to the nature of the Mexican Lake, appear to me to be more worthy of consideration:—"The bottom of this lake is shallow, and one passes imperceptibly from the lake into extensive marshy regions before reaching solid ground; perhaps this circumstance makes the Axolotl incapable of reaching dry land, and prevents the transformation."

In any case the Lake of Mexico offers very

peculiar conditions for Amphibian life. My esteemed friend Dr. v. Frantzius has called my attention to the fact that this lake—as well as many other Mexican lakes—is slightly saline. At the time of the conquest of Mexico by Ferdinand Cortez, this circumstance led to the final surrender of the city, as the Spaniards cut off the supply of water to the besieged, and the water of the lake is undrinkable. The ancient Mexicans had laid down water-conduits from the distant mountains, and the city is still supplied with water brought through conduits.

Now this saltness cannot in itself be the cause of the degeneration to the perennibranchiate form, but it may well be so in combination with other pecularities of the lake. The narrowest part of the lake is the eastern, and it is only in this part that the Axolotl lives. Now in winter, violent easterly gales rush down from the mountains and blow continuously, driving the water before them to such an extent that it becomes heaped up in the western portion of the lake, where it frequently causes floods, whilst 2000 feet of the shallow eastern shore are often laid completely dry.[36]

Now if we consider these two peculiarities, viz. salineness and periodical drying up of a part of the bottom of the lake through continuous gales, we certainly have for the Axolotl, conditions of life

[36] Mühlenpfordt, "Versuch einer getreuen Schilderung der Republik Mejico," Hanover, 1844, vol. ii. p. 252.

which are only to be found in few species. One might certainly attempt to apply these facts in a quite opposite sense, and to regard them as unfavourable to my theory, since the retreat of the water from a great portion of the bottom of the lake would—so one might think—rather facilitate transition to a life upon land, and indeed compel the adoption of such a mode of existence. But we should thus forget that the exposed bottom of the lake is a sterile surface without food or place of concealment, and, above all, without vegetation; and further, that owing to the considerable saline-ness of the water (specific gravity = 1.0215),[37] the whole of the exposed surface must be incrusted with salt, a circumstance which would render it quite impossible for the creatures to feed upon land. Sodic chloride and carbonate are dissolved in the water in such considerable quantities, that they are regularly deposited upon the shores of the lake as a crust, which is collected during the dry season of the year and sent into the market under the name of "tequisquite" (Mühlenpfordt).[38]

Thus the supposition is not wanting in support, that peculiar conditions make it more difficult for the creature to obtain its food upon land than in

[37] [The specific gravity of sea water (Atlantic), according to the determinations of Mr. Buchanan on board the "Challenger," at 15.56° C. varies from 1.0278 to 1.0240. That of the water of the Dead Sea is 1.17205.—Watts' "Dict. of Chemistry," vol. v., table, p. 1017. R.M.]

[38] *Loc. cit.* p. 252.

the water, and this alone may have been sufficient to have induced it to acquire the habits of a purely aquatic existence, and thus to revert to the perennibranchiate or Ichthyodeous form.

But enough of supposition. We must not complain that we are unable from afar to discover with precision the causes which compelled the Axolotl to abandon the Amblystoma stage, as long as we are not able to explain the much nearer cases of reversion in Filippi's and Jullien's Tritons; nevertheless, in these cases also, the causes affecting the whole colony of Tritons must be general, since—at least in the case noticed by Filippi—the greater majority of the individuals remained in the larval condition. Experiments with Triton larvæ could throw greater light upon this subject; it would have in the first place to be established whether reversion could be artificially induced, and if so, by what influences.

From the previously mentioned experiments with butterflies, as well as from the results obtained with Axolotls, we should expect that in Tritons, reversion to the Ichthyodeous form would take place if we allowed the inciting cause, viz. the bathing of the gills and of the whole body with water, to act persistently, and at the same time withheld that in-influence under whose action the salamander form became developed, viz. the bathing of the gills, the skin, and the surfaces of the lungs with air.

Old experiments of this kind are to be met with,

but they were never carried on for a sufficient time to entirely allay the suspicion, that the specimens concerned would perhaps have undergone the ordinary metamorphosis if their existence had been prolonged.

Thus, Schreibers [39] relates that "by confining tadpoles of the salamander found at large in their last stage of growth, under water by means of an arrangement (net ?), and feeding them with finely chopped earthworms, he was able to keep them for several months—and indeed throughout the winter—in this condition, and in this way to forcibly defer their final change, and their transition from the tadpole stage to that of the perfected creature during this period." It is not stated whether the animals finally underwent transformation, so that it cannot be decided whether we have here a case of reversion or simply one of retarded development. That metamorphosis may occur after a long period of time, is shown by experiments upon the tadpole of *Pelobates* conducted by Professor Langer in Vienna.[40] The creatures were kept in deep water in such a manner that they were not able to land, and by this means three out of a large number

[39] "Über die specifische Verschiedenheit des gefleckten und des schwarzen Erdsalamanders oder Molchs, und der höchst merkwürdigen, ganz eigenthümlichen Fortpflanzungsweise des Letzteren." Isis, Jahrg. 1833, p. 527.

[40] The experiments referred to have not been made known; I am indebted for them to a written communication kindly furnished by an esteemed colleague.

of individuals had their metamorphosis delayed till the second summer; notwithstanding this, transformation then occurred.

It cannot be objected to my reversion hypothesis, that it opposes on the one side what on the other it postulates, viz. a *per saltum* change of structure. Reversion is characterized by the sudden acquisition of an older, *i.e.* a formerly existing phyletic stage. That reversion occurs is a fact, whilst nobody has hitherto been able to prove, or even to make probable, that a stage of the future (*sit venia verbo*) has been attained at once (*per saltum*).

Now if it is possible to find influences in the present conditions of life of the Axolotl which make it difficult or quite impossible for it to live upon land, and which therefore appear as incentives to the reversion to the Ichthyodeous form, the other portion of my hypothesis—the assumption that the Axolotl had become an Amblystoma at a former period—can also be supported by facts.

We know from Humboldt[41] that the level of the Lake of Mexico at a comparatively recent period was considerably higher than at present. We know further that the Mexican plateau was covered with forest, which has now been destroyed wherever there are human, and especially Spanish settlements. Now if we suppose that at some post-glacial period the mountain forests extended

[41] See Mühlenpfordt's work already quoted, vol. i.

to the borders of the lake, at that time deep, with precipitous sides and much less saline, not only should we thus have presented different conditions of life to those at present existing, but also such as would be most favourable for the development of a species of salamander.

On the whole, I believe that my attempt to explain the exceptional metamorphosis of the Axolotl of the Mexican lake cannot be objected to as being a too airy phantasy. In any case it is the only possible explanation which can be opposed to that which supposes that the occasional transformation of the Axolotl is not reversion, but an attempt at advancement. This last assumption must, in my judgment, be rejected on purely theoretical grounds by those who hold that a sudden transformation of a species, when connected with adaptation to new conditions of life, is inconceivable—by those who regard adaptation, not as the sudden work of a magic power, but as the end result of a long series of natural, although minute and imperceptible causes.

If my interpretation of the facts be correct, there arises certain consequences which I may here briefly mention in conclusion.

First, with regard to more obvious results. If *Siredon Mexicanus*, Shaw, only by occasional reversion assumes the Amblystoma form, and never, or only exceptionally, propagates as such, but only as *Siredon*, the more recent systematists are not justified in striking out the genus *Siredon* and in placing *S. Mexicanus* as an undeveloped form in the genus *Amblystoma*. So long as there exists not one only, but several species of *Siredon* which as such regularly propagate themselves, the genus exists; and although we would not deprive systematists of all hope of these species of *Siredon* being one day re-elevated to *Amblystomæ*, it nevertheless better accords with the actually existing state of affairs if we allow the genus *Siredon* to remain as before among the genera of *Salamandrina*, and to include therein all those species which, like the Paris Axolotl, *S. Mexicanus*, Shaw, and probably also *S. Lichenoides*, Baird, only exceptionally, or through artificial influences, assume the Amblystoma form, but without propagating regularly in this condition. On the other hand, we should correctly comprise under the genus *Amblystoma* all those species which propagate in this state regularly, and in which the perennibranchiate stage occurs only as a larval condition.

To arrive at a decision in single cases would chiefly concern the American naturalists, whose ever increasing activity may lead us to hope soon for a closer investigation of the reproduction of the numerous species of *Amblystoma* of their native country. I should rejoice if the facts and arguments which I have here offered should give an impetus to such researches.

The second consequence to which I may refer,

is of a purely theoretical nature, and concerns a corollary to the "fundamental biogenetic law" first enunciated by Fritz Müller and Haeckel. This, as is well known, consists of the following law:—The ontogeny comprises the phylogeny, more or less compressed and more or less modified.

Now according to this law, each step in phyletic development when replaced by a later one, must remain preserved in the ontogeny, and must therefore appear at the present time as an ontogenetic stage in the development of each individual. But my interpretation of the transformation of the Axolotl appears to stand in contradiction to this, since the Axolotl, which at a former period was an Amblystoma, retains nothing of the latter in its ontogeny. The contradiction is, however, only apparent. As long as we are concerned with an actual advance in development, and therefore with the attainment of a new step never formerly reached, the older stages will be found in the ontogeny. But this is not the case when the new stage is not an actual novelty, but formerly represented the final stage of the individual development; or, in other words, when we are concerned with the reversion, not of single individuals, but of the species as such, to the preceding phyletic stage, *i.e.* with a phyletic degeneration of the species. In this case the former end-stage of the ontogeny would be simply eliminated, and we should then only be able to recognize its former existence by its occasional appearance in a reversion form. Thus, under certain conditions the Triton sinks back to the perennibranchiate stage; not in such a manner that the individual first becomes a Triton and then undergoes perennibranchiate re-modification, but simply, as I have already shown above, by its remaining at the Ichthyodeous stage and no longer attaining to the Salamander form. So also, according to my hypothesis, the salamandrine *Amblystoma Mexicanum*, formerly inhabiting the shores of the Lake of Mexico, has degenerated to the perennibranchiate stage, and the only trace that remains to us of its former developmental status is the tendency, more or less retained in each individual, to again ascend to the salamander stage under favourable conditions.

The third and last consequence which my interpretation of the facts entails, is the change in the part played by reversion in organic nature. Whilst atavistic forms have hitherto been known only as isolated and exceptional cases, interesting indeed in the highest degree, but devoid of significance in the course of the development of organic nature, a real importance in this last respect must now be attached to them.

I may assume that reversion can in two ways be effectual for the preservation or re-establishment of a living form. In the first place, where, as in Axolotl, the new and organically higher form becomes untenable through external influences, instead of simply perishing—since advancement in another direction does not appear to be possible—a reversion of the species to the older and more lowly organized stage occurs. In the second place, the older phyletic form may not be abandoned while a newer form is being developed therefrom, but the former may alternate with the latter, as we see in the case of seasonally dimorphic butterflies. It can hardly be objected if I regard the alternation of the summer and winter form in this case as a periodic reversion to the phyletically older (winter) form.

Although the reversion of an entire species, such as I suppose to have been the case with the Axolotl, may be of rare occurrence, this is certainly not the case with periodic or cyclical reversion; the latter plays a very important part in the development of the various forms of alternating or cyclical propagation.[42]

POSTSCRIPT.

In the previous portion of this essay it was pointed out that the causes to which I attributed the reversion of the hypothetical *Amblystoma Mexicanum* to the existing Axolotl, did not appear to me to amount to a complete explanation

[42] In the province of botany such a case has already been made known by Fritz Müller (Botan. Zeitung, 1869, p. 226;

of the phenomenon. In the first place these seemed to me too local, since they could only be applied with any certainty to the Axolotl of the lake of the Mexican capital, whilst the Paris Axolotls obtained from other parts of Mexico still required an explanation. On the other hand, these causes did not appear to me sufficiently cogent. Should we even learn subsequently that the Paris Axolotl is also derived from a salt lake which is exposed to similar winds to the Lake of Mexico, we still have in this peculiarity of the lakes only a cause tending to make it difficult for the larva to undergo metamorphosis, and to reach a suitable new habitat on the land. The impossibility of doing this, or the complete absence of such habitat, does not however follow as a necessary consequence.

It would obviously be a much more solid support for my hypothesis if it were possible to point to some physical conditions of the land which

1870, p. 149). I may be here permitted to quote a passage from the letter in which Dr. Müller calls attention to this interesting discovery. "As a proof of the possibility that a reversion form can again become a persistent character in a species or in the allied form of a particular district, I may refer you to an *Epidendrum* of the island of Santa Catharina. In all Orchids (with the exception of *Cypripedium*) only one anther is developed; in very rare cases well-formed anthers appear as reversions among the aborted lateral anthers of the inner whorl. In the *Epidendrum* mentioned, these are however *always present*."

there precluded the possibility of the existence of Amblystomas.

For a long time I was indeed unable to discover such causes, and I therefore concluded the previous portion of this essay and went to press. Afterwards, when residing in one of the highest valleys of our Alps in the Upper Engadine, an idea accidentally occurred to me, which I do not now hesitate to regard as correct after having tested it by known facts.

It happens that in the Upper Engadine there live only such Amphibia as persistently, or at least frequently resort to the water. I found frogs up to nearly 7000 feet above the sea, and Tritons at 6000 feet (Pontresina and Upper Samaden). On the other hand, the land-living mountain salamander, *S. Atra*,[43] was absent, although suitable stations for this species were everywhere present, and it would have wanted for food as little as do its allies the water-newts. Neither would the great elevation above the sea offer any obstacle to its occurrence, since it occa-

[43] [This species is interesting as being ovoviviparous, the young passing through the branchiate stage within the body of the mother. Some experiments, which were partially successful, were made by Fräulein v. Chauvin with a view to solve the question whether the branchiate stage could be prolonged by taking the larvæ directly from the mother before birth and keeping them in water. See "Zeit. für wissen. Zoo." vol. xxix., p. 324. R.M.]

sionally ascends to a height of 3000 metres (Fatiot).[44]

Now it is well known that the atmosphere of the Upper Engadine,[45] like that of other elevated Alpine valleys enclosed by extensive glaciers, is often extraordinarily dry for a long period, a condition which appears to me to explain why the black land-salamander is there absent,[46] whilst its near water-living ally occurs in large numbers. The skin of the naked Amphibia generally requires moisture, or else it dries up, and the creature is deprived of a necessary breathing apparatus, and often dies as rapidly as though some important internal organ had been removed. Decapitated frogs hop about for a long time, but a frog which escapes from a conservatory and wanders about for one night in the dry air of a room, is found the following day with dry and dusty skin half dead in some nook, and perhaps perishes in the course of another day if left without moisture.

All that we know of the biology of the Amphibia is in accordance with this. Thus, all the land-salamanders of southern Italy avoid the hot and

[44] See Fatiot, "Les Reptiles et les Batraciens de la haute Engadine." Geneva, 1873.

[45] I can remember at Upper Engadine a peculiar kind of preserved beef, prepared by simply drying in the air; also the mummification of entire human bodies by drying in the open air, as is practised at Great St. Bernard.

[46] "Faune des Vertébrés de la Suisse," vol. iii. "Histoire Naturelle des Reptiles et des Batraciens." Geneva, 1873.

dry air of summer by burying in the ground, where they undergo a summer sleep. This is the case with the interesting *Salamandrina Perspicillata*,[47] and with the land-living Sardinian Triton, the remarkable *Euproctus Rusconii*, Gené,[48] (*Triton Platycephalus*, Schreiber). With respect to *Geotriton Fuscus* I learn from Dr. Wiedersheim, who has studied the life conditions of this, the lowest European Urodelan, in its own habitat, that in Sardinia it sleeps uninterruptedly from June till the winter; whilst on the coast of Spezia and at Carrara, where it also occurs, it avoids the summer sleep in a very peculiar manner. It makes use of the numerous holes in the calcareous formation of that region, and for some months in the year becomes a cave-dweller. As soon as the great heat occurs, often in May, it withdraws into the holes, and again emerges in November during the wet weather. In these lurking holes it does not fall into a sleep, but is found quite active, and its stomach, filled chiefly with scorpions, shows that it goes successfully in search of food; the moist air of the holes makes it unnecessary for it to bury in the earth.

In the same sense it appears to me must be conceived the fact that the solitary species of frog

[47] See Wiedersheim, "Versuch einer gleichenden Anatomie der Salamandrinen." Würzburg, 1875.

[48] See Gené, "Memorie della Reale Acad. di Torino," vol. i.

of the Upper Engadine, *Rana Temporaria*,[49] the brown grass frog, is there much more a frequenter of the water than in the plains. It is true that I can find no remark to this effect in the excellent work of Fatiot, already referred to above, and I am therefore obliged to resort to my own observations, which, although often repeated, have always been carried on for only a short time. I was much struck with the circumstance that the Engadine frogs were to be found in numbers in the water long after the pairing season, which, according to Fatiot, lasts at most to the end of June. In the numerous pools around Samaden I found them in July and August, whilst in the plains they only take to the water at the time of reproduction, and seek winter quarters in the mud on the first arrival of this season. (Fatiot, p. 321.) In the Engadine they have therefore in some measure adopted the mode of life of the aquatic frogs, but this of course does not prevent them from returning in damp weather to their old habits and roving through meadows and woods.

After these considerations had made it appear to me very probable that the dry air of the Upper Engadine accounted for the absence of the black land-salamander, the question at once arose whether the absence of Amblystomas from the

[49] *Rana esculenta* never reaches Alpine regions, this species not having been found higher than 1100 meters. (Fatiot, *loc. cit.*, p. 318).

Mexican plateau might not perhaps be due to the same cause, *i.e.* whether such a dryness of the atmosphere might not perhaps prevail also in that region, so that Amphibia, or at least salamander-like Amphibia, could not long exist on the land. The height above the sea is still greater (7000 to 8000 feet), and the tropical sun would more rapidly dessicate everything in a country poor in water.

As I was at the time without any books that might have enlightened me on the meteorological conditions of Mexico, I wrote to Dr. v. Frantzius, who, by many years residence in Central America was familiar with the climate of this region, and solicited his opinion. I received the reply that on the high plains of Mexico an extraordinary dryness of the atmosphere certainly prevails. "The main cause of the dryness of the high plains is to be found in the geographical position, the configuration of the land, and the physical structure. The north-eastern trade-wind drives the clouds against the mountains, on the summits of which they deposit their moisture, so that no vapour is carried over; as long as the north-east trade-wind blows, the streams feeding the rivers flowing into the Atlantic Ocean are abundantly fed with water, whilst on the western slopes, and especially on the high plains, the clouds give no precipitation. In the second half of the year also, during our summer, the so-called rainy season brings but little rain[50]—little in comparison with the more southern regions, where the heavy tropical thunderstorms daily deluge the earth with water. Mexico lies much too northerly, and does not reach the zone of calms, within which region these tropical rains are met with."

Thus, in the high degree of dryness of the air lasting throughout the year, I do not doubt that we have the chief cause why no Amblystomas occur on these elevated plains; they simply cannot exist, and would become dried up if taken there, supposing them not to be able to change their mode of life and to take to the water. If therefore in former times Amblystomas inhabited Mexico, the coming on of the existing climatic conditions left them only the alternative of becoming extinct, or of again taking to the aquatic life of their Ichthyodeous ancestors. That this was not *directly* possible—that the Amblystoma form was not able to become aquatic without a change of structure, is shown by the fact that even in the Lake of Mexico no Amblystoma occurs. A retreat to an aqueous existence could, as it appears, only be effected by complete reversion to the Ichthyodeous form, which then also took place.

But my hypothesis of the transformation of the

[50] See also the excellent work upon Mexico by Mühlenpfordt already quoted, vol. i., pp. 69—76.

Axolotl not only requires the proof that Amblystomas cannot exist under present conditions in Mexico, but also the further demonstration that at a former period other conditions prevailed there, and these of such a nature as to make the existence of land-salamanders possible.

With respect to my question, whether we might not perhaps assume that at some post-glacial period the conditions of atmospheric moisture on the high plains of Mexico were essentially different from those at present prevailing, I recollected Dr. v. Frantzius and the above-quoted observation of Humboldt's,[51] who discovered in the neighbourhood of the Lake of Tezenco (Mexico) distinct evidence of a much higher former level of the water. "All such elevated plains were certainly at a former period so many extensive water-basins, which gradually became filled, and are still filling up with detritus. The evaporation from such large surfaces of water must at that time have caused a very moist atmosphere, favourable to vegetation and adapted for the life of naked Amphibia."

From this side also my hypothesis thus receives support, and we may assume with some certainty that at the beginning of the diluvial period[52] the

[51] "Essai politique sur le Royaume de la Nouvelle Espagne," 1805, p. 291.

[52] [The expression made use of by the author, viz. "Dilu-vialzeit," would perhaps be more in harmony with the views of English geologists if rendered as the "pluvial period,"

woods surrounding the Mexican lakes were inhabited by Amblystomas, which, as the lakes subsequently became more and more dried up and the air continually lost moisture, found it more difficult to exist on the land. They would at length have completely died out, had they not again become aquatic by reversion to the Ichthyodeous form. It may perhaps be supposed that the above-mentioned physical conditions—desolate, salt-incrusted shores—co-operated in the production of the reversion, by making it difficult for the larvæ to quit the water; but we can only judge with certainty upon this point when, by means of experiment, we have discovered the causes which produce reversion in the Amphibia.

Addendum.

I have lately met with another interesting notice on the reproduction of the native North American Amblystomas. Professor Spence F. Baird, of Washington, has often observed the

thereby indicating the period of excessive rainfall which, according to Mr. Alfred Tylor, succeeded to and was a consequence of the thawing of the great glaciers which accumulated during the last glacial epoch. There is abundant evidence to show that during the latter period glacial action extended in North America at least as far south as Nicaragua. See Belt on "The Glacial Period in North America," Trans. Nova Scotian Inst. of Nat. Sci. 1866, p. 93, and "The Naturalist in Nicaragua," pp. 259—265. R.M.]

development from the egg of various species, and especially of *Amblystoma Punctatum* and *A. Fasciatum.* His observations do not appear to be as yet published, so that I was unable to discover any account of the development of *Amblystoma* in existing literature.[53] I am authorized to extract the following brief data from a letter addressed to Dr. v. Frantzius.

In order to deposit their eggs the Amblystomas go into the water, where the eggs are laid enclosed in a jelly-like mass, but never more than fifteen to twenty together. The spherical eggs are very large, perhaps a quarter of an inch in diameter. They soon develop into a *Siredon*-like larva, which remains several months in this condition. The gills then shrivel up, the creature begins to crawl, and gradually passes through the different transformations to the complete Amblystoma form.

It appears from this communication that the Amblystomas lay much larger and much fewer eggs than the Axolotl, and that their development throughout resembles that of our salamanders.

In concluding I may mention an anatomical fact which most strongly supports my view that the Mexican Axolotl is a reverted Amblystoma. I learn from Dr. Wiedersheim that the Axolotl possesses the "intermaxillary gland" which occurs

[53] [Eng. ed. A memoir by Samuel Clarke has since been published upon the embryonic development of *Amblystoma punctatum*, Baird. Baltimore, 1879.]

in all the land Amphibia. This organ, lying in the intermaxillary cavity, appears, whenever it occurs, to produce a kind of birdlime, *i. e.* a very glutinous secretion, which serves to attach the prey to the rapidly protrusible tongue. Although this secretion may perhaps also have another function, from the absence of the intermaxillary gland in all exclusively aquatic Amphibia, it follows that it must be devoid of importance for, and inapplicable to feeding in the water. The intermaxillary gland is absent in all *Perennibranchiata* and *Derotremata* which Wiedersheim has hitherto investigated, viz. in *Menobranchus*, *Proteus*, *Siren*, *Cryptobranchus*, *Amphiuma*, and *Menopoma*, all of which are indeed without the cavity in which the gland is situated in the *Salamandrina*, *i. e.* the *cavum intermaxillare*.

Now in the *Salamandrina* the gland appears at an early stage. It is possessed in a well-developed state by the larvæ both of species of *Triton* and of *Amblystoma*, where indeed the glandular structure completely fills the *cavum intermaxillare*.

Were the Axolotl a species retarded in phyletic development, the presence of a gland which does not occur in any other *Perennibranchiata*, and which is only of use for life upon land, would be quite inexplicable.

The matter becomes still more enigmatical through the fact that the gland, although present, is quite rudimentary. Whilst in the *Salamandrina* the capacious intermaxillary cavity is entirely filled by the tubes of the gland in question, in Axolotl this cavity is almost completely filled with a closely woven connective tissue, in which there can only be found a small number of gland-tubes—in the extreme front, and at the base immediately over the intermaxillary teeth—these tubes agreeing in the details of their histological structure with the elements of the same gland in the *Salamandridæ*.

I give these anatomical details from Dr. Wiedersheim's verbal communication. An amplified account will subsequently appear in another place.[54]

An explanation of this rudimentary intermaxillary gland in the Axolotl only appears to me possible on the supposition that the latter is an atavistic form. From this point of view it is evident that the gland already present in all *Amblystoma*-larvæ must have been taken over by the perennibranchiate form of the existing Axolotl, through the reversion of the hypothetical *Amblystoma Mexicanum* of the "diluvial period."[55] It can also be easily understood that this organ would become more and more rudimentary in the course of time, since it has no further use in the water,

[54] [Eng. ed. See this author's work, "Das Kopfskelet der Urodelen." Leipzig, 1877, p. 149.]

[55] [See preceding note 52. R.M.]

and the gap thus arising in the formerly present *cavum intermaxillare* would become filled with connective tissue.

While the German edition of this work was going through the press I obtained, through the kindness of my friend Dr. Emil Bessels of Washington, the Mexican memoir upon the new Axolotl,[56] which even in Mexico regularly, or at least in many cases, becomes developed into the Amblystoma form.

The facts are briefly as follows:—The small Lake of Santa Isabel is some hours' journey from the Mexican capital. In this lake there lives a species of Axolotl which had hitherto remained unknown, and was described by Señor Velasco as *Siredon Tigrinus*. This species propagates itself indeed in the Axolotl state, but in many cases it becomes transformed into Amblystoma and takes to the land. Although propagation in the Amblystoma condition was not observed, it can hardly be doubted that it also propagates in this form.

At first sight these facts appear to refute my hypothesis, that the extreme dryness of the air of the Mexican plateau precludes the existence of land Amphibia. Nevertheless I do not abandon this hypothesis for the former one, since a closer study of the data furnished by Velasco confirms rather than refutes my supposition.

Velasco expressly corroborates the statement

[56] See note 2, p. 566.

that the Axolotl hitherto known from the great Mexican lake which never dries up (Lake of Xochimilco and Chalco), is only met with in its native habitat in the *Siredon* form, *i.e.* as *Siredon Humboldtii*. According to Velasco the cause of the frequent assumption of the Amblystoma form by the new *Siredon Tigrinus*, is to be found in the local conditions of life of this species. The Lake Santa Isabel is shallow, its greatest depth amounting to three meters, and it is liable to a periodical drying up, which is so complete that one can pass dry-shod through it in several places. The species must therefore have long since died out had it not been able to adapt itself periodically to a land life. Now it could have become transformed into a land Amphibian—as Señor Velasco observed—at various stages of growth; and indeed this author believes that "the Creator has implanted an instinct in this creature," which enables it to always undergo metamorphosis at the right time.

This last assumption may or may not be taken as correct, but this much is established, viz. that numerous individuals of this species take to the land, and remain there during a period of many months.

But does this contain the proof that salamander-like animals are actually able to lead a land life in Mexico—that the dry air is advantageous, or at least supportable to them? It does not appear so to me, but rather that all which has been reported of this Amblystoma by Señor Velasco goes to show that the animal does not, properly speaking, live upon land like the North American Amblystomas, or like our land-salamanders, but that *it only experiences a summer sleep lasting over the period of drought.* These Amblystomas were observed as they left the dried-up lake at night in order to seek some moist lurking-place in the neighbourhood, where they might remain concealed. They are only known in the villages situated near the lake, and were only seen there at large just when they were wandering from the lake to their place of concealment. At other times they were mostly found in the earth, buried under walls, the pavement of the market-place, &c. When laying down a line of railway, a workman found in the earth a whole nest of twelve Amblystomas lying close together. All these are not mere lurking-holes which could be abandoned at any moment; it would rather appear that we have here places of refuge for the entire duration of the period of drought, and that these would only be forsaken when the water of the rainy season penetrated the soil. I am not myself in a favourable position for investigating these suppositions more closely, but this could be done by Señor Velasco, who lives in Mexico, and science would be much indebted to him if he would examine as precisely as possible into the habits and conditions of life of this, and of the other species of Mexican Axolotls. Unfortunately this gentleman can, it would appear, have seen only the French publications upon the transformation of the Axolotl, and could not therefore have asked himself questions arising from my conception of the facts; otherwise many of his observations would have led to more definite results. The above conclusion can however be still further supported by Señor Velasco's data.

One might indeed insist that with us also the land-salamanders conceal themselves in moist places during dry weather, and often lie hidden, as in Mexico, in a hole, in a cluster of as many as ten together; but with us they leave their lurking-place from time to time and go in search of food. Señor Velasco mentions nothing with respect to this. What especially struck me was the statement that the Mexican Amblystomas *were also to be found in the water.*[57] When Lake Santa Isabel

[57] [Prof. Semper also remarks ("Animal Life," note 47, p. 430) with reference to the Axolotl of Lake Como in the Rocky Mountains, which he states always becomes transformed into *Amblystoma Mavortium*, that this metamorphosis "takes place in the water, and the Amblystomas, so long as they are little, actually live exclusively in the water, as I know by my own experience. A young Amblystoma which I kept alive for a long time, never went out of the water of its own free will, while one nearly twice as large lives entirely on land and only takes a bath now and then. It always goes into the water when the temperature of the air in the cellar, in which my aquaria stand, falls below that of the water—down to about 6° or 8° C." This statement appears to suggest that the effect of tempera-

is drained, the fishermen stretch large nets across the exit channels, and in these they not only find ordinary Axolotls, but also some "*sin arcos*," which they also designate "*mochos*," *i.e.* hornless Axolotls, because they have no gills, but have already reached the Amblystoma stage. Our land-salamanders live in the water only as larvæ, but they also love and require moisture. Only the female enters the water when she wants to deposit her young (eggs with mature larvæ), and then only at the margin of shallow pools or small brooks. The Mexican Amblystoma thus much more resembles in its habits our water-salamanders

ture may be a factor in some way concerned in these interesting cases of transformation, and would in any case be well worthy of experimental investigation. Some further details concerning the *Siredon Lichenoides* of Lake Como have been recently published by Mr. W. E. Carlin (Proc. U.S. National Museum, June, 1881). The lake, which is shallow, is fed by a constant stream of fresh water, but the water of the lake is intensely saline. The Siredon never enter the fresh water stream, but congregate in large numbers in the alkaline waters of the lake. "When about one hundred and fifty were placed in fresh water they seemed to suffer no inconvenience, but it had a remarkable effect in hastening their metamorphosis into the Amblystoma form. Of an equal number kept in fresh water and in the lake water, quite a change occurred with the former after twenty-four hours, while the latter showed no change after several days of captivity. Those that were kept well fed in jars usually began to show a slight change in from two to three weeks, and all of them completed the change into the Amblystoma inside of six weeks, while in some kept, but not specially fed, there were but three changes in three months." (Nature, Aug. 25th, 1881, p. 388). R.M.]

(Tritons), which remain in the water at least during the whole period of reproduction. These also leave the water later, and, like the land-salamander, seek concealment in the earth. They have this habit also in those districts which possess a very dry atmosphere; and especially in the Engadine, where I first conceived the idea of taking into account the dryness of the air, I found in the pools at the end of August and the beginning of September only larvæ of Tritons. The older Amphibians must therefore have been on the land, presumably in their places of winter concealment.

From what we have hitherto learnt from Señor Velasco, the mode of life of *Amblystoma Tigrinum* must resemble that of our Tritons, although its structure is that of a land-salamander. I would thus offer the following explanation of the facts at present known:—Owing to the periodic drying up of the lake of Santa Isabel, the *Siredon Tigrinus* would be again compelled to undergo metamorphosis. Whether this was formerly entirely abandoned, or whether it always occurred in solitary individuals, is almost immaterial; in any case the habit of metamorphosis must have been very rapidly acquired through natural selection, and must have again become general, if the faculty was only present in the species, although latent. Through the dryness of the air, the Amblystomas that had taken to the land would be compelled to bury themselves at once, and to remain asleep till the recurrence of the rainy season, when they would hasten back into the water and would there live as a species of Triton.

Now one might feel inclined to ask why the species of the great Mexican lake has not also taken to this mode of life. To this it may be simply replied that the water of this lake never dries up, and that the Axolotls have thus never been reduced to the alternative of undergoing metamorphosis or of perishing. If therefore the conditions of existence in water were more favourable than on land, the tendency to abandon metamorphosis would increase from generation to generation, and the deportment at present observed would finally result, *i.e.* propagation would take place exclusively in the Axolotl state. As has already been mentioned above, the latest observations of Velasco furnish further confirmation that the Axolotl of the great lake is never met with in the Amblystoma condition, "although it (the Axolotl) is brought daily from Mexico into the market throughout the whole year." I should not however regard it as a refutation of my view if prolonged investigation should show that this species also (*Siredon Humboldtii*) occasionally developed into an Amblystoma; on the contrary, it would not at all surprise me if such cases of reversion occurred in Mexico as well as in Europe. The fact that an immense majority of the Amphibians propagate in the Axolotl state would not be thereby affected, and would still require an explanation : this I am still inclined to see in the dryness of the air of the high plains, which is so unfavourably adapted for a life passed entirely on land.

IV.

ON THE MECHANICAL CONCEPTION OF NATURE.

INTRODUCTION.

In the first of the three preceding essays it was attempted to solve the question whether the transformations of a given complex of characters in a certain systematic group could be completely explained by tbe sole aid of Darwinian principles. It was attempted to trace the origin of the marking and colouring of the Sphinx-caterpillars to individual variability, to the influences of the environment, and to the laws of correlation acting within the organism. These principles as applied to the origin of a certain well-defined, although narrowly restricted range of forms, were tested in order to see whether they were alone sufficient to explain the transformation of the forms.

It appeared that this was certainly the case. In all instances, or at least where the facts necessary to obtain a complete insight were available, the transformations could be traced to these known factors; there remained no inexplicable residual phenomena, and we therefore had no reason for inferring the existence of some still unknown modifying cause lying concealed in the organism. In this region of the marking and colouring of caterpillars, the assumption of a phyletic vital force had to be abandoned, as being superfluous for the explanation of the facts.

In the second essay the attempt was next made with reference to double form-relationship, as presented for observation in metamorphic insects, to draw conclusions as to the causes of the transformations. It appeared here that form- and blood-relationship do not always coincide, since the larvæ of a species, genus, or family, &c., may show quite different form-relationships to their imagines. These facts alone told very decisively against the existence of an internal developmental power, so that the latter had likewise to be set aside by the method of elimination, since the observed incongruences as well as the congruences of form-relationship, found sufficient explanation in the action of the environment on the organism.

This investigation thus also led to the denial of a phyletic vital force.

In the third essay I finally sought to prove that the only case of transformation of one species into another at present actually observed[1], could

[1] [Some experiments on the transformation of the Crustacean *Artemia Salina* into *A. Milhausenii* by gradually increasing the saltness of the water, and conversely, the transformation of

not without further evidence be interpreted as the result of the action of a phyletic vital force, but that more probably we had here only an apparent case of new formation, which was in reality but a reversion to a stage formerly in existence.

If this last investigation removes the only certain observation which could have been adduced in favour of the hypothesis of a phyletic vital force, so also do the two former essays show that this hypothesis, at least in the case of insects, must be abandoned as inadeqate.

The question now arises whether this conclusion, based on such a limited range of inquiry, can also be applied to the other groups of the organic world without further evidence.

The supporters of a principle of organic development will deny this in each individual case, and will demand special proof for each group of organisms; I believe this position, however, to be incorrect. Here, if anywhere, it appears to me justifiable to apply the conclusions inductively from special cases to general ones, since I cannot at all see why a power of such pre-eminent and fundamental importance as a phyletic vital force should have its activity limited to solitary groups

A. Milhausenii into *A. Salina* by diminishing the saltness of the water, have been made by Schmankewitsch (Zeitschrift f. wiss. Zool. xxv. Suppl. 103 and xxix. 429), but the changes which occur here are much less considerable than in the case of the Axolotl. R. M.]

in the organic world. If such a power exists it must be the inciting cause of organic development in general, and must be equally necessary in every part of creation, as no advancement could take place without it. In this case, however, the force would be recognizable and demonstrable at every point; the phenomena should nowhere stand in opposition to its admission, and should in no case be explicable or comprehensible without it. The same laws and forces which caused the development of one group of forms must underlie the development of the whole organic world.

I therefore believe that we are correct in applying to the whole living world the results furnished by the investigation of insects, and in thus denying the existence of an innate metaphysical developmental force.

There is, however, a quite distinct method which leads to the same results, and to the preliminary, if not to the complete and definitive rejection of such a principle; *the admission of this power is directly opposed to the laws of natural science*, which forbid the assumption of *unknown* forces as long as it is not demonstrated that *known* forces are insufficient for the explanation of the phenomena. Now nobody will assert that this has in any case been proved; the test of applying the known factors of transformation has only just commenced, and wherever it has been made they have proved sufficient as causal forces.

Thus, even without the foregoing special investigations we should deny a phyletic vital force ; the more so as its admission is fraught with the greatest consequences, since it involves a renunciation of the possibility of comprehending the organic world. We should, on this assumption, at once cut ourselves off from all possible mechanical explanation of organic nature, *i.e.* from all explanation conformable to law. But this signifies no less than the renunciation of all further inquiry ; for what is investigation in natural science but the attempt to indicate the mechanism through which the phenomena of the world are brought about? Where this mechanism ceases science is no longer possible, and transcendental philosophy alone has a voice.

This conception represents very precisely the well-known decision of Kant:—" Since we cannot in any case know *à priori* to what extent the mechanism of Nature serves as a means to every final purpose in the latter, or how far the mechanical explanation possible to us reaches," natural science must everywhere press the attempt at mechanical explanation as far as possible. This obligation of natural science will be conceded even by those who lay great stress upon the necessity for assuming a designing principle. Thus, Karl Ernst von Baer states that we have no right "to assert of the individual processes of Nature, even when these evidently lead to a definite result, that some Mind has originated them designedly. The naturalist must always commence with details, and may then afterwards ask whether the totality of details leads him to a general and final basis of intentional design."[2]

But even if we are precluded on these grounds only from assuming the existence of a directive power, *i. e.* a phyletic vital force, for explaining detailed phenomena, and are at the same time debarred from the possibility of arriving at a physical or mechanical explanation—which amounts to no less than the abandoning of the scientific position—it certainly cannot be asserted that the development of the organic world is already conceived of as a mechanical process. We rather acquiesce in the belief that the processes both of organic and of inorganic nature depend most probably upon purely causal powers, and that the attempt to refer these to mechanical principles should not therefore be abandoned. There is no ground for renouncing the possibility of a mechanical explanation, and the naturalist *must not* therefore resign this possibility; for this reason he cannot be permitted to assume a phyletic power so long as it is not demonstrated that the phenomena can never be understood without such an assumption.

[2] "Reden und kleinere Aufsätze, Th. II.: Studien aus dem Gebiete der Naturwissenschaften." St. Petersburg, 1876. P. 81.

It cannot be raised as an objection that even for the explanation of individual life a vital power was long ago admitted, as there was not then sufficient material at hand to enable the phenomena of life to be traced to physical forces. It is now no longer questionable that this assumption was a useless error—a false method—at the time when made certainly very excusable, since the aspect of the question was then, owing to the imperfect basis of facts, very different to the present analogous question as to the causes of derivative development. Thus, although it is now easy to prove this assumption to be erroneous, it was in the former sense correct, as it was in accordance with the existing state of knowledge. At that time there was hardly one of the numerous bridges which now connect inorganic with organic nature, so that the supposition that life depended upon forces which had no existence outside living beings was sufficiently near.

In any case the philosophers of that period cannot be blamed for filling up the gaps in the existing knowledge by unknown powers, and in this manner seeking to establish a finished system. The task of philosophy is different to that of natural science; the former strives at every period to set up a completely finished representation of the universe in accordance with the existing state of knowledge. Natural science on the other hand is only concerned in collecting this knowledge;

she need not therefore always finish off, and indeed can never close her account, since she will never be in a position to solve all problems.[3] But science must not for this reason pronounce any question to be insoluble simply because it has not yet been completely solved; this she does, however, as soon as she renounces the possibility of a mechanical explanation by invoking the aid of a metaphysical principle.

That this is the correct mode of scientific investigation is seen by the abandoning of the (ontogenetic) vital force. The latter is no longer admitted by anybody, now that we have turned from mere speculation to the investigation of Nature's processes; nevertheless its non-existence has not been demonstrated, nor are we yet in a position to prove that all the phenomena of life must be traced to purely physico-chemical processes, to say nothing of our being actually able to thus trace them. Von Baer also states "that the abolishment of the vital force is an important advance; it is the reduction of the phenomena of life to physico-chemical processes, although these

[3] This obviously does not imply that the naturalist should not investigate Nature's processes, and not only correlate these, but also work them up into a universal conception; this is indeed both desirable and necessary if natural knowledge is to be regarded in its true value. The naturalist by this means becomes a philosopher, and the vitality of the so-called "natural philosopher" has been inspired, not by the necessity for investigation, but by philosophy proper.

indeed still contain many gaps." He points out how very far we are still removed from being able to reduce to physical causes, the processes through which the fertilized yelk of an egg becomes developed into a chicken.

How comes it therefore that we all have a conviction that such a complete reduction will in time become possible, or if not this, that the development of the individual depends entirely upon the same forces which are in operation without the organism? For what reason have we rejected the "vital force"?

Simply because we see no reason for assuming that known forces are insufficient for explaining the phenomena, and because we are not justified in admitting directive forces as long as we have any hope of one day furnishing a mechanical explanation.

But if it is not only permissible, but even necessary, to explain the *ontogenetic* vital power by known forces, and to commence to indicate the mechanism which produces the individual life, why should it not be equally necessary to abandon that assumption of a *phyletic* vital force which stifles any deeper inquiry, and to attempt to point out that here also the co-operation of mechanical forces has brought about the multitudinous and wonderful phenomena of the organic world?

The renunciation of the old vital force was certainly an immediate consequence of the acquisition of new facts—of the knowledge that the same compounds which compose organic bodies can be produced without the latter. This discovery, due to Wöhler and his followers, showed that organic products could be prepared artificially.[4] In brief, the decline of the vital force followed from the knowledge that at least one portion of the processes of life was governed by known forces.

But in the domain of the development of the organic world have we not quite analogous proofs of the efficacy of known forces? Is not the *variability* of all types of forms a fact? and must not this under the action of natural selection and heredity lead to permanent changes? Has not the problem of explaining the subserviency of all organic form to law as a *result* without invoking its aid as a *principle* been thus successfully solved? It is true that we have not directly observed the process of natural selection from beginning to end; neither has anybody directly

[4] [The discovery here referred to is the synthesis of urea by Wöhler in 1828 (Pogg. Ann. xii., 253; xv. 619), by the molecular transformation of ammonium cyanate. Since that period large numbers of organic syntheses have been effected by chemists, and many of the compounds formerly supposed to be essential products of life have been built up in the laboratory from their inorganic elements. The division of chemistry into "organic" and "inorganic" is thus purely artificial, and is merely retained as a matter of convenience, the former division of the science being defined as the chemistry of the carbon compounds. R. M.]

observed the mode in which the heat of the animal body is generated by the processes of combustion going on in the blood and in the tissues; nevertheless, this is believed as a certainty, and a "vital force" is not invoked.

Now the above-mentioned Darwinian principles of transmutation are certainly not simple forces of nature like those underlying the development of the individual, *i. e.* chemico-physical forces, and it cannot be said *à priori* whether in one of these principles—perhaps in variability or in correlation—there may not lie concealed a metaphysical principle in addition to the physical forces. In fact it has lately been asserted by Edward von Hartmann[5] that the theory of selection is not a mechanical explanation, since it combines forces which are only partly mechanical and in part directive.

It must therefore be next investigated whether this assertion is tenable.

[5] "Wahreit und Irrthum im Darwinismus." Berlin, 1875.

I.

Are the Principles of the Selection Theory Mechanical?

Edward von Hartmann may justly claim that his views should be considered and tested by naturalists.[1] He would be correctly classed with

[1] [Eng. ed. I have been reproached by competent authorities for having clothed my ideas upon the theory of selection in the form of a reply to Von Hartmann. I willingly admit that this author cannot be considered as the leader of existing philosophical views upon the theory of descent in Germany; Frederick Albert Lange has certainly a much greater claim to this position. Lange does not however combat this theory; he accepts and develops it most beautifully and lucidly on a sound philosophical basis in such a manner as has never been done before from this point of view ("Geschichte des Materialismus," 3rd. ed., 1877, vol. ii. pp. 253—277). On most points I can but agree with Lange. Von Hartmann, however, whose objections appeared to me to be supported by a wide scientific knowledge, afforded me a suitable opportunity of developing my own ideas upon some essential points in the theory of selection. In this sense only have I attempted to interfere with this author, the refutation of his views, as such, having been with me a secondary consideration.] [The chief exponent of the doctrine of organic evolution in this country is Mr. Herbert Spencer, in whose "Principles of Biology," vol. i. chap. xii., will be found a masterly treatment of the theory of descent from a "mechanical" point of view. R. M.]

those philosophers who have approached this question with a many-sided scientific preparation. It can nevertheless be perceived in his case how difficult, and indeed how impossible, it is to estimate the true value of the facts furnished by the investigation of nature, when we attempt to take up only the results themselves, without being practised in the methods by which these are reached, *i. e.* without being completely at home in one of the scientific subjects concerned through one's own investigations. It appears to me that the denial of the purely mechanical value of the Darwinian factors of transformation arises in most part from an erroneous classification of the scientific facts with which we have to deal. There can certainly be no mistake that the entire philosophical conception of the universe, as laid down by Von Hartmann in his "Philosophy of the Unconscious," is unfavourable to an unprejudiced estimate of scientific facts and to their mechanical valuation.

Variability, heredity, and above all correlation, would not be regarded by Von Hartmann as purely mechanical principles, but he would therein assume a metaphysical directive principle.

In the first place, as regards variability, Von Hartmann endeavours to show that it is only a quite unlimited variability which suffices for the explanation of necessary and useful adaptations by means of selection and the struggle for existence. But this does not exist—variation rather takes place in a fixed direction only (in Askenasy's sense), and this can be nothing else than the expression of an innate law of development, *i.e.* a phyletic vital force.

This deduction appears to me in two ways erroneous. In the first place it is incorrect that a quite unlimited variability is a postulate of the theory of selection, and in the next place the admission of variability, which is in a certain sense "fixed in direction," does not necessitate the assumption of a phyletic vital force.

A mere unsettled variability, uniform in all possible directions, is, according to Von Hartmann, necessary for the theory of selection, because only then does the variability offer a certain guarantee "that under given conditions of life the variations necessary for complete adaptation will not be wanting." But it is hereby overlooked that the new life conditions to which the adaptation must take place are as little fixed and unchangeable as the organism itself. In such a case of transformation we have not to deal with a type of organization which was before fixed and immutable, and which has to be squeezed into new life-conditions as into a mould. The adaptation is not one-sided, but mutual; a species in some measure selects its new conditions of life, corresponding with those possible to its organization, *i. e.* with the variations actually occurring. I will choose an instance which will even be conceded by Von Hartmann as being only explicable by natural selection, viz., a case of mimicry.

Supposing that among the South American *Heliconiidæ* there occurred a species of *Pieris* which had no resemblance to these protected butterflies, either in form, marking, or colouring; who can deny that it would be most useful to this species to acquire the form and colouring of a Heliconide, and thus, by taking to new conditions of life, to avoid the persecutions of its foes? But if the physical nature of the Pieride concerned precluded the occurrence of Heliconoid variations, would this incapability of insinuating itself into these new conditions necessitate the decline of the species? Could not its existence be secured in some other manner? could not the destruction of numerous individuals by foes be compensated for by increased fertility? to say nothing of the numerous other means through which the number of surviving individuals might become increased, and the existence of the species secured. This case is not arbitrarily chosen; in the districts where the *Heliconiidæ* occur there are actually a large number of Whites which do not possess the protective colours of the former nauseous family. In the adoption of these new life conditions we have not to deal therefore with survival or extermination, but only with amelioration. It is not every species of "White" that can become adapted to these conditions, because every species does not give rise to the necessary colour variations; those that do, become in this way modified, because they are thus better protected than before. And so it is throughout; wherever we find protected insects enjoying immunity from foes we see also mimickers, sometimes only single, sometimes several, and generally from very diverse groups of insects, according to the general resemblance which existed before the commencement of the process of adaptation, and to the variations made possible by the physical nature of the species concerned.

In the first essay of the second part of this work it was shown that in certain Lepidopterous larvæ a process of adaptation is at the present time still in progress, this depending upon the fact that while the young caterpillar is very well protected by the leaf-green colour of its body, this colour becomes insufficient to conceal the insect as soon as it exceeds the leaf in size. All such caterpillars—and there is a whole series of species—as they increase in size acquire the habit of concealing themselves on the earth by day, and of feeding only at night. New conditions of life are thus imposed, and these are even compulsory, *i. e.* they could not be abandoned without risking the existence of the species. Now in accordance with these new conditions, some individuals in these species have lost the green colouring of the young stages, and have acquired the brown coloration of the dark surroundings of the insects which conceal themselves by day. In one species this change has now occurred in almost all individuals, in others in only a larger or smaller proportion of them. Now supposing that among these species there occurred one, the physical nature of which did not admit of the production of brown shades of colour, would the species for this reason succumb? Is it not conceivable that the want of colour adaptation might be compensated for by better concealment, *i. e.* by burrowing into the earth, or by a greater fertility of the species, or by the development of warning signals—supposing the species to be unpalatable—or finally, by the acquisition of a terrifying marking? In other words, could not the caterpillar itself modify the new condition of life—that of being concealed by day—in accordance with variations made possible by its physical nature?

As a matter of fact in one of these species the green colour remains unchanged in spite of the altered mode of life, and this species, wherever it occurs, notwithstanding the persecution of entomologists, is always common (*Deilephila Hippophaës*); it conceals itself better and deeper however than those other species which, like *Sphinx Convolvuli*, are difficult to detect on account of their brown colour. In another species the striking yellowish green colouring is likewise retained in the majority of individuals, but this species buries itself by day in the loose soil (*Acherontia Atropos*).

To this it may be objected that there are also compulsory changes in the conditions of life from which the species cannot withdraw itself, but in which adaptation must necessarily follow, or extermination would take place.

Such compulsory conditions of life do most assuredly occur, and there is indeed no doubt that many living forms have perished through not becoming transformed. I believe, however, that such conditions occur much more rarely than one is inclined to admit at first sight. As a rule the alternative of immediate change or of extermination is offered only by such changes in the conditions of life as occur very rapidly. The sudden appearance of a new and dominant enemy, such as man, has already caused the extinction of the Dodo (*Didus ineptus*), and of Steller's Sea Cow (*Rhytina Stelleri*), and of other vertebrate animals, and constantly leads to the extermination of many other species of different classes. When in America hundreds of thousands of acres of primeval forest are annually destroyed, the conditions of life of a numerous fauna and flora must be thereby suddenly changed, leaving no choice but extermination.

Such abrupt changes in the conditions of life occur, however, but seldom in nature unless caused by man, and must therefore have very rarely happened in former epochs of the earth's history. Even climatic changes, which we might at first regard as of this character, and which produce a modification in one fixed direction, occur always so gradually that the species has time either to adapt itself to the conditions in this or that direction, according to the variations possible to its physical nature, or else to emigrate.

It thus appears to me erroneous to suppose that variability must be "merely undetermined" in order to complete its part in Darwin's theory of selection, and its "illimitedness" seems to me also as little necessary for this purpose. Von Hartmann imagines that it is only unlimited variability that furnishes a guarantee that any type, to whatever extent diverging from its point of departure, will be reached by the Darwinian method of gradual transmutation by means of selection and the struggle for existence.

But who has ever asserted that *any* type can be reached from any point? Or if anybody has said such nonsense, who can prove that its admission is necessary for the theory of selection? Nowhere in systemy do we see any point of support for such an assumption. But when Von Hartmann imagines that the "unlimited" variability which he postulates for Darwin "is in itself unlimited, the limits of its divergence in a given direction being found, not in itself, but only in external obstacles," he conceives variability to be something independent of, and in some way added to, the animal body, and not a mere expression for the fluctuations in the type of the organism. If, however, we conceive variability in this latter, the true scientific sense, it is in no way "quantitatively unlimited," nor are its limits even determined by *external* influences, but essentially by *internal* influences, *i. e.* by the underlying physical nature of the organism. Darwin has indeed already shown this in a most beautiful manner in his investigations upon the correlations of organs and systems of organs of the body. To make use of a metaphor, the forces acting within the body are in equilibrium; if one organ becomes changed this causes a disturbance in the forces, and the equilibrium must be restored by changes in other parts, and these again entail other modifications, and so forth. Herein lies the reason why the primary change cannot exceed a certain amount if the restoration of the equilibrium is not to be quite impossible. This is but a metaphor, and I do not wish to assert that we are at present in a position to formulate and demonstrate mathematically for any particular case, how much an organ can become changed in any one species before an interruption of the internal harmony of the body takes place. But such impossibility of demonstration does not appear to me to furnish a sufficient reason for regarding variability as the expression of a directive power—as an "innate tendency to variation conformable to law."[2] On the contrary, it is to me easily conceivable that we only learn to analyse the processes of nature in detail very slowly, because of their necessary complexity. It thus appears to me quite useless when in this sense Wigand makes use of the objection, that "the gooseberry has not undergone any enlargement since 1852, although it is inconceivable why it should not attain the size of a pumpkin if variability was not internally limited." It may well be that this is for the present "inconceivable;" nevertheless, this does not justify us in setting up a hypothetical "force of variation" which will not admit of the gooseberry surpassing the pumpkin in size. We are bound to maintain that it is the action and reaction of known forces which sets a limit to the enlargement of this fruit.

In more simple instances the causes of such limitations to growth can be well perceived. Several decades have passed since Leuckart proved in how exact a relation the proportion of volume and surface stood to the degree of organi-

[2] [The above views on the nature of variability, which were also broadly expressed in the first essay "On the Seasonal Dimorphism of Butterflies" (pp. 114, 115), are fully confirmed by Herbert Spencer (*loc. cit.* chaps. ix. and x.), and more recently by A. R. Wallace in an article on "The Origin of Species and Genera" (*Nineteenth Century*, vol. vii., 1880, p. 93). See also some remarks by Oscar Schmidt in his "Doctrine of Descent and Darwinism," Internat. Scien. Ser. 3rd. ed. 1876, p. 173. R. M.]

zation of an animal. In animals of a spherical form the surface is quite sufficient for respiration, so long as they are of microscopic size. But such an organism cannot become enlarged at pleasure, because the ratio of the surface to the volume would become quite different. The surface increases as the square, whilst the volume increases as the cube, so that very soon the surface of the more rapidly increasing bodily mass can no longer suffice for respiration.[3] This sort of limitation is in no way equivalent to that purely external kind which, for instance, manifests itself in such a manner as to prevent the indefinite lengthening of the tail feathers of the Bird of Paradise. In this case feathers that were too long would hinder flight, and such individuals would accordingly be eliminated by natural selection. The cause is in the former case purely internal, depending upon the equilibrium of the forces governing the organism.

Von Hartmann is entirely in the right when he asserts that variability is neither qualitatively nor quantitatively unlimited. In both senses it is limited (in direction as well as in amount) by the physico-chemical forces acting in some contrary way in each specific organism—by the physical

[3] [This law has been beautifully applied by Herbert Spencer in order to explain why, with an unlimited supply of food, an organism does not indefinitely increase in size. "Principles of Biology," vol. i. p. 121—126. R. M.]

nature of each living form. He errs, however, both in making absolute illimitability a necessary postulate of the theory of selection, as also in inferring the existence of a directive principle from that limitation of variability which is certainly present. "Tendencies to variation" do however exist, not in the sense of a directive power, but as expressions of the different physical constitutions of species, which necessarily cause unequal reactions to the same external actions, as will be more clearly proved below.[4]

This is, of course, a modification of Darwin's original assumption of an unbounded variability not limited in direction; but Darwin himself has later coincided in the view that the quality of the variations is essentially determined by the nature of the organism.[5]

[4] [Eng. ed. This idea, formerly expressed by me, occurs also in Lange ("Geschichte des Materialismus," ii. 265), and is there exemplified in a very beautiful manner by illustrations from modern chemistry. Lange compares what I have termed above the "physical constitution" of the organism to the chemical constitution of one of those organic acids which by substitution of single elements may become transformed into more complicated acids, but which, as it were, always undergo "further development" in only one determined and narrowly restricted course. Here, as with the organism, the number of possible variations is very great, but is nevertheless limited, since "what can or cannot arise is determined beforehand by certain hypothetical properties of the molecule."]

[5] "Origin of Species." 4th German ed., p. 19; 5th English ed., p. 6.

I now turn to the consideration of the second factor of the theory of selection—heredity. This also, according to Von Hartmann is not a mechanical principle. Darwin himself has now become convinced how great is the probability against the hereditary retention of modifications which, whether feebly or strongly pronounced, appear *only in single individuals*, *i. e.* of those so-called "fortuitous" variations which are not the expression of a directive developmental principle. "But as among the numberless possible directions of an indefinite variability, useful modifications can only occur in single cases, Darwin has by this supplementary admission himself retracted an inadmissible assumption of his theory of selection," and so forth. A "regular, designed tendency to variation, acting from within and contemporaneously affecting a large number of individuals," must therefore be assumed "in order to insure the by itself improbable inheritance."

But even from the unbounded variability laid down by the author, it by no means follows that useful variations can only occur in single individuals. In the whole category of *quantitative* variations the reverse is always the case. Is it the lengthening of some part that is concerned; so would a large number of individuals always possess the useful variation, since we are not dealing with an *absolute* enlargement, but only with the fact that the part concerned is longer than in other individuals.[6]

But if *qualitative* variations come into consideration, it may be asked whether Darwin's "supplementary admission" does not go too far. Such calculations as those quoted by Darwin from the article in the *North British Review* of March 1867 are extremely deceptive, since we have no means of measuring the amount of protection afforded by a useful variation, and we can therefore hardly compute with any certainty, in how great a percentage of individuals a change must contemporaneously occur in order to have a chance of becoming transferred to the following generation. If our blue rock-pigeon could exist in a polar climate, and if we had the power of introducing it gradually, but not suddenly, into these regions in a wild state, who can doubt that it would assume the white colour of all polar animals? Nevertheless, among wild rock-pigeons white varieties do not occur more frequently than among swallows, crows, or magpies. Or must the white colour of polar animals, the yellow colour of desert species, and the green

[6] [Mr. A. R. Wallace, in his article last referred to, quotes some most valuable measurements of mammals and birds, showing the amount of variation of the different parts. These observations were published by J. A. Allen, in a memoir "On the Mammals and Winter Birds of East Florida," &c. (Bulletin of the Museum of Comparative Zoology at Harvard College, Cambridge, Mass., vol. ii. No. 3.) R. M.]

colour of leaf-frequenting forms, be always referred to a "regular, designed, fixed tendency to variation acting from within," and causing a "large number of individuals" to vary in a similar manner?

There is, however, a grain of truth in the foregoing; variations which occur singly have but little chance of becoming predominant characters, and this is obviously what Darwin concedes. But this is by no means equivalent to the assumption that only those variations which from the first occur in numerous individuals have a chance of being perpetuated. Let us keep to the facts. We have not the slightest reason either for regarding the white colour of polar animals as the direct action of cold, or for considering that the green colour of foliage-living caterpillars depends upon direct action arising from the habit of resting upon the leaves;[7] both these characters are explicable only by natural selection, and there is nothing to favour the assumption (which Von Hartmann postulates as necessary for success) that many individuals varied into white at the same time. We know no single extra-polar species of a dark colour which frequently, *i. e.* in many individuals of every generation, varies into white, but we know many species which from time to time produce single white individuals. Now

[7] [See note 2, p. 310. R. M.]

when, on the other hand, we find that all polar animals to which the white coloration is advantageous, and indeed none but species of which the nearest allies vary only individually into white, possess this colour, must we not conclude from this alone that *single* variations can, under favourable conditions, become predominant characters?

It appears to me that in this question one weighty factor has been too little regarded, even by the supporters of the selection theory, viz., the slowness of most, and especially of climatic changes, which I have already insisted upon. If the transformation of a temperate into an arctic climate occurred so rapidly that the species exposed to it had the alternative either of becoming white in ten or twenty generations or of being unable to exist, then the hasty intervention of a directive power could alone save them from extermination by causing hundreds of thousands of individuals to become similarly coloured with all speed. But it is quite different if the change of climate takes place only in the course of several thousand generations; and this, according to the geological evidence, must have been the true state of the case.

Let us take a definite example—the well-known one of the hare. With us this animal remains brown in the winter and but seldom produces white varieties, whilst its ally the Alpine hare is white during seven months of the year, the Norwegian hare during nine months, and the Greenland hare throughout the whole year. If our climate became transformed into an arctic one, after a given time there would arrive a period when the older coloration no longer possessed any advantage over the occasional and singly-appearing white variations; the winter days during which the ground was covered with snow would have become so numerous, that the protection afforded to the white animals would be equal to the protection enjoyed by the brown individuals on the equally numerous days free from snow. From this time forth the hares that were white in winter would not be subjected to a greater decimation by foxes, &c., than the brown individuals. This period must however be represented as consisting of one or more centuries, and it would be strange if from the individual white hares, which now had an equal chance of existing, some white families did not become established. But the state of affairs would gradually become reversed —the brown hares would experience greater decimation, and wherever there were white families these would possess an advantage in the struggle for existence. It does not follow that the dark individuals would be forthwith extirpated; on the contrary, the advantage in favour of the white would be but small throughout a long period of time, and these individuals would only gradually increase to a higher percentage of the total population; nevertheless their numbers would constantly but very slowly augment. In the course of time this increase would become more rapid for two reasons—first, because even a very small advantage in favour of the increasing number of individuals would always leave a greater number of these victorious; and secondly, because on the whole as the climate became more arctic, the advantage of being white would continually become more decisive in determining which should live and which should succumb.

Thus I see no reason why individual variations which do not appear only once, but which frequently recur in the course of generations, should not acquire predominance under favourable conditions. All facts are in accord with this. Even the common hare shows us that it would be quite capable of becoming coloured in a similar manner. In the museum of Stuttgart there are three specimens of *Lepus timidus*, killed in Wurtemburg, which are completely white, and several others which are silver-grey or spotted with white. In eastern Russia the common hare possesses a light grey, almost white, winter coat, and Seidlitz[8] makes known the interesting observation that such light specimens occur *singly* in Livonia, where "the common hare has become naturalized since the commencement of the century."

[8] "Die Darwin'sche Theorie," Dorpat, 1875.

As I have already insisted upon above, from the point of view of the conditions of life there is no reason for assuming rapid transformations; the change of conditions is almost always extremely slow; and indeed in numerous instances no objective change occurs, but simply a subjective one, if we may thus designate those cases in which the alteration in the conditions of life depends upon a change in the animal form which is undergoing transformation, and not in that of the environment. This is the case in the above-mentioned instances of mimicry, where the whole change in the conditions of life arises from one species becoming similar to another. The process of natural selection has here as long a period of time as it requires to perfect its results. It is quite similar in all cases of special protective adaptations of form and colour. In all these it is always *improvement* that is concerned, and not the question "to be or not to be" with which we have to deal.

It is just cases of this last kind, however, which are best fitted for exposing the improbability and insufficiency of the assumption of a variational tendency as a distinct directive power. We have only to fix our attention upon some particular case of sympathetic colouring, or, still better, of mimicry. A "tendency to variation" implies that a large number of individuals produce varieties resembling the model to be imitated, and this—at least according to Von Hartmann—must take place in each of the successive generations, so that by this means, combined with heredity, the useful variation becomes increased. But how comes it that this "tendency to variation" coincides with the existence of the model both in time and place? Can this be due to accident if the two have not a common cause? The upholders of a directive power will certainly not admit this; so that there remains only Leibnitz's assumption of a pre-established harmony contained in the first organic germ, which, after innumerable transformations of the organic form and after millions of years, gave rise in the midst of the Amazonian region to an inedible Heliconide with certain yellow, black, and white markings on the wings, and at precisely the same time developed the tendency in a Pieride at the same spot on the globe to imitate this Heliconide as a model!

In adddition to this assumption, which is certainly but little worthy of consideration, there is perhaps one other remaining, viz., that all or many Pierides and other species of butterflies possessed the same tendency to a Heliconoid variation and were always everywhere striving to develop this type, but succeeded only where they accidentally coincided in time and place with the model, the "tendency" being thus furthered by natural selection. But the facts negative this assumption, since such imitative variations have never been observed to a perceptible extent in other species.[9]

All variations which are demonstrably useful can be similarly dealt with if their origin is explained by variational tendencies.

We perceive that the objection which Von Hartmann brings against heredity is only valid on the ground that this process affords no security for the preservation of variations which occur singly. That heredity itself is a mechanical process is not directly disputed; it is simply assumed that new characters can be transferred by inheritance only when they are produced by the metaphysical "developmental principle," and not when they arise "accidentally." This critic does not therefore direct his attack against heredity, but rather against the mechanical origin of variability.

Von Hartmann might have said here that a reference of the phenomenon of heredity to purely mechanical causes, *i. e.* a mechanical theory of

[9] [A certain number of instances of mimicry are known to occur between species both of which are apparently nauseous. A most able discussion of this difficult problem is given by Fritz Müller, in the case of the two butterflies *Ituna Ilione* and *Thyridia Megisto,* in a paper published in *Kosmos,* May, 1879 (p. 100). The author shows by mathematical reasoning that such resemblances between protected species can be accounted for by natural selection if we suppose that young birds and other insect persecutors have to learn by experience which species are distasteful and which can be safely devoured. See also Proc. Ent. Soc. 1879, pp. xx—xxix. R. M.]

heredity, is up to the present time wanting. That he has not done so proves on the one hand that he despised the dialectical art, but, on the other hand, that he himself has not overlooked the subserviency of the total phenomenon to law, and that he grants the possibility of finding a mechanical explanation therefor. If, in fact, the power of inheritance does not depend upon mechanical principles, I know not what organic processes we are entitled to regard as mechanical, since they are all dependent in essence upon heredity, with which process they are at one, and from which they cannot be thought of as isolated. Haeckel correctly designates reproduction as surplus individual growth, and accordingly refers the phenomena of heredity to those of growth. Conversely, growth may also be designated reproduction, since it depends upon a continuous process of multiplication of the cells composing the organism, from the germ-cell to the innumerable congeries of variously differentiated cells of the highly developed animal body. Who can fail to see that these two processes, the reproduction of the germ-cell and its offspring in the economy of the individual, and the reproduction of individuals and species in the economy of the organic world, show an exact and by no means simply superficial analogy?[10] But whoso grants this must also conceive both processes to depend upon the

[10] See Haeckel's "Generelle Morphologie," ii. 107.

same cause—he cannot assume for the one a causal power and for the other a directive principle. If nutrition and cell-multiplication are purely mechanical processes, so also is heredity. Although it has not yet been possible to demonstrate the mechanism of this phenomenon, it can nevertheless be seen broadly that by means of a minimum of living organic matter (*e.g.* the protoplasm of the sperm and germ-cell) certain motions are transferred, and these can be regarded as directions of development, as I have already briefly laid down in a former work.[11] The power of organisms to transmit their properties to their offspring appears to me to be only conceivable in such a manner "that the germ of the organism by its chemico-physical composition together with its molecular structure, has communicated to it a fixed direction of development—the same direction of development as that originally possessed by the parental organism . . . . . ." (*loc. cit.* p. 24). This is confessedly nothing more than a hint, and we do not learn therefrom the means by which developmental direction can be possibly transferred to another organism.

Recently Haeckel, that indefatigable pioneer to whom we are indebted for such a rich store of new ideas, has attempted to bridge over this gap in his essay on "The Perigenesis of the Plasti-

[11] "Über die Berechtigung der Darwin'schen Theorie," Leipzig, 1868.

dule," Berlin, 1876. The basic idea, that heredity depends upon the transference of motion, and variability upon a change of this motion, completely corresponds with the conviction gained in the province of physical science, that "all laws must finally be merged in laws of motion" (Helmholtz [12]). I hold this view to be the more completely justifiable—although certainly not in the remotest degree as proved—because I formerly designated the acquired individual variations as the "diversion of the inherited direction of development." Haeckel's hypothesis in so far accomplishes more than Darwin's pangenesis, in which a transference of matter, and not of a species of motion peculiar to this matter, is assumed. But although the germ of a mechanical theory of heredity may be contained in Haeckel's hypothesis, this nevertheless appears to me to be somewhat remote from completely solving the problem. It brings well into prominence one portion of the process of inheritance; under the image of a molecular motion of the plastidule, which motion is modifiable by external influences, we can well understand the fact of a change gradually taking place in the course of generations. On the other hand, the assumption of consciousness in the plastidule,—however admissible philosophically—although only as a

[12] "Populäre wissenschaftl. Vorträge," vol. ii., Brunswick, 1871, p. 208.

formula, scarcely furnishes any deeper knowledge. In the light of a theory, detailed instances which were formerly obscure should become comprehensible. I fail to see, however, how the various forms of atavism, *e.g.* the reversions which so commonly occur by crossing different races, become more comprehensible by assuming consciousness in the plastidule. If in both parents the plastidule long ago acquired different molecular motions, why, in its rencounters in the germ, does it recollect past times and reassume the older and long abandoned motion? That it does acquire the latter is indeed a fact if we once refer the directional development of the individual to molecular motion of the plastidule; the wherefore does not appear to me, however, to become clearer by assuming consciousness in the plastidule. A mechanical theory of heredity must rather be able to show that the plastidule movements of the male and female germ-cells, in their rencounter in the case of the crossing of widely divergent forms, become mutually modified in such a manner that the motion of the common ancestral form must occur as the resultant. To such demonstration there is however as yet a long step. Haeckel himself moreover points out that his hypothesis is by no means a "mechanical theory of heredity," but only an introduction to this theory, which he hopes "will be capable of being elevated to the rank of a genetic molecular theory" (*loc. cit.* p. 17). But although we must also confess with the critic of the "Philosophy of the Unconscious," that "the facts of heredity have hitherto defied every scientific explanation,"[13] this furnishes us with no excuse for flying to a metaphysical explanation, "which is here certainly least able to satisfy the inability to understand the connection arising from natural laws."

It is not to be wondered at that Von Hartmann, on the ground of the "Unconscious" on which he takes his stand, speaks of the law of correlation as an unconscious acknowledgment of a "non-mechanical universal principle on the side of Darwinism." By "correlation" he understands something quite different to the idea which we attach to this expression. He supposes that "Darwinism sees itself compelled to acknowledge through empirical facts the uniform correlation of characters pertaining to the specific type; but it thereby contradicts its mechanical principles of explanation, all of which amount to the same thing as conceiving the type as a mosaic, chequered, superficial, and accidental aggregate of characters, which have been singly acquired, contemporaneously or successively, by selection or habit." I do not believe, however, that any such conception has ever been admitted either by

[13] "Das Unbewusste vom Standpunkte der Physiologie u. Descendenztheorie," Berlin, 1872, p. 89. The second edition appeared in 1877, in Von Hartmann's own name.

Darwin or any one else. The admission that not all, but only every deep-seated *physiological* detailed modification, is or may be bound up with a system of correlated changes, indeed implies that we on our side also acknowledge an internal harmony of parts—an equilibrium, as I have above expressed it.

But does this include the admission of a teleological principle, or exclude a mechanical explanation? Do we thereby acknowledge a "specific type" in the sense of an inseparably connected complex of characters, none of which can be taken away without all the others becoming modified? Does such a view agree generally with the empirical facts?

Neither of these views appears to me to represent the case.

I will first answer the second question. On all possible sides the earlier view of the absolute nature of species is contradicted; there is no boundary between species and varieties. But when Von Hartmann assumes that by the transformation of one species "into another" the "whole uniformly connected complex must become changed," he falls back into the old doctrine of the absolute nature of species, which is sharply contradicted by multitudes of facts. We not unfrequently observe varieties which differ from the parent-form by only a single character, whilst others show numerous differences, and again others may be seen in which the differences predominate. This last deviation would then be designated by many systematists as a new species, but not so by others.

The "specific type" is thus indeed a kind of mosaic-work, but it is a structure to which all the single characters—the stones of the mosaic—belong and build up one harmonious whole, and not a meaningless confusion. Some of the stones or groups of stones can be taken away and replaced by others differently coloured without the structure being thereby necessarily distorted, *i. e.* destroyed as a structure; but the larger the stones which are exchanged the more necessary will corrections in the other parts of the structure become, in order that the harmony of the whole may be preserved.

Still more weighty than those insensible transitions which in various groups of animals so frequently connect species with species, appear to me, however, the facts made known in the second essay of the second part of this volume, which prove that the two forms in which one species appears can change entirely independently of one another. The caterpillar changes and becomes a new variety or even species (according to the form-value of the change), whilst the butterfly remains unaltered. How could this occur if some other law than that of physiological equilibrium linked together the parts or characters and permitted them to become severed? Must not the two stages become changed with and through one another, like the parts of one body, since they first together constitute the specific type? Is not the fact of this not happening a proof that the whole "uniformly connected complex" of the specific type is not bound and held together by a metaphysical principle, but simply by natural laws?

Now when Von Hartmann comprises the relations of different species to one another under the idea of correlation, such for instance as the relation of dependence in which orchidaceous flowers stand with respect to the insects which visit them, he completely abandons the scientific conception which should be associated with this expression, and compares together two heterogeneous things which have nothing in common excepting that they are both considered by him as a result of the "Unconscious." The consequence which is then deduced from this correlation of his own construction, viz., that an organic law of correlation is only another expression for a "law of organic development" in the sense of a metaphysical power, obviously cannot be admitted.

By correlation we understand nothing more than the dependence of one part of the organism upon the others and the mutual inter-relations of these parts, which depend entirely upon a "physiological relation of dependence," as Von Hartmann himself has correctly designated it. Herein is evidently comprised the total morphology of the organism—the structure as a whole, the length, thickness and weight of the single parts, as well as the histological structure of the tissues, since upon all these depends the performance of the single parts. But when, under correlation, Von Hartmann comprises "also a morphological, systematic, inter-action of all the elements of the organism with reference both to the typical ground-plan of the organization as well as to the microscopic anatomical structure of the tissues," he drags into the idea something foreign to it, not on the ground of facts, but actually in opposition to them, and supported only by a supposed "innate developmental principle" which "is not of a mechanical nature."

The living organism has already been often compared with a crystal, and the comparison is, *mutatis mutandis*, justifiable. As in the growing crystal the single molecules cannot become joined together at pleasure, but only in a fixed manner, so are the parts of an organism governed in their respective distribution. In the crystal where nothing but homogeneous parts become grouped together their resulting combination is likewise homogeneous, and it is obvious that they offer but very little possibility of modification, so that the governing laws thus appear restricted and immutable. In the organism, whether regarded microscopically or macroscopically, various parts become combined, and these therefore offer numerous possibilities of modification, so that the governing laws are more complex, and appear less restricted and unchangeable. In neither instance do we know the final causes which always lead to a given state of equilibrium; in the case of a crystal it has not occurred to anybody to ascribe the harmonious disposition of the parts to a teleological power; why then should we assume such a force in the organism, and thus discontinue the attempt, which has already been commenced, to refer to its natural causes that harmony of parts which is here certainly present and equally conformable to law?

On these grounds the assertion that the theory of selection is not an attempt at a "mechanical" explanation of organic development appears to me to be incorrect. Variability and heredity, as well as correlation, admit of being conceived as purely mechanical, and must be thus regarded so long as no more cogent reasons can be adduced for believing that some force other than physico-chemical lies concealed therein.

But we certainly cannot remain at the purely empirical conception as laid down by Darwin in his admirable work on the "Origin of Species." If the theory of selection is to furnish a method of mechanical explanation, it is essential that its factors should be formulated in a precise mechanical sense. But as soon as we attempt to do this it is seen that, in the first enthusiasm over the newly discovered principle of selection, the one factor of transformation contained in this principle itself has been unduly pushed into the background, to make way for the other more apparent and better known factors.

I have for many years insisted that the first, and perhaps most important, or in any case the most indispensable, factor in every transformation, is *the physical nature of the organism itself.*[14]

It would be an error to believe that it is entirely the external conditions which determine what changes shall appear in a given species; the nature of these changes depends essentially upon the physical constitution of the species itself, and a modification actually arising can obviously be only regarded as the resultant of this constitution and of the external influences acting thereon.

But if an essential or perhaps even a preponderating share in determining new characters is to be undoubtedly ascribed to the organism itself, for a mechanical representation of organic develop-

[14] "Über die Berechtigung," &c., Leipzig, 1868. In this work will be found briefly laid down the theoretical conception of variability here propounded somewhat more broadly. [In the last edition of the "Origin of Species" Darwin states, with respect to the direct action of the conditions of life as producing variability, that in every case there are two factors, "the nature of the organism and the nature of the conditions." 6th ed. p. 6. R. M.]

mental processes everything depends upon our being able to conceive this most important factor in a definite theoretical manner, and to comprise under one common point of view its apparently contradictory manifestations of constancy and variability.

Now every change of considerable extent is certainly considered by Darwin to be the direct or indirect consequence of external actions; but indirect action always presupposes a certain small variability (individual variability), without which larger modifications cannot be brought about. Empirically this small amount of variability is doubtless present, but the question arises, upon what does it depend? Can it be conceived as arising mechanically, or is it perhaps just at this point that the metaphysical principle steps in and offers those minute variations which make possible that course of development which, according to this view, is immutably pre-determined? It is certainly the absence of a theoretical definition of variability which always leaves open a door for smuggling in a teleological power. A mechanical explanation of variability must form the basis of this side of the theory of selection.

This explanation is not difficult to find. All dissimilarities of organisms must depend upon the individuals having been affected by dissimilar external influences during the course of the development of organic nature. If we ascribe to the organism the power of giving rise by multiplication only to exact copies of itself, or, more correctly, the power of transmitting unaltered to its successors the motion of its own course of development, each "individual variation" must depend upon the power of the organism to react upon external influences, *i. e.* to respond by changes of form and of function, and consequently to modify its original (inherited) developmental direction.

It has sometimes been insisted upon, that the "individuals of the same species" or the offspring of one mother cannot be absolutely equal, because, from the commencement of their existence, they have been subjected to dissimilar actions of the environment. But this implies that by perfectly equal influences they would become equal, *i.e.* it supposes that variability is not inseparably bound up with the essence of the organism, but is only the consequence of developmental tendencies which are in themselves equal being unequally influenced. As a matter of fact the first germs of an individual certainly cannot be supposed to be perfectly equal, because the individual differences of the ancestors must be contained therein in different degrees according to their constitution, and we should have to go back to the primordial organism of the earth in order to find a perfectly homogeneous root, a *tabula rasa* from which the descendants would commence their development.

Whether such a homogeneous root ever existed is however doubtful; it is much more probable that *numerous* organisms first arose spontaneously,[15] and these cannot be presumed to have been absolutely equal, since the conditions under which they came into life cannot have been perfectly identical. Let us, however, for the sake of simplicity assume a single primordial organism; the first generation which took its rise from this by reproduction could only have possessed such individual differences as were produced by the action of dissimilar external influences. But the third generation, together with self-acquired, would also have shown *inherited*, dissimilarities, and in each succeeding generation the number of tendencies to individual difference imparted to the germ by heredity must have increased to a certain degree, so that it may be said that all germs,

[15] [Although hardly necessary to the evolutionist, it may perhaps be well to remind the general reader, that all experiments upon spontaneous generation, or abiogenesis, have hitherto yielded negative results; no life is produced when the proper precautions are taken for excluding atmospheric germs. But although we have so far failed to reproduce in our laboratories the peculiar combination of conditions necessary to endow colloidal organic matter with the property of "vitality," the consistent evolutionist is bound to believe, from the analogy of the whole of the processes of nature, that at some period of the earth's history the necessary physical and chemical conditions obtained, and that some simple form or forms of life arose "spontaneously," *i. e.* by the operation of natural causes. R. M.]

from their first origination, bear in themselves a tendency to show individual peculiarities, and would develop these even if they should not be again affected by dissimilar influences. This is obviously the case, since the youngest egg-cells in the ovary of an animal are, as can be demonstrated, always exposed to unequal external conditions with respect to nutrition and pressure.[16] Hence, if it were possible that two germs were exactly equal with respect to the direction of development imparted to them by heredity, they would nevertheless furnish two incongruent individuals; and if, conversely, it were possible that two individuals could be exposed to absolutely the same external influences from the formation of the embryo, these also could not be identical, because the individual differences of the ancestors would entail small differences, even in asexual reproduction, in the direction of development transmitted to the egg. The differences between individuals of similar origin thus finally depend entirely upon the dissimilarity of external influences—on the one side upon those which divert the development of the progenitors, and on the other side upon those which divert the individual itself from its course, *i. e.* from the developmental direction transmitted hereditarily. Although I thus essentially agree with Darwin and Haeckel

[16] See Haeckel's "Generelle Morphologie," vol. ii. p. 203, and Seidlitz, "Die Darwin'sche Theorie," 1875, p. 92 *et. seq.*

in so far as these authors refer the "universal individual dissimilarity" to dissimilar external actions, I differ from Darwin in this, that I do not see an essential distinction between the direct and indirect production of individual differences, if by the latter is meant only the unequal influencing of the germ in the parental organism. Haeckel is certainly correct in referring the "primitive differences of the germs produced by the parents" to the inequalities of nutrition to which the single germs must inevitably have been exposed in the parent organism; but another dissimilarity of the germs must evidently be added —a dissimilarity which has nothing to do with unequal nutrition, but which depends upon unequal inheritance of the individual differences of the ancestors, a source of dissimilarity which must arise to a greater extent in sexual than in asexual reproduction. Just as in sexual propagation there occurs a blending of the characters (or more precisely, developmental directions) of two *contemporaneous individuals* in *one* germ, so in every mode of reproduction there meet together in the same germ the characters of a whole *succession of individuals* (the ancestral series), of which the most remote certainly make themselves but seldom felt in a marked degree.

The fact of individual variability can in this way be well understood; the living organism contains in itself no principle of variability—it is the *statical element* in the developmental processes of the organic world, and would always reproduce exact copies of itself if the inequality of the external influences did not affect the developmental course of each new individual; these influences are therefore the *dynamical elements* of the process.

From this conception of variability two important empirically established facts can be theoretically deduced, viz. the limitability of variation with respect to quality, which has already been previously mentioned, and the origination of transformations by the direct action of external conditions of life.

If the differences in individuals of the same origin depend upon the action of unequal influences, variation itself is nothing else than the reaction of the organism to a definite external inciting cause, the quality of the variation being determined by the quality of the inciting cause and by that of the organism. In the cases of individual variation hitherto considered, the quality of the organism is equal but that of the inciting cause is unequal, and in this way there arise minute differences in organisms of an equal physical constitution—variations of a different quality.

The same result, viz., different qualities of variation, may also arise in a reverse manner by organisms of a different physical nature being affected by equal external influences. The response of the organism to the cause inciting change would be different according to its nature, or, in other words, organisms of different natures react differently when affected by equal modifying influences. The physical nature of the organism plays the chief part with respect to the quality of the variations; each specific organism can thus give rise to extremely numerous, but not to all conceivable, variations; that is, only to such variations as are made possible by its physical composition. From this it follows further that the possibilities of variation in two species are more widely different, the wider they diverge in physical constitution (including bodily morphology)—that a cycle of variation is peculiar to every species. In this manner we are led to the knowledge that there must certainly exist a "fixed direction of variation," but not in the sense of Askenasy and Von Hartmann, as the result of an unknown internal principle of development, but as the necessary, *i. e.* mechanical, consequence of the unequal physical nature of the species, which must respond even to the same inciting cause by unequal variations.

The facts, as far as we know them, agree very well with this conclusion. Allied species vary in a similar manner, whilst species which are more distantly related vary in a different manner, even when acted upon by the same external influences. Thus, in the first part of these "Studies" I have remarked that many butterflies under the influence of a warm climate acquire an almost black coloration (*Polyommatus Phlœas*), whilst on the other hand others become lighter (*Papilio Podalirius*).

We can thus understand why always certain courses of development are followed, a fact which cannot be completely explained by the nature of the conditions of life which induce the variations. But as soon as we clearly perceive that the quality of the changes essentially depends upon the physical nature of the organism itself, we arrive at the conclusion that species of widely diverging constitutions must give rise to different variations, whilst those of allied constitutions would produce similar variations. But definite courses of development are thus traced out, and we perceive that from any point of the organic developmental series, it is impossible that any other point can be attained at pleasure. Variation in a definite direction thus by no means necessitates the acknowledgment of a metaphysical developmental principle, but can be well conceived as the mechanical result of the physical constitution of the organism.

The manner in which the dissimilar physical constitution of organisms must arise can also be easily shown, although the first commencement of the whole developmental series, *i. e.* the oldest living forms must be assumed to have been almost homogeneous in their physical constitution. The quality of the variation is, as said before, not merely the product of the physical constitution, but the resultant of this and of the quality of the changing external conditions. Thus from the first "species" there proceeded, through the dissimilar influence of external conditions of life, several new "species," and as this took place the former physical nature of the organism at the same time became changed, necessitating also a new mode of reacting upon external influences, *i. e.* another direction of variation. The difference from the primary "species" must certainly be conceived as having been very minute, but it must have increased with each new transformation, and must have proceeded exactly parallel with the degree of physical change connected with each transformation. Thus, hand in hand with the modifications, the power of modification, or mode of reaction of the organism to changing influences, must have continually become re-modified, and we finally obtain an endless number of differently constituted living forms, of which the variational tendencies are different in exact proportion to their physical divergence, so that nearly allied forms respond similarly, and widely divergent forms very differently, to the same inciting causes.

Individual variation arises, as I have attempted to show, by each individual having been continually affected by different, and indeed by constantly changing, influences. Let us, however, imagine on the contrary, that a large group of individuals is affected by the same influences—in fact by such influences as the remaining individuals of the species are not exposed to: this group of individuals would then vary in a nearly similar manner, since both factors of variation, viz. the external influence and the physical constitution, are equal or nearly so. Such local variations would first become prominent when the same external influence had acted upon a series of generations, and the minima of variation produced in the individual by the once-exerted action of the cause inciting change had become augmented by heredity. Transformations of some importance (up to the form-value of species) can thus arise simply by the direct action of the environment, in the same way as that in which individual differences are produced—only the latter fluctuate from generation to generation, since the inciting influences continually change; whilst, in the former, the constant external cause inciting modification always reproduces the same variation, so that an accumulation of the latter can take place. Climatic varieties can be thus explained.

A more efficacious augmentation of the variations arising in the single individual is certainly brought about by the *indirect* action of the environment upon the organism. It is not here my intention to explain once more the processes of natural selection; I mention this only in order to point out that in these cases transformation depends upon a *double action* of the environment, since the latter first induces small deviations in the organism by direct action, and then accumulates by selection the variations thus produced.

By regarding variability in this manner—by considering each variation as the reaction of the organism to an external action, as a diversion of the inherited developmental direction, it follows that without a change in the environment no advance in the development of organic forms can take place. If we imagine that from any period in the earth's history the conditions of life remain completely unchanged, the species present on the earth at this period would not, according to our view, undergo any further modification. Herein is clearly expressed the difference of this view from that other one according to which the inciting principle of modification is not in the environment, but lies in the organism itself in the form of a phyletic vital force.

I cannot here refrain from once more returning to the old (ontogenetic) vital force of the natural philosophers, since the parallel between this and its younger sister, the "phyletic vital force" which appears in so many disguises, is indeed striking. Were the inciting principle of the development of the individual actually an independent vital force acting within the organism, the birth and growth of the individual would be able to take place without the continuous encroachment of the environment, such as occurs in nutrition and respiration. Now this is known to be impossible, so that those who support the existence of such a force, if any still exist, would be driven to the obscure idea of a co-operation between the designing power and the influences of the environment, just in the same manner as such a co-operation is at present postulated by the defenders of the phyletic vital force. I shall further on take the opportunity of pointing out that this last idea is quite untenable; with respect to the (ontogenetic) vital force any clearer proof cannot well be adduced, but it will be admitted that the confused notion of the co-operation and inter-action of teleological and causal powers is, from our point of view, opposed to those very simple and clear ideas which are in harmony with the views on phyletic development. As in racial development each change of the organic type is entirely dependent upon the action of the environment upon the organism, so in the development of the individual, the totality of the phenomena of the personal life must depend upon similar actions. Physiology, as is known, herein entirely supports our view, since this shows that without the continual alternating action of the environment and of the organism there can be no life, and that vital phenomena are nothing but the reactions of the organism to the influences of the environment.

It will be immediately perceived how exactly the processes of phyletic and of ontogenetic development coincide, not merely in their external phenomena but in their nature, if we trace the consequences of the existing knowledge of the structure of the animal body. Although we may not entirely agree with Haeckel's doctrine of individuality in its details, its correctness must on the whole be conceded, since it cannot be disputed that the notion of individuality is a relative one, and that several categories of morphological individuals exist, which appear not only *singly* as *physiological individuals, i. e.* as independent living beings of lowest grade, but which can also *combine* to form beings of a higher order.

But if we admit this, we should see with Haeckel nothing but reproduction in the origination of a high organism from a single cell, the egg; this reproduction being at the same time combined with various differentiations of the offspring, *i.e.* with adaptations of the latter to various conditions of life. Not even in the fact that the tissues and organs of a single physiological individual stand in great dependence upon one another through physical causes,[17] is there any

[17] [In a recently published work by Dr. Wilhelm Roux this author has attempted to work out the idea of an analogy

striking difference between this view and the phyletic composition of the animal (and vegetable) kingdom out of physiological individuals (Haeckel's "*Bionten*"), since contemporaneous animals (individuals and species) are known to influence one another in the most active manner.

Now if we further consider that the same units (cells) which, by their reproduction and division of labour, at present compose the body of the highest organism, must at one time have constituted as independent beings the beginning of the whole of organic creation, and that consequently the same processes (division of cells) which now lead to the formation of a mammal, at that time led only to a long series of different independent beings, it will be admitted that both developmental series must depend upon the same inciting powers, and that with reference to the causes of the phenomena it is not possible that any great gap can exist between ontogeny and phylogeny, *i. e.* between the life-phenomena of the individual and those of the type. According to our view both depend upon that co-operation of the same material physical forces which admits of being

between the struggle for existence and survival of the fittest in individuals and species, and the struggle for existence and survival of the parts in the individual organism. See "Der Kampf der Theile im Organismus: ein Beitrag zur Vervollständigung der mechanischen Zweckmässigkeitslehre," Leipzig, 1881. R. M.]

briefly summarized as the reaction of organized living matter to influences of the environment.

Our opponents either cannot boast of such harmony in their conception of nature, or else they must, together with the phyletic vital force, re-admit into their theory the old ontogenetic vital force. I know not indeed why they should not do so. Whoever inclines to the view that organic nature is governed not merely by causal, but at the same time by teleological, forces, may admit that the latter are as effective as inciting causes of individual, as they are of phyletic, development. According to my idea they are even bound to admit this, since it cannot be perceived why the adaptations of the ontogeny should not depend upon the same metaphysical principle assumed for each individual, as the adaptations of the phylogeny; the latter are indeed only brought about by the former. I believe therefore that the vital force (ontogenetic) of the ancients stands or falls with the modern (phyletic) vital force. We must admit both or neither, since they both rest on the same basis, and are supported or opposed by the same arguments. Whoever feels justified in setting up a metaphysical principle where complete proof that *known forces* are sufficient for the explanation of the phenomena has not yet been adduced, must do the same with respect to individual, as he does to phyletic, development, since this proof is in both cases very far from being complete, and still contains large and numerous gaps.[18]

The theoretical conception of variation as the reaction of the organism to external influences has also not yet been experimentally shown to be correct. Our experiments are still too coarse as compared with the fine distinctions which separate one individual from another; and the difficulty of obtaining clear results is greatly increased by the circumstance that a portion of the individual deviations always depends upon heredity, so that it is frequently not only difficult, but absolutely impossible, to separate those which are inherited from those which are acquired. Still further are we removed from being able to refer variation to its final mechanical causes, *i.e.* from a mechanical theory of reproduction, which would bring within the range of mathematical calculation both the phenomena of stability (heredity) and of change (variability).

But although sufficient proofs of the correctness of the views here advocated cannot at present be adduced, these views are not contradicted by any known facts—they are, on the contrary, supported by many facts which they in turn make comprehensible (local forms, different

[18] [Eng. ed. Meanwhile it has been shown by Oscar Schmidt that Von Hartmann, under the name of "the Unconscious," reinvests the old vital force with some portion of its former power. "Die naturwissenschaftlichen Grundlagen der Philosophie des Unbewussten," Leipzig, 1877, p. 41.]

cycles of variation in heterogeneous species). These views are finally completely justified by their furnishing the only possible theoretical formulation of variability on which a mechanical conception of organic development can be based. That such a conception is not only admissible, but is unavoidable, at least to the naturalist, I have already attempted to prove.

II.

Mechanism and Teleology.

In the third volume of his smaller works Karl Ernst von Baer submits the theory of selection to a most searching examination. Without actually calling in question its scientific admissibility, he believes that this theory is dependent upon its satisfying one condition, viz. that it should connect the teleological with the mechanical principle.

"The Darwinian hypothesis, as stated by its supporters, always ends in denying to the processes of nature any relation to a future, *i.e.* any relation of aim or design. Since such relations appear to me *quite evident*," &c. And further:—"If the scientific correctness of the Darwinian hypothesis is to be admitted, it must accommodate itself to this universal striving after a purpose. If it cannot do this we should have to deny its value."

These words appear almost equivalent to passing a sentence of doom upon the theory of selection and the mechanical conception of nature, for how can one and the same process be effected simultaneously by necessity and by designing powers? The one excludes the other, and we must—so it appears—take our stand either on one side or the other.

Nevertheless we cannot set aside Von Baer's proposition without further examination simply because it is apparently incapable of being fulfilled, since it contains a truth which should not be overlooked, even by those who uphold the mechanical theory of nature. It is the same truth which is also made use of by the philosophical opponents of this theory, viz. that the universe as a whole cannot be conceived as having arisen from blind necessity—that the endless harmony revealed in every nook and corner by all the phenomena of organic and of inorganic nature cannot possibly be regarded as the work of chance, but rather as the result of a "vast designed process of development." It is also quite correct when, in reply to the supposed objection that the mechanical theory of nature is not concerned with chances but with necessities, Von Baer answers that the operations of a series of necessities which "are not connected together" can only be termed accidents in their opposing relations. He illustrates this by instancing a target. If I hit the latter by a well-aimed shot, nobody would explain this as the result of an accident, but if "a horseman is riding along a gravelly road past this target, and one of the pebbles thrown up by the hoof of the galloping horse hits the mark, this would be termed an accident of extremely rare occurrence. My target was not the mark for the pebble, therefore the hit was purely accidental, although the projection of the stone in this precise direction with the velocity which it had acquired, was sufficiently explained by the kick given by the horse. But the hit was accidental because the kick of the galloping horse, although it necessarily projected the pebble, had no relation at all to my target. For the same reason we must regard the universe as an immense accident if the forces which move it are not designedly regulated—the more immense because it is not a single motion of projection that acts here, but a large number of heterogeneous powers, *i.e.* a large number of variously acting necessities which are, as a whole, devoid of purpose, but which nevertheless accomplish this purpose, not only at any single moment, but constantly. A truly admirable series of desirable accidents!" [1]

The same idea is expressed, although in a very different manner, by Von Hartmann, in the concluding chapter of his work already quoted. He thinks that "design is a necessary and certain consequence of the mechanical laws of nature."

[1] *Loc. cit* p. 175.

"Were the mechanism of natural laws not teleological there would be no mechanically regulated laws, but a weak chaos of obstinate and capricious powers. Not until the causality of the laws of inorganic nature had superseded the expression "dead" nature, and had shown itself as the mainspring of life and of a conformability to design visible on all sides, did it deserve the name of mechanical lawfulness; just as a complication of wheels and machinery made by man, which move in some definite manner with respect to one another, only acquires the name of a mechanism or of a machine when the immanent teleology of the combination and of the various movements of the parts is revealed."[2]

Against the correctness of the idea underlying these statements scarcely anything can in my opinion be said. The harmony of the universe and of that portion of it which we designate organic nature, cannot be explained by chance, *i.e.* without a common ground for co-operating necessities; by the side of mere mechanism it is impossible not to acknowledge a teleological principle—the only question is, in what manner can we conceive this as acting without abandoning the purely mechanical conception of nature?

This is obviously effected if, with Von Baer and Von Hartmann, we permit the metaphysical principle to interrupt the course of the mechanism

[2] *Loc. cit.* p. 156.

of nature, and if we consider both the former and the latter to work together with equal power. Von Hartmann expressly makes such an admission under the name of an "internal principle of development," to which he attributes such an important share that one cannot understand why it should have any need for the employment of causal powers, and why it does not simply do everything itself. Von Baer expresses himself much less decisively, and even in many places insists upon the purely mechanical connection of organic natural phenomena; but that with him also the idea of interruption by a metaphysical principle is present, is principally shown by his assuming, at least partly, the *per saltum* development of species. This necessarily involves an actively internal power of development.

Although I have already brought forward many arguments against the existence of such a power, and although in refuting it every form of development by directive powers is at the same time overthrown, it nevertheless appears to me not to be superfluous in such a deeply important question to show that a *per saltum* development, and especially the so-called heterogeneous generation, is inconceivable, not only on the ground of the arguments formerly employed against the phyletic vital force in general, but quite independently of these.

In the first place it must be said that the

positive basis of this hypothesis is insecure. Cases of sudden transformation of the whole organism with subsequent inheritance are as yet quite unknown. It has been shown that the occasional transformation of the Axolotl must most probably be regarded in a different light. Another case, taken for heterogeneous generation, viz. the budding of twelve-rayed Medusæ in the gastric cavity of an eight-rayed species, has lately been shown by Franz Eilhard Schulze[3] to be a kind of parasitism or commensalism. The buds of the *Cuninæ* do not spring, as was supposed, from the *Geryonia*, but are developed from a *Cunina* egg. But even if we recall here the cases of alternation of generation and heterogenesis, this would not be of any value by way of proof; it would only be thus indicated how one might picture to oneself a sudden transformation. That in alternation of generation, or generally, in every mode of cyclical reproduction, we have not to deal with the abandonment of one type of organization and the transition to some other, is proved by the continual return to the type of departure—by the cyclical character of the entire transformation. That two quite heterogeneous types can belong to one cycle of development is, however, capable of a far better and more correct

[3] "Über die Cuninen-Knospenähren im Magen von Geryonien." Reprint from "Mittheil. des naturwiss. Vereines," Graz, 1875.

explanation than would be given by the supporters of *per saltum* development. If we trace cyclical reproduction to the adaptation of different developmental stages or generations to deviating conditions of life, we thus not only explain the exact and often striking agreement between form and mode of life—we not only bridge over the gap between metamorphosis and alternation of generation, but we can also understand how, within one and the same family of Hydrozoa, species can occur with or without alternation of generation, and further how other species can exist in which the alternation of generation (the production of free Medusæ) is limited to the one sex; we can understand in general how one continuous series of forms may lead from the simple sexual organ of the Polypes to the independent and free swimming sexual form of the Medusæ, and how hand in hand with this the simple reproduction becomes gradually cyclical. It is just these intermediate steps between the two kinds of reproduction that make quite untenable the idea that the heterogeneous forms in cyclical propagation arise through so-called "heterogeneous generation," *i. e.* through sudden *per saltum* transformation. It is excusable if philosophers to whom these facts are strange, or who have to take the trouble of working them up, should adduce alternation of generation as an instance of "heterogeneous generation," but by naturalists this should be once and for ever abandoned.

All other facts which have hitherto been referred to "heterogeneous generation" are still less explicable as such, inasmuch as they always relate to changes in single parts of an organism, such as the sudden change of fruit or flower in cultivated plants. The notion of *per saltum* development, however, demands a total transformation—it comprises (as Von Hartmann quite correctly and logically admits) the idea of a *fixed specific type* which can only be re-modelled *as a whole*, and cannot become modified piecemeal. It must further be added, that the observed variations which have arisen abruptly in single parts are not as a rule inherited:[4] fruit-trees are only propagated by grafting, *i.e.* by perpetuating the individual, and not by ordinary reproduction by seeds. Now, if we nowhere see sudden variations of large amount perpetuated by heredity, whilst we everywhere observe small variations which can all be inherited, must it not be concluded that *per saltum* modification is not the means which Nature employs in transforming species, but that an accumulation of small variations takes place, these leading in time to large differences? Is it logical to reject the latter conclusion because our period of observation is too

[4] [See Darwin's "Origin of Species," 6th ed. pp. 33, 34, and 201—204. R. M.]

brief to enable us to directly follow long series of accumulations, whilst *per saltum* variation is admitted, although unsupported by a single observation? As long as there remains any prospect of tracing large deviations to the continually observed phenomenon of small variations, I believe we have no right to resort to the purely hypothetical explanation afforded by *per saltum* variations.

But the hypothesis of "heterogeneous generation" is not only without a basis of facts—it can also be directly shown to be untenable. Since the operation of an internal power of transformation does not explain adaptation to the conditions of life, the claims of natural selection to explain these transformations must be admitted; but the co-operation of a phyletic vital force and natural selection is inconceivable if we imagine the modifications to occur *per saltum.*

The supposed "heterogeneous generation" is always illustrated by the example of alternation of generation. The origination of a new animal form is thus conceived to take place in the same manner as we now see, in the cyclical reproduction of the Medusæ, free swimming, bell-shaped Medusoids, produced from fixed polypites, or *Cercariœ* from Trematode worms by internal budding; in brief, it is imagined that one animal form suddenly gives rise to another widely deviating form by purely internal causes. Now on this theory it would be an unavoidable postulate, that by such a process of *per saltum* development there arises not merely a new type of some species, but at the same time individuals capable of living and of persisting under, and fitted to, given conditions of life. But every naturalist who has attempted to completely explain the relation between structure and mode of life knows that even the small differences which separate one species from another, always comprise a number of minute structural deviations which are related to well defined conditions of life—he knows that in every species of animal the whole structure is adapted in the most exact manner *in every detail* to special conditions of life. It is not an exaggeration when I say in every detail, since the so-called "purely morphological parts" could not be other than they are without causing changes in other parts which exercise a definite function. I will not indeed assert that in the most closely related species all the parts of the body must in some manner differ from one another, if only to a small extent; it seems to me not improbable, however, that an exact comparison would very frequently give this result. That animals which are so widely removed in their morphological relations as Medusæ and Polypes, or Trematoda and their "nurses," are differently constructed in each of their parts can, however, be stated with certainty.

Now if this wide deviation in every part were in itself no obstacle to the assumption of a designing and re-modelling power, it would become so by the circumstance that all the parts of the organism must stand in the most precise relation to the external conditions of life, if the organism is to be capable of existing—all the parts must be exactly adapted to certain conditions of life. How can this be brought about by a transforming force acting spasmodically? Von Hartmann—who, in spite of his clear perception and widely extended scientific knowledge, cannot possibly possess a strong conviction of that harmony between structure and life-conditions prevailing throughout the whole system of the organism, and which personal research and contemplation are alone able to give—simply bridges over the difficulty by permitting natural selection to come to his aid as an "auxiliary principle" of the re-modelling power. It would not be supposed that naturalists would resort to the same device—nevertheless those who support the phyletic force and *per saltum* development generally invoke natural selection as the principle which governs adaptation. But when does this agency come into operation? When by germinal metamorphosis a new form has arisen, this, from the first moment of its existence, must be adapted to the new conditions of life or it must perish. No time is allowed for it to continue in an unadapted state throughout a series of generations until adaptation is luckily reached through natural selection. Let us have either natural selection or a phyletic force—both together are inconceivable. If there exists a phyletic force, then it must itself bring about adaptation.

It might perhaps be here suggested that the same objection applies to that process of modification which is effected by small steps, but that it does so only when the change occurs suddenly. This, however, as I have already attempted to show, but very rarely takes place; in many cases (mimicry) the conditions even change in the first place through the change in form and therefore, as is evident, as gradually as the latter. It must be the same in all other cases where transformation of the existing form and not merely extinction of the species concerned takes place. The transmutation must always keep pace with the change in the conditions of life, since if the latter change more rapidly the species could not compete with rival species—it would become extinct.

The abrupt transformation of species implies sudden change in the conditions of life, since a Medusa does not live like a Polype, nor a Trematode like its "nurse." For this reason it is impossible that natural selection can be an aiding principle of "heterogeneous generation." If such abrupt transformation takes place it must produce the new form instantly equipped for the struggle for existence, and adapted in all its organs and systems of organs to the special conditions of its new life. But would not this be "pure magic"? It is not thereby even taken into consideration that here—as in the cases of mimicry—time and place must agree. The requirements of a pre-established harmony ("*prästabilirte Harmonie*") further demand that an animal fitted for special conditions of life should only make its appearance at that precise period of the earth's history when these special conditions are all fulfilled, and so forth.

But he who has learnt to perceive the numerous and fine relations which, in every species of animal, bring the details of structure into harmony with function, and who keeps in view the impelling power of these conditions, cannot possibly hold to the idea of a *per saltum* development of animal forms. If development has taken place, it must have occurred gradually and by minute steps—in such a manner indeed that each modification had time to become equilibrated to the other parts, and in this way a succession of modifications gradually brought about the total transformation of the organism, and at the same time secured complete adaptation to new conditions of life.

Not only abrupt modification however, but every transformation is to be rejected when based upon the interference of a metaphysical principle of development. Those to whom the arguments already advanced against such a principle appear insufficient may once more be asked, how and where should this principle properly interfere? I am of opinion that one effect can have but one sufficient cause; if this suffices to produce it, no second cause is required. The hand of a watch necessarily turns once round in a circle in a given time as soon as the spring which sets the mechanism in movement is wound up; in an unwound watch a skilful finger can perhaps give the same movement to the hand, but it is impossible that the latter can receive both from the operator and from the spring at *the same time*, *the same motion* as that which it would receive through either of these two powers *alone.* In the same manner it appears to me that the variations which lead to transformation cannot be at the same time determined by physical and by metaphysical causes, but must depend upon either one or the other.

On no side will it be disputed that at least one portion of the processes of organic life depends upon the mechanical co-operation of physical forces. How is it conceivable that sudden pauses should occur in the course of these causal forces, and that a directive power should be substituted therefor, the latter subsequently making way again for the physical forces? To me this is as inconceivable as the idea that lightning is the electric discharge of a thunder-cloud, of which the formation and electrical tension depends upon causal forces, and of which the time and place are purely determined by such forces, but that Jupiter has it nevertheless in his power to direct the lightning flash according to his will on to the head of the guilty.

Now although I deny the possibility or conceivability of the contemporaneous co-operation of teleological and of causal forces in producing any effect, and although I maintain that a purely mechanical conception of the processes of nature is alone justifiable, I nevertheless believe that there is no occasion for this reason to renounce the existence of, or to disown, a directive power; only we must not imagine this to interfere directly in the mechanism of the universe, but to be rather behind the latter as the final cause of this mechanism.

Von Baer himself points this out to us, although he does not follow up the complete consequences of his arguments. He especially insists in his book, which abounds in beautiful and grand ideas, that the notions of necessity (causality) and of purpose by no means necessarily exclude one another, but rather that they can be connected together in a certain manner. Thus, the watchmaker attains his end, the watch, by combining the elastic force of a spring with wheel-work, *i. e.*

by utilizing physical necessities; the farmer accomplishes his purpose, that of obtaining a crop of corn, by sowing the seed in suitable land, but the seed must germinate as an absolute necessity when exposed to the influences of warmth, soil, moisture, &c. Thus, in these instances a chain of necessities is undoubtedly connected with a teleological force, the human will; and it directly follows from such cases that wherever we see an aim or result attained through necessities, the directive force does not interrupt the course of the series of necessities which have already commenced, but is active before the first commencement of these necessities, since it combines and sets the latter in movement. From the moment when the mechanism of the watch is combined harmoniously and the spring wound up, it goes without the further interference of the watchmaker, just as the corn-seed when once placed in the earth develops into a plant without assistance from the farmer.

If we apply this argument to the development of the organic world, those who defend mechanical development will not be compelled to deny a teleological power, only they would have with Kant[5] to think of the latter in the only way in which it can be conceived, viz. *as a Final Cause.*

In the region of inorganic nature nobody any

[5] [Eng. ed. See Kant's "Allgemeine Naturgeschichte und Theorie des Himmels."]

longer doubts the purely mechanical connection of the phenomena. Sunshine and rain do not now appear to us to be whims of a deity, but divine natural laws. As the knowledge of the processes of nature advances, the point where the divine power designedly interrupts these processes must be removed further back; or, as the author of the criticism of the philosophy of the Unconscious[6] expresses it, all advance in the knowledge of natural processes depends "upon the continual elimination of the idea of the miraculous." We now believe that organic nature must be conceived as mechanical. But does it thereby follow that we must totally deny a final Universal Cause? Certainly not; it would be a great delusion if any one were to believe that he had arrived at a comprehension of the universe by tracing the phenomena of nature to mechanical principles. He would thereby forget that the assumption of eternal matter with its eternal laws by no means satisfies our intellectual need for causality. We require before everything an explanation of the fact that relationships everywhere exist between the parts of the universe—that atoms everywhere act upon one another.[7] He who can content himself with the assumption

[6] "Das Unbewusste vom Standpunke der Physiologie und Descendenz-Theorie," Berlin, 1872, p. 16.

[7] [Eng. ed. See Lotze's "Mikrokosmos," 1st ed., vol. iii. pp. 477—483.]

of matter may do so, but he will not be able to show that the assumption of a Universal Cause underlying the laws of nature is erroneous.

It will not be said that there is no advantage in assuming such a Final Cause, because we cannot conceive it, and indeed cannot so much as demonstrate it with certainty. It certainly lies beyond our power of conception, in the obscure region of metaphysics, and all attempts to approach it have never led to anything but an image or a formula. Nevertheless there is an advance in knowledge in the assumption of this Cause which well admits of comparison with those advances which have been led to by certain results of the new physiology of the senses. We now know that the images which give us our sense of the external world are not "actual representations having any degree of resemblance,"[8] but are only signs for certain qualities of the outer world, which do not exist as such in the latter, but belong entirely to our consciousness. Thus we know for certain that the world is not as we perceive it—that we cannot perceive "things in their essence"—and that the reality will always remain transcendental to us. But who will deny that in this knowledge there is a considerable advance, in spite of its being for the most part of a negative character? But just as we must assume behind the phenomenal world of our

[8] See Helmholtz's "Populäre wissenschaftl. Vorträge," vol. ii., Brunswick, 1872.

senses an actual world of the true nature of which we receive only an incomplete knowledge (*i.e.* a knowledge corresponding only in reality with the relations of time and space), so behind the co-operating forces of nature which "aim at a purpose" must we admit a Cause, which is no less inconceivable in its nature, and of which we can only say one thing with certainty, viz., that it must be teleological. Just as the former first leads us to perceive the true value of our sensual impressions, so does the latter knowledge lead us to foresee the true significance of the mechanism of the universe.

It is true that in neither case do we learn more than that there is something present which we do not perceive, but in both instances this knowledge is of the greatest value. The consciousness that behind that mechanism of the universe which is alone comprehensible to us there still lies an incomprehensible teleological Universal Cause, necessitates quite a different conception of the universe—a conception absolutely opposed to that of the materialist. Most correctly and beautifully does Von Baer say that "a purpose cannot be otherwise conceived by us than as proceeding from a will and consciousness; in this would the 'aiming at a purpose,' which appears to us as reasonable as it is necessary, have its deepest root." If we conceive in this world a divine Universal Power exercising volition as the ultimate basis of matter and of the natural laws resident therein, we thus reconcile the apparent contradiction between the mechanical conception and teleology. In the same way that Von Hartmann, somewhere speaks of the immanent teleology of a machine, we might speak of the immanent teleology of the universe, because the single forces of matter are so exactly adjusted that they must give rise to the projected world, just as the wheels and levers of a machine bring forth a required manufactured article. I admit that these are grossly anthropomorphic ideas. But as mortals can we have any other ideas? Is not the notion of purpose in itself an equally anthropomorphic one? and is there any certainty that the idea of causality is less so? Do we know that causality is unlimited, or that it is universally valid? In the absence of this knowledge, should it not be permissible to satisfy as far as we can the craving of the human mind for a spiritual First Cause of the universe, by speaking of it in terms conceivable to human understanding? We can take up such a final position and still be conscious that we thereby form no certain conception, and indeed come no nearer to the reality. The materialist still makes use of the notion of "eternity," and frequently handles it as though it were a perfectly known quantity. We nevertheless do not seriously believe that by the expression "eternal matter," any true idea resulting from human experience is gained.

If it is asked, however, how that which in ourselves and in the remainder of the animal world is *intellectual* and *perceptive*, which *thinks* and *wills*, is ascribable to a mechanical process of organic development—whether the development of the mind can be conceived as resulting from purely mechanical laws? I answer unhesitatingly in the affirmative with the pure materialist, although I do not agree with him as to the manner in which he derives these phenomena from matter, since thinking and extension are heterogeneous things, and one cannot be considered as a product of the other. But why should not the ancient notion of "conscious matter" given out by Maupertuis and Robinet, not be again entertained, as pointed out in recent times by Fechner?[9] Would there not thus be found a useful formula for explaining phenomena hitherto quite incomprehensible?

Von Hartmann in criticizing himself, designates the sensibility of atoms as an "almost inevitable hypothesis" (p. 62), "inevitable because if sensibility were not a general and original property of the constituent elements of matter, it would be absolutely incomprehensible how through its potentiality and integration that sensibility known to us as being possessed by the organism could

[9] See also Fr. Vischer's "Studien über den Traum. Beilage zur Augsburger Allgem. Zeitung," April 14th, 1876. Haeckel also includes this idea in his recent essay already quoted, "Die Perigenesis der Plastidule," Berlin, 1876, p. 38 *et seq.*

have arisen." "It is impossible that from purely external elements devoid of all internality (*Innerlichkeit*) there should suddenly appear, by a certain mode of combination, an internality which becomes more and more richly developed. The more certainly science becomes convinced that in the sphere of externality (*Äusserlichkeit*) the higher (organic) phenomena are only results of combination, or are the aggregate phenomena of the elementary atomic forces, the more surely, when she once seriously concerns herself with this other question, will she not fail to be convinced that the sensibility possessed by higher stages of consciousness can be only combination-results, or the aggregate phenomena of the elementary sensations of atoms, although these atomic sensations as such always remain below the level of the higher combinations of consciousness." In confusing this double-sided nature of the objective phenomenon "lies the main error of all materialism and of all subjective idealism. Just as the attempt of the latter (subjective idealism) to construct the external phenomena of existence in space out of functions of internality and their combinations is impossible, so is the endeavour of the former (materialism) to build up internal sensation out of any combinations of force acting externally in space equally impossible."

I have no intention of going any deeper into these questions. I mention them only in order to point out that even from this side there appears to me no obstacle in the way of a purely mechanical conception of the processes of the universe. The naturalist may be excused if he attempts to penetrate into the region of philosophy; it arises from the wish to be able to contribute a little towards the reconciliation of the latest knowledge of the naturalist with the religious wants of the human mind—towards the aim striven for by both sides, viz. a satisfactory and harmonious view of the universe, according with the state of knowledge of our time.

I believe that I have shown that the theory of selection by no means leads—as is always assumed—to the denial of a teleological Universal Cause and to materialism, and I thereby hope that I have cleared the way for this doctrine, the importance of which it is scarcely possible to overestimate. Many, and not the most ill-informed, do not get so far as to make an unbiassed examination into the facts, because they are at the outset alarmed by the to them inevitable consequence of the materialistic conception of the universe. Mechanism and teleology do not exclude one another, they are rather in mutual agreement. Without teleology there would be no mechanism, but only a confusion of crude forces; and without mechanism there would be no teleology, for how could the latter otherwise effect its purpose?[10]

Von Hartmann correctly says:—"The most

[10] See Von Hartmann, *loc. cit.* p. 158.

complete mechanism conceivable is likewise the most completely conceivable teleology." We may thus represent the phenomenal universe as such a completely conceivable mechanism. With this conception vanish all apprehensions that the new views would cause man to lose the best that he possesses—morality and purely human spiritual culture. He who, with Von Baer, considers the laws of nature as the "permanent expressions of the will of a creative principle," will clearly perceive that a further advance in the knowledge of these laws need not divert man from the path of increasing improvement, but must further him in this course—that the knowledge of truth, whatever may be its purport, cannot possibly be considered a backward step. Let us take our stand boldly on the ground of new knowledge, and accept the direct consequences thereof, and we shall not be obliged to give up either morality or the comforting conviction of being part of an harmonious world, as a necessary member capable of development and perfection.

Any other mode of interference by a directive teleological power in the processes of the universe than by the appointment of the forces producing them, is however, at least to the naturalist, inadmissible. We are still far removed from completely understanding the mechanism by means of which the organic world is evoked—we still find ourselves at the very beginning of knowledge.

We are, however, already convinced that both the organic and the inorganic worlds are dependent only upon mechanical forces, for to this conclusion we are led, not only by the results of investigators who have restricted themselves to limited provinces, but also by the most general considerations. But although the force of these arguments may not be acknowledged, and although one might maintain that the inductional proofs against the existence of a "phyletic vital force" have been directed only against points of detail, or have never been completely demonstrated, *i. e.* for all points, it must nevertheless be conceded, that for the naturalist the mechanical conception of Nature is the only one possible—that he is not at all justified in abandoning this view so long as the interference of teleological forces *in the course* of the processes of organic development has not been demonstrated to him. Thus, it will not be immaterial whether a conception of Nature which to many seems inevitable is consistent with the idea of universal design, or a final directive universal principle, since the value which we may attach to our own lives and aims, essentially depends thereon. The final and main result of this essay will thus be found in the attempted demonstration that the mechanical conception of Nature very well admits of being united with a teleological conception of the Universe.

THE END.

INDEX.

B.

C.

D.

N.

O.

P.

Q.

R.

S.

ERRATA.

Page 81, line 8 from top, and throughout essay, for "*Daphnidæ*" read "*Daphniidæ*."

Page 95, line 3 from bottom, for "Daphnoidea" read "Daphniacea."

Page 166, line 7 from bottom (note), for "p. 438" read "p. 433."

Page 245, line 17 from bottom (note), for "Ståll" read "Stoll."

Page 263, after the word "insects" (bottom line of note), add, "but the whole marking is suggestive of distastefulness."

Page 296, line 3 from bottom, for "Stähelina—collector" read "Stähelin—a collector."

Page 305, line 5 from bottom (note), for "In 1869" read "In 1865."

Page 434, bottom line of note, for "Geometræ" read "Bombycidæ."

Page 494, line 2 from top, for "from which a larval form" read "from a larval form which."

Page 542, line 12 from top, for "*Dione Vanilla*" read "*Dione Vanillæ*."

Page 544, line 15 from bottom, for "*Siderome*" read "*Siderone*."